James R. Davis
Muskingum Area Technical College

LABORATORY MANUAL

to accompany

MODERN INDUSTRIAL ELECTRONICS

Fourth Edition

Timothy J. Maloney

Upper Saddle River, New Jersey
Columbus, Ohio

Editor in Chief: Stephen Helba
Assistant Vice President and Publisher: Charles E. Stewart, Jr.
Production Editor: Alexandrina Benedicto Wolf
Design Coordinator: Robin G. Chukes
Production Manager: Matthew Ottenweller
Marketing Manager: Barbara Rose

This book was set by James R. Davis. It was printed and bound by The Banta Company. The cover was printed by Phoenix Color Corp.

Copyright © 2001 by Prentice-Hall, Inc., Upper Saddle River, New Jersey 07458. All rights reserved. Printed in the United States of America. This publication is protected by Copyright and permission should be obtained from the publisher prior to any prohibited reproduction, storage in a retrieval system, or transmission in any form or by any means, electronic, mechanical, photocopying, recording, or likewise. For information regarding permission(s), write to: Rights and Permissions Department.

10 9 8 7 6 5 4 3 2 1
ISBN 0-13-032332-2

To my children--Janna, Samantha, and Taylor
My grandchild, Matthew Vincent

This work would not have been possible without the relationship I have with my lord and savior Jesus Christ.

In remembrance of Bo and Blackey

Preface

This *Laboratory Manual to accompany Modern Industrial Electronics, Fourth Edition,* was written for students enrolled in industrial electronics courses in electronics engineering technology programs or industrial electronics apprenticeship programs.

To meet the goals of individual programs and differing student learning styles, a typical experiment in this manual has several different components. Materials were developed to meet the needs of various programs through diverse experiments, while emphasizing the basics. Each experiment begins with an introduction, which provides essential background information to support the actual hands-on phase of the students' work. Encourage your students to read this material before coming to class. In most cases, I have also included troubleshooting problems, practice problems, design problems, study questions, and data manual usage. As students complete each experiment, encourage them to construct their circuits modularly, testing each stage as they go. Emphasize that it will be easier to troubleshoot their circuits that way.

Where possible, I have attempted to use readily available components for experiments. You may use replacement parts of equal or superior quality if the exact device is not available. Each experiment contains a list of all materials required to construct each circuit. The complete listing of all materials and equipment is found in the Appendix.

Lastly, it is expected that the lab instructor who supervises the students during lab will have completed each experiment before it is introduced to the class. This ensures that the instructor is aware of the experiment's subtleties, and can make any changes necessary to accommodate each program's parts and equipment inventory. This also represents good practice and is an important element of a good lab safety program. I have included comments on safety at the beginning of each experiment, but this is not intended as a substitute for the instructor's responsibility. Rather, this is intended to remind the students that safety, when working in the lab at work or at school, is their number-one. Safety precautions should be reviewed with students before each experiment. Students should also wear appropriate apparel such as safety glasses and follow established lab rules and policies at all times.

I have added a number of web sites for student and faculty use in this edition. These are web sites of for major companies whose products are widely used in industrial electronics applications in the United States.

Acknowledgments

I would like to thank the staff at Prentice Hall, especially Delia Uherec and Alex Wolf for their contribution to the production of this lab manual. I also want to acknowledge William Martin for his help with the schematic diagrams and the revision of the circuits contained in this manual. I would like to thank Jason Griffith for his editorial comments, Jeff Wilbur for his help with programmable controllers, Chris Morrison for circuit testing, Angie Merckle for her help with software and the scanner, and Ky Davis for her review of the manuscript.

Table of Contents

1	Switches and Schematic Symbols	1
2	Transistor Switches	9
3	Basic Control Relay Circuits	23
4	Transformer Basics	39
5	The Half-Wave Rectifier	57
6	The Full-Wave Rectifier	75
7	Power Supply Filtering	97
8	Voltage Regulator Principles	111
9	I-C Voltage Regulators	127
10	Current Sources and Regulators	141
11	Introduction to the Wattmeter	151
12	Delta and Wye Circuits	159
13	Measuring Three-Phase Power	169
14	Three-Phase, Half-Wave Rectification	183
15	Three-Phase, Full-Wave Rectification	199
16	Fundamental Op-Amp Circuits	217
17	Instrumentation Amplifier	233
18	Op-Amp Summers	243
19	SCRs: D.C. Characteristics	255
20	SCRs: A.C. Characteristics	265
21	Triacs and Diacs	279
22	Unijunction Transistors	293
23	Photoelectric Devices	303
24	Optoelectronic Devices	315
25	Light-Activated Thyristors	331
26	The Hall-Effect Switch	347
27	Thermistors	353
28	I. C. Temperature Sensors	365
29	Thermocouples	375
30	The D/A Converter	385
31	The A/D Converter	409
32	The 555 Timer	429
33	Frequency-to-Voltage Converters	441
34	Solid-State Relays	457
35	Solid-State Timers	467
36	Programmable Controllers	477
37	Magnetic Proximity Switch	489
38	The D.C. Shunt Motor	495
39	Stepper Motors	505
40	Three-Phase Motors	515
APPENDIX: Materials and Equipment Lists		527

Switches and Schematic Symbols: 1

INTRODUCTION:
Fundamental to the field of industrial electronics is the principle that electricity must be controlled. For our purposes the term *control* will have two possible meanings. Electricity can be controlled either by turning it on or off, or by controlling the magnitude of a current or voltage source (making it larger or smaller). At work or in your home, you can turn a light on or off by simply flipping a switch from one position to another. In a factory, the speed of a D.C. motor can be varied (controlled) by varying its field current. Electrical maintenance technicians are responsible for installing, repairing, and maintaining the electrical/electronics equipment that performs these two basic tasks. In addition to their hands-on skills, these technicians must also be able to read and interpret schematics, wiring diagrams, and blueprints. With these thoughts in mind, your first experiment will introduce you to basic switching devices, their current and voltage ratings, and schematic symbols.

SAFETY NOTE:
Have you ever had an accident? Can you remember what caused it? Now, be honest with yourself. There are two primary causes of accidents involving electronics technicians. Can you guess what they are? Haste and lack of knowledge. Accidents occur when you are in too big of a hurry to observe well-established safety precautions. Accidents also happen when you don't know what the safety precautions are or you get in too big of a hurry to ask for help. Take your time and observe all safety precautions when wiring electronic circuits. When performing lab experiments, ask your instructor if you are in doubt about *anything*. At your place of work, ask your supervisor. I challenge you to have an accident-free career as a professional in the field of electronics. Always wear safety glasses, and remove all jewelry from your hands and fingers before the start of any experiment.

OBJECT:
Upon successful completion of this experiment and all reading assignments, the student should be able to:
- recognize and draw the schematic symbols for several common switching and control devices
- determine the current and voltage ratings for a given switch
- obtain engineering data for switches and control devices from manufacturer web sites

REFERENCES:
www.eaton.com
www.ab.com
www.nema.org
www.honeywell.com
www.squared.com
www.cutlerhammer.com

MATERIALS:
    1 - light bulb, 120 volts, 60 or 75 watts
    1 - light bulb socket
    2 - switches, SPST, rated for 120 VAC applications or higher
      - miscellaneous switches, control devices, lead wires, and connectors

EQUIPMENT:
1 - digital multimeter

PART 1: Schematic Symbols and Switch Terminology
Background:
Before going into the many different types of switches available to the circuit designer, it would be appropriate to define a few common terms associated with switches. Let's first consider the term *single-pole, single-throw* (SPST). This type of switch can be used to turn on or off an electrical load. Put another way, this type of switch will either start or stop the flow of current by closing a set of contacts or opening them. An example is the wall switch in your home that you use to turn on or off a ceiling light. The schematic symbol of the SPST switch is shown in the circuit of Figure 1.1 a). A single-pole, double-throw (SPDT) switch is shown in Figure 1.1 b) controlling two different loads. The SPDT switch then can control two branches of a series-parallel circuit; whereas, the SPST switch can control only one single closed-loop circuit. Figure 1.1 c) shows a double-pole, single-throw switch. It can control two different circuits. In the case of the SPST and SPDT switches, notice that there is one common hinge point or pole. The DPST switch has two hinge or pivot points. Hence the term *double-pole*. The SPST switch can throw in or out (connect or disconnect) one load. It is therefore referred to as a single-throw switch. The SPDT switch can throw in or out two loads. Yes, that's why it's called double-throw. The DPST is referred to as a single-throw switch because <u>each</u> pole can throw in or out <u>only one</u> load. Now, what about the double-pole, double throw (DPDT) switch? I will let you think about that one!

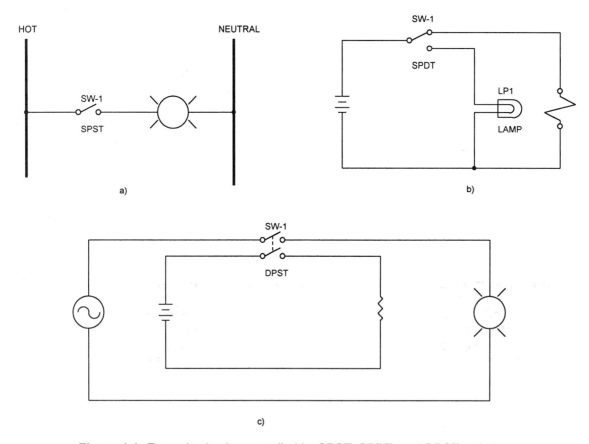

Figure 1.1: Example circuits controlled by SPST, SPDT, and DPST switches.

The switches discussed so far will stay in one position or the other when turning on or off a load. However, some switches are spring-loaded and will return to their original position after you release them. Just consider all of the push button switches that you use every day: the channel select buttons on your car radio, the eject button on a computer's disk drive, the on/off switch on your computer or monitor, the date button on your wrist watch, and who could forget all of those buttons on your TV and VCR remotes, etc. Figure 1.2 shows two push button switches connected in series. One is a normally-open (N.O.) switch; the other is a normally-closed (N.C.) switch. In most cases (not all, but most) the schematic symbols for switches and other electronic devices should be drawn in what is called their *normal* state: de-energized, or de-activated, or at rest state. Pressing SW-1 will activate it, closing its *contacts*. Contacts are the connections (usually strips of copper) that come in contact with each other, completing the circuit. Pressing SW-2 will activate it, opening its contacts. Releasing either switch will allow it to return to its normal state as shown on the schematic diagram.

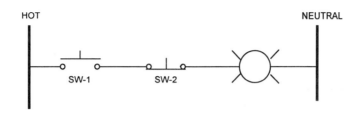

Figure 1.2: Circuit controlled by N.O. and N.C. push button switches.

Figures 1.3 a), b), and c) show the schematic symbols for limit switches. Limit switches operate very much like push button switches, except that they are operated (activated) by some part of a machine such as a cam coming into contact with the lever or arm of the limit switch rather than being activated by a human. Figure 1.3 a) is the schematic symbol for a normally-open (N.O.) limit switch. Figure 1.3 b) is the schematic symbol for a normally-closed (N.C.) limit switch. Figure 1.3 c) is the schematic symbol for a limit switch having both normally-open and normally-closed contacts. Note the one similarity between this particular limit switch and the SPDT switch. Each has a common connection or common pole. On Figure 1.3 c) this common connection has been labeled with a C (for common).

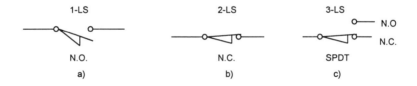

Figure 1.3: Schematic symbols for limit switches.

I mentioned earlier that the schematic symbol of an electronic device or switch is usually drawn in its normal or deactivated state. There are some exceptions. When a machine comes to a stop during an operating cycle, part of it may come into contact with a limit switch. If that represents the normal or at rest state of the machine, then it is acceptable to draw the symbol for the switch showing it in the activated state. For example, Figures 1.4 a) and b) show the activated states of the normally-open and normally-closed limit switches. The normally-open limit switch is shown being held closed, and the normally-closed limit switch is shown being held open.

Figure 1.4: Schematic symbols showing a normally-open limit switch held closed and a normally-closed limit switch held open.

Procedure:

1.1 Obtain several different types of switches from your instructor. Try to identify the terminals or connection points of each switch. If you think a switch has both normally-open and normally-closed contacts, use an ohmmeter to measure the resistance from one terminal to another as shown in Figure 1.5. As you do, move the switch actuator from one position to another and see if there is a difference.

Figure 1.5: Schematic diagrams showing an ohmmeter connected to the common and normally-closed terminals of a limit switch. In figure a) the meter will read as a short. In figure b) the meter will indicate an open.

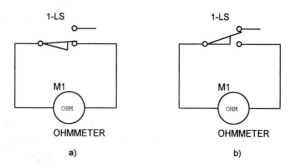

1.2 In the space provided below, make a sketch of the physical (package) outline (front or side view) of the switches given to you by your instructor. Beside each write the name of the device: SPST, SPDT, limit switch, etc. Where appropriate, label the terminals as C, N.O., or N.C. If necessary, draw both a front and bottom view of the switch to completely describe it.

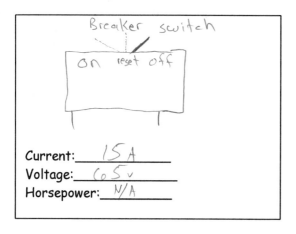
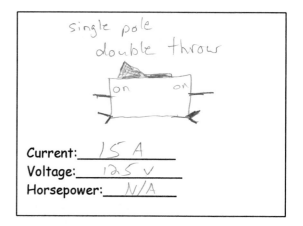
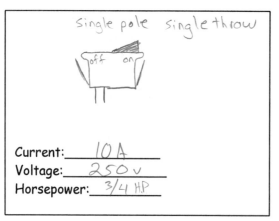
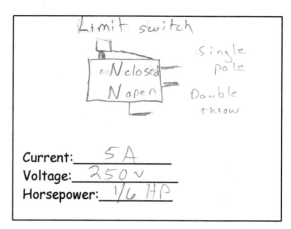

Figure 1.6: Student sketches of several different switch types.

PART 2: Switch Ratings

Background:

When you design and build a circuit with devices such as resistors and capacitors, you must specify the resistance and wattage rating of each resistor as well as the capacitance and working voltage D.C. of each capacitor. Failure to correctly specify the ratings of these devices would likely result in circuit failure, or worse, human injury. In a similar fashion, it is very important to correctly specify the current and voltage ratings of a switching device before installing it in a circuit. Of course, you will have to do a little research before you can specify the current and voltage ratings. For example, what will be the open circuit voltage across a certain switch? In your home, light switches must be rated for 120 volts RMS. In a manufacturing setting a switch, control relay, or magnetic contactor may be required to control voltages as high as 208 VAC, 480 VAC, or even higher. Now don't go thinking that selecting a switch with a voltage rating equal to the line voltage or power supply voltage for a circuit will be adequate to make it live. You must remember your knowledge of inductors, specifically their response when powered by a D.C. supply and the inductor circuit is opened. A large voltage spike due to a phenomenon referred to as *inductive kick* will occur. The resulting voltage spike can be many times larger than the supply voltage.

Hopefully you are starting to see that specifying the voltage and current ratings for a switch are not necessarily a simple matter. When determining the current rating for a switch, the switch must be able to withstand (without melting its contacts together) the current flowing through it. Now this may also appear to be deceptively easy. Just use Ohm's Law to calculate it, right? That would be true if all loads were purely resistive. However, the current to inductive loads such as A.C. motors can vary greatly. If a motor is operating at rated or full load, it will draw more current than when operated under no load. You might be saying to yourself, "Well, why not just determine the full-load current specified by the manufacturer on the nameplate data of the motor?" It's true, that particular approach would take care of the current rating required by the switch when the motor is operating at rated load. But what if the motor is overloaded, even for a short period of time? A factor of safety should always be applied. If a circuit will continuously draw a current of 2 amperes, you may wish to select a switch rated for 3 amperes or more. Your circuit should be designed to operate under all anticipated and unanticipated load and environmental conditions, and if any of these are exceeded, then protective devices such as fuses and circuit breakers should shut off the circuit. Most switches designed for industrial applications are rated for a certain maximum continuous current and a maximum current that they are capable of interrupting--turning off. Circuit breakers, fuses, and thermal overload relays should be sized according to the National Electric Code.

Let's get back to our discussion about current ratings for switches controlling inductive devices. When a motor is rotating, it produces a counter EMF that opposes or limits the amount of current that flows through it. However, when the motor shaft is not rotating, such as at start up or when stalled due to a heavy load, the current is limited only by the nominal resistance of the windings. So, if a motor with an armature resistance of 2 Ohms is powered by a 480 VAC source, it can draw as much as 240 amperes at start up. Fortunately, designers of motor control circuits know this and provide measures at motor start up to limit the current to a much lower level. In any case though, the current drawn by a motor at start up is much greater than when it is rotating at rated RPM. The point is this, switches controlling inductive loads require higher current ratings than non-inductive loads of equal power rating. Sometimes manufacturers of switches, control relays, and magnetic contactors specify the horsepower or wattage rating that their switch can control in addition to the load voltage and line current. This makes life a little easier for the circuit designer. The current and voltage ratings of switches are usually stamped, molded, or painted onto the body of the switch. If you can't find these ratings when examining a switch, look up its part number in the manufacturer's catalog to obtain its specifications.

On a final note, the current rating for a switch at contact closure is usually much higher than its current at opening or "break" rating. And, the A.C. voltage or current rating for a switch's contacts is usually greater than its D.C. voltage or current rating. In other words, do not exceed *any* of the manufacturer's ratings.

Procedure:

    2.1 Again, examine the switches given to you by your instructor. Look closely at the switches, and try to find the manufacturer's current and voltage rating for each switch.

    2.2 Record the voltage, current, and horsepower or wattage (if included) ratings for each switch beside your sketch of that device in Figure 1.6.

PART 3: Circuit Design

3.1 Design a circuit using two single-pole, double-throw (SPDT) switches that will perform the following function: Either switch can be used to turn on or off a light. This is the same function performed by a light switch in your house and another in your garage. You can turn on or off a light in the garage from either of two locations. Neatly draw your schematic diagram in the space provided below. Have your instructor check your design. If time permits, construct this circuit in your lab and demonstrate it to your instructor.

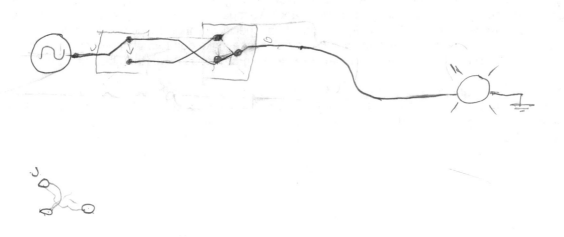

Figure 1.7: Sketch of student design.

PART 4: Questions

4.1 Use one of the web sites provided in the list of References or one provided by your instructor to find a limit switch that is rated for 120 volts RMS and 25 HP or greater and has an explosion-proof enclosure. Write the manufacturer's name and part number in the space provided below. From the web site, print a copy of the manufacturer's data sheet for this switch and attach it to your report.

Manufacturer's name:_____

Part Number:_____

4.2 To answer the following question, you will need to search the web site of the National Electrical Manufacturer's Association (NEMA), www.nema.org. What NEMA document (standard) deals with the requirements for "control relays, limit switches, proximity switches, pushbutton selector switches, indicating, and pushbutton stations"?

_____

4.3 Beside each of the schematic symbols shown in Figure 1.8, write the name of the device corresponding to that symbol. In addition to the schematic symbols of several switch types, some symbols of other commonly occurring devices comprising industrial control circuits are also included. Where appropriate, label the contacts as N.C. or N.O.

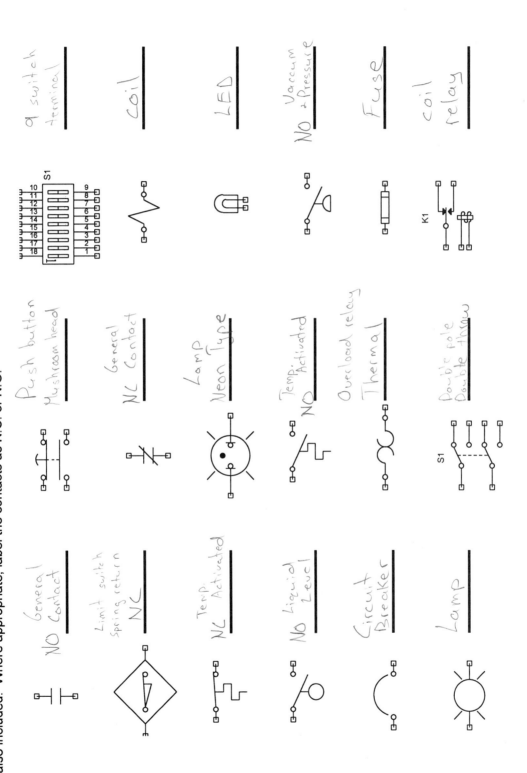

Figure 1.8: Schematic symbols of several different switch types and control devices.

4.4 In the space provided below, draw the schematic symbol for each of the devices listed.

a) center-off SPDT

b) rotary selector switch

c) time-delay opening contacts, TDO

d) time-delay closing contacts, TDC

e) normally-open, flow-activated switch

f) normally-closed float switch

g) normally-open, foot-operated switch

h) safety interlock

i) normally-open proximity switch

j) normally-closed pressure switch

k) normally-closed, mushroom-head push button switch

l) SPDT slide switch

Transistor Switches: 2

INTRODUCTION:
In Experiment 1 you worked with several types of common mechanical or manually-operated switches. In Experiment 3 you will work with an electromagnetic switch--the control relay. In fact, throughout this lab manual, many different types of electrical and electronic switches are incorporated into the electronic circuits that you will construct. Fundamental to understanding the role of electronic devices in industrial control electronics is a study of how transistors can be used as a switch. While many electronic switching applications are being performed by complex integrated circuits, these devices still rely on the transistor as their basic building block. These devices incorporate hundreds of transistors on a chip; yet, understanding how one works leads to an understanding of how the entire unit works. For example, in Experiment 39 you will build a stepper-motor controller circuit using a total of eight transistors. Four of them—optoisolators--will be used to electrically isolate the low-power digital control circuitry from the higher-power field windings of the motor. The remaining four transistors will be used to provide current gain to switch voltage on and off to the motor's four field windings. Now, all of this could be performed by a single chip. For the electrical maintenance technician to be able to install, troubleshoot, and repair complex integrated circuits, he/she must understand the basic operations such chips perform. In this experiment you will study how bipolar junction transistors (BJTs) and metal-oxide semiconductor field-effect transistors (MOSFETs) can be configured to perform basic switching operations.

SAFETY NOTE:
MOSFETs are static sensitive devices! Therefore, it is suggested that your instructor review the proper steps to follow when handling such devices in order to avoid component failure due to electrostatic discharge (ESD). <u>Always</u> familiarize yourself with the manufacturer's ratings for any electronic component before designing or constructing a circuit utilizing it. As always, wear safety glasses, and remove all jewelry from your hands and fingers before the start of this or any experiment.

OBJECT:
Upon successful completion of this experiment and all reading assignments, the student should be able to:
- construct a low-power electronic switch using one BJT
- construct a high-power electronic switch using a low-power BJT and a high-power BJT
- construct an electronic switch using an enhancement-mode MOSFET

REFERENCES:
Chapter 1 of Maloney's <u>Modern Industrial Electronics</u>
www.motorola.com

MATERIALS:
1 - 14 volt lamp rated for 200 mA
1 - bayonet-style lamp holder
1 - NPN transistor, 2N3904 or the equivalent
1 - PNP transistor, 2N3905 or the equivalent
1 - NPN transistor, TIP29 or the equivalent
1 - N-channel, enhancement-mode MOSFET, SK9155 or a MOSFET of equal or higher power

(if you do not have a low-power MOSFET, you may substitute a high-power MOSFET such as the SK9502)
1 - 560 Ohms resistor, 1/2 watt
1 - 5600 Ohms resistor, 1/4 watt
1 - 10 Kohms resistor, 1/4 watt
1 - 50 Kohms ten-turn trim potentiometer
1 - SPST switch
1 - solderless breadboard
- miscellaneous lead wires and connectors

EQUIPMENT:
1 - D.C. power supply
1 - digital multimeter

PART 1: NPN and PNP Transistor Switches
Background:
The basic microstructure of NPN and PNP transistors is shown in Figures 2.1 a) and b), respectively, alongside their schematic symbols. The three terminals of these devices have been identified as the collector, base, and emitter. Transistors are current devices. A small amount of current (conventional current) flowing into the base of an NPN transistor will cause a large amount of current to flow into its collector. This is depicted in Figure 2.2. Transistors are current amplifiers--small current in, large current out. Transistors can operate in one of three modes. These three modes are referred to as the *active region*, *cutoff*, and *saturation*. When operated in the active region, a transistor is being used as an amplifier. When used as a switch, a transistor is operated in either cutoff or saturation. When a transistor is in cutoff, it is basically acting like an open (as in open switch) from the collector to the emitter. When a transistor is operating in the saturation region, it is acting like a short (closed switch) from the collector to the emitter. However, the transistor is not a perfect switch. When closed, a perfect switch has a voltage drop of 0.0 volt across it. When a transistor is in saturation, it may have as little as 0.1 volt to as much as 0.6 volt drop from the collector to emitter depending upon the amount of current flowing through it. Well, how do we cause a transistor to operate in cutoff or saturation?

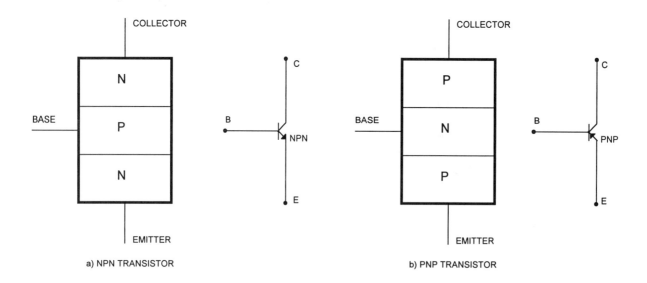

Figure 2.1: Basic microstructure and schematic symbols for NPN and PNP transistors.

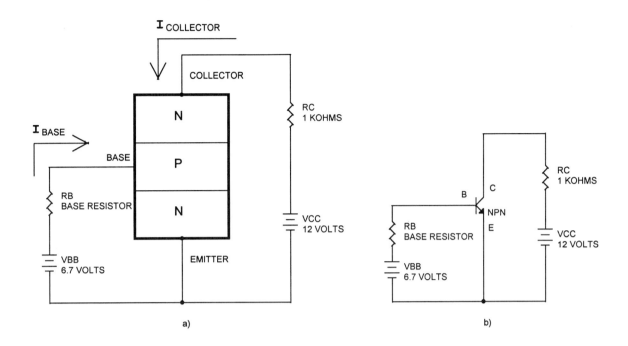

Figure 2.2: NPN transistor with forward-biased base-emitter junction showing direction of base and collector currents.

There are two basic ways to put a transistor in cutoff. If there is no bias voltage applied to the base-emitter junction as shown in Figure 2.3 a), then the transistor will be in cutoff. Reverse biasing the base-emitter junction of a transistor will also cause it to operate in cutoff. This is shown in Figure 2.3 b). Compare the polarity of V_{BB} shown in Figure 2.3 b) to the polarity of V_{BB} shown in Figure 2.2. Putting a transistor in saturation requires the base-emitter junction to be forward biased as shown in Figures 2.2 a) and b). However, a transistor can also be in the active region when forward biased.

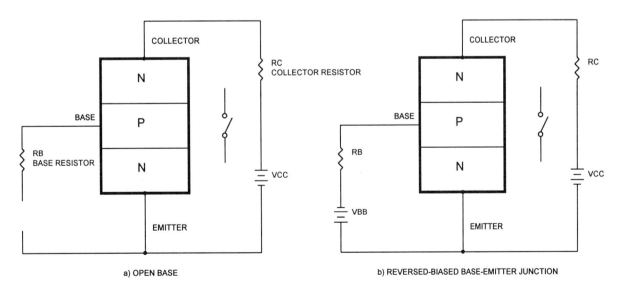

Figure 2.3: Two methods to put NPN transistors in cutoff. A transistor in cutoff acts like an open switch.

So, how do we design a transistor switch to operate in saturation and guarantee that it is not operating in the active region? First, let's consider the relationship between the base current, I_B, and the collector current, I_C. The D.C. current gain of a transistor, β_{DC}, is the ratio of I_C to I_B. Written as a formula we have:

$$\beta_{DC} = I_C/I_B \qquad \textbf{2.1}$$

The value of β_{DC} depends on the transistor being used and at what point on the transistor's I-V curves it is operating. To cause a transistor to go into saturation, the amount of I_B must be great enough to cause the collector-to-emitter voltage, V_{CE}, to be as close to 0.0 volt as possible. As I pointed out before, it will not be possible to achieve a value of 0.0 volt for V_{CE}. Rather, a transistor deep in saturation will have a voltage drop of 0.1 to 0.2 volt from collector to emitter. It is not necessary to have a set of transistor I-V curves to determine how much I_B must flow to cause V_{CE} to approach 0.0 volt (although it would help us be more precise). If we let $\beta_{DC} = 10$, this will allow us to compute a value of I_B that will cause saturation for a given value of I_C. Consider for the moment the circuit shown in Figure 2.2. If we assume that the transistor acts as a short when in saturation, then approximately all 12 volts from the power supply will appear across the load resistor, R_C. Applying Ohm's Law we would then find the saturation current, I_{Csat}, as follows:

$$I_{Csat} = V_{CC}/R_C \qquad \textbf{2.2}$$

$$= 12\ V/1\ K\Omega$$

$$= 12\ mA$$

Applying Formula 2.1 for $\beta_{DC} = 10$, we can find the value of base current that will cause saturation, I_{Bsat}, as follows:

$$I_{Bsat} = I_{Csat}/\beta_{DC} \qquad \textbf{2.3}$$

$$= 12\ mA/10$$

$$= 1.2\ mA$$

Next we must find the value of R_B that will result in an I_B of 1.2 mA for the given value of 6.7 volts for V_{BB}. If the transistor is made of silicon, we can assume a voltage drop across the base-emitter junction, V_{BE}, of approximately 0.7 volt. When a transistor is operated in deep saturation, or if working with a power transistor, it is possible that the value for V_{BE} will be greater than 0.7 volt. Assuming a value of 0.7 volt will yield satisfactory results in most cases. Applying Kirchhoff's Voltage Law for the transistor's input stage, we have:

$$V_{BB} = V_{RB} + V_{BE} \qquad \textbf{2.4}$$

Rearranging, we have

$$V_{RB} = V_{BB} - V_{BE} \qquad \textbf{2.5}$$

Solving for V_{RB}, we have

$$V_{RB} = 6.7\ V - 0.7\ V$$

= 6.0 V

Applying Ohm's Law we can find R_B as follows:

$$R_B = V_{RB}/I_{Bsat} \qquad \textbf{2.6}$$

$$= 6.0 \text{ V}/1.2 \text{ mA}$$

$$= 5000 \text{ Ohms}$$

A standard 4.7 Kohms resistor would be an appropriate choice for this particular circuit. Thus far our analysis has been for an NPN transistor. The same analysis would apply for a PNP transistor. The important thing for you to remember is that the base-emitter junction must be forward-biased to produce saturation whether it is an NPN transistor or a PNP transistor.

Procedure:

1.1 Obtain all materials and equipment required to complete this experiment.
1.2 The first circuit you are to construct is shown in Figure 2.4. The lamp described in the parts list is rated for an operating voltage in the range 12 to 14 volts with an operating current of 200 mA. If you are using a different lamp, obtain its voltage and current ratings from your instructor.
1.3 Assuming an I_{Csat} of 200 mA (or the current rating for your lamp if different than this), a β_{DC} of 10, and $V_{BE} = 0.7$ volt, calculate the value of R_B that will cause transistor saturation. Select the closest standard resistor value having a wattage rating of 1/2 watt for your circuit.

Calculations:

1.4 Have your instructor check your calculations. Construct the circuit shown in Figure 2.4 using your final value for R_B.

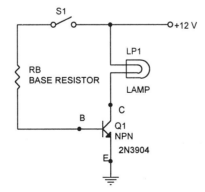

Figure 2.4: NPN transistor switch circuit.

1.5 With switch S1 open, use your voltmeter to measure the voltages specified in Table 2.1. Record your results.
1.6 Close switch S1. Repeat your voltage measurements. Record your results in Table 2.1.

Table 2.1: Experimental data for the circuit shown in Figure 2.4.

	LAMP OFF	LAMP ON
V_{LAMP}, VOLTS		
V_{CE}, VOLTS		
V_{RB}, VOLTS		

1.7 Construct the circuit shown in Figure 2.5. Use the same resistor for R_B in this circuit as you used in the circuit shown in Figure 2.4. Note, the load (lamp) is connected to the collector.

Figure 2.5: PNP transistor switch circuit.

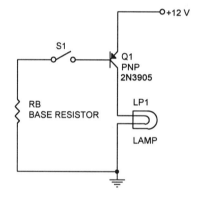

1.8 With switch S1 open, use your voltmeter to measure the voltages specified in Table 2.2. Record your results.
1.9 Close switch S1. Repeat your voltage measurements. Record your results in Table 2.2.

Table 2.2: Experimental data for the circuit shown in Figure 2.5.

	LAMP OFF	LAMP ON
V_{LAMP}, VOLTS		
V_{CE}, VOLTS		
V_{RB}, VOLTS		

PART 2: The Two-Stage Transistor Switch

Background:

In some control applications the signal source may not be able to provide the amount of base current required to cause transistor saturation. Consider for example the circuit shown in Figure 2.6. When the output of the 74LS26 goes low, the base-emitter junction of the PNP transistor will be forward-biased; however, the maximum current that the 74LS26 can sink, I_{OL}, according to the manufacturer's specifications is 8 mA. According to our previous calculations, 20 mA of base current is required to cause saturation for this lamp circuit.

For your information, the 74LS26 is an open-collector NAND gate with a V_{OH} rated for 15 volts. What this means is that when the 74LS26 is in its open state, it can withstand up to 15 volts at its output without breaking down. This will allow the 74LS26 to turn off our lamp circuit, which is powered by 12 volts, but what do we do to enable the 74LS26 or some other low power device to turn our lamp on? One solution is to cascade two transistors as shown in the circuit of Figure 2.7. In this circuit two transistors are being used to obtain the current gain necessary to lower the base current required by the 2N3904 transistor to turn the lamp off. After constructing this circuit and analyzing it mathematically, you should be able to explain how it operates and solve for all voltages and currents when the lamp is on or off.

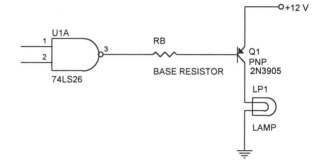

Figure 2.6: Open-collector NAND gate being used to control the turn-on of a transistor switch.

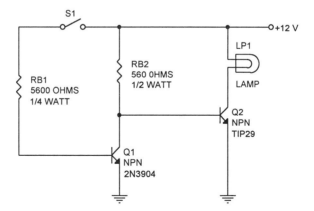

Figure 2.7: Two-stage transistor switch circuit.

Procedure:

2.1 Construct the circuit shown in Figure 2.7. If you are using a different lamp than the one specified in the parts list, then you should use the same base resistor for R_{B2} that you used in Figures 2.4 and 2.5. The resistor value for R_{B1} should be ten times the value of R_{B2}. If you don't yet know why this factor of ten is being used, then take a moment to think about it.

Before constructing the circuit, measure the actual resistance values for R_{B1} and R_{B2}. Record these values in Table 2.3.

Table 2.3: Nominal and measured base resistor values for the circuit shown in Figure 2.7.

	NOMINAL RESISTANCE, OHMS	MEASURED RESISTANCE, OHMS
R_{B1}		
R_{B2}		

2.2 Turn on your power supply. With switch S1 open, measure and record the voltages shown in Table 2.4.

2.3 Close switch S1. Repeat your voltage measurements. Record your results in Table 2.4.

2.4 Assuming a base-emitter voltage of 0.7 volt for each transistor, use Kirchhoff's Voltage Law to calculate the theoretical lamp, transistor, and resistor voltages for the circuit shown in Figure 2.7. You may also assume that each transistor is either in complete saturation or in complete cutoff depending upon whether the lamp is on or off. Record your results in Table 2.4. How do your measured voltages compare to your calculated values? Reconcile any <u>major</u> differences before proceeding with the experiment. Turn off the power supply.

Calculations:

Table 2.4: Experimental and theoretical data for the circuit shown in Figure 2.7.

	MEASURED VOLTAGES		CALCULATED VOLTAGES	
	LAMP OFF	LAMP ON	LAMP OFF	LAMP ON
V_{LAMP}, VOLTS				
$V_{CE(Q1)}$, VOLTS				
V_{RB1}, VOLTS				
$V_{CE(Q2)}$, VOLTS				
V_{RB2}, VOLTS				

PART 3: The MOSFET Switch
Background:
Because of their high input impedance, MOSFETs have become popular in a number of different amplifier and switching applications. From your earlier study of electronic devices, you may recall that junction field-effect transistors (J-FETs) are available in the N-channel and P-channel configurations. The conductivity of these two devices is controlled by varying the amount of reverse-bias voltage applied to their gate-source junction. This is in contrast to NPN and PNP transistors, which have their conductivity varied by varying the amount that the base-emitter junction is forward-biased. MOSFETs have been designed in several different semiconductor configurations. However, the two basic configurations of MOSFETs are the enhancement-mode MOSFET (E-MOSFET) and the depletion-mode MOSFET (D-MOSFET). In this part of your experiment you will work with an E-MOSFET.

We will not go into the details of the construction nor biasing of MOSFETs. That is best left to an in-depth electronic devices class. For more details on this topic, consult an electronic devices text such as Paynter's Introductory Electronic Devices and Circuits: Conventional Flow Version, 5th edition, published by Prentice-Hall, Copyright 2000 or Boylestad and Nashelsky's Electronic Devices and Circuit Theory, 7th edition, published by Prentice-Hall, Copyright 1999. While they can come in different configurations, MOSFETs are basically three-terminal devices having a drain, source, and gate, which are analogous to the collector, emitter, and base of a BJT. However, this is where any similarities between the two families of devices ends. Unlike N-channel and P-channel J-FETs, whose gate-source junction is reverse-biased, the gate-source junction of E-MOSFETs is forward-biased to control its conduction. In the first two parts of this experiment you controlled the amount of collector current flowing through the NPN and PNP transistors by varying the amount of base current, I_B. Specifically, if you increased I_B, there was a linear increase in I_C. Conversely, if you decreased I_B, the amount of I_C decreased. In the case of the E-MOSFET, the amount of drain current, I_D, that flows through it will increase with an increase in the value of the gate-to-source voltage, V_{GS}. In fact, the relationship between I_D and V_{GS} can be compared to that of a forward-biased diode's I-V curve. Recall that the shape of a diode's I-V curve is parabolic. So is the I_D-V_{GS} curve for an E-MOSFET. Remember that a silicon diode does not conduct heavily until it reaches a certain threshold voltage, typically 0.6 to 0.7 volt. In a similar fashion, the E-MOSFET will not start to conduct (turn on) until the value for V_{GS} reaches a certain minimum referred to as the threshold voltage, $V_{GS(th)}$. The value for $V_{GS(th)}$ depends on the particular E-MOSFET that you are working with. You will have to consult the manufacturer's data sheet for this parameter. Figure 2.8 shows the circuit you are to build. It is a voltage-divider bias circuit that will allow you to vary the value of V_{GS} in order to experimentally determine the value of $V_{GS(th)}$. You will also determine the value of V_{GS} that will cause the lamp to reach full intensity.

Procedure:

3.1 Identify the drain, source, and gate leads of your E-MOSFET, then construct the circuit shown in Figure 2.8. Before installing your E-MOSFET, construct the voltage-divider bias portion of the circuit and adjust the voltage across the potentiometer for 0.0 volt. Turn off the power supply, and complete construction of your circuit.

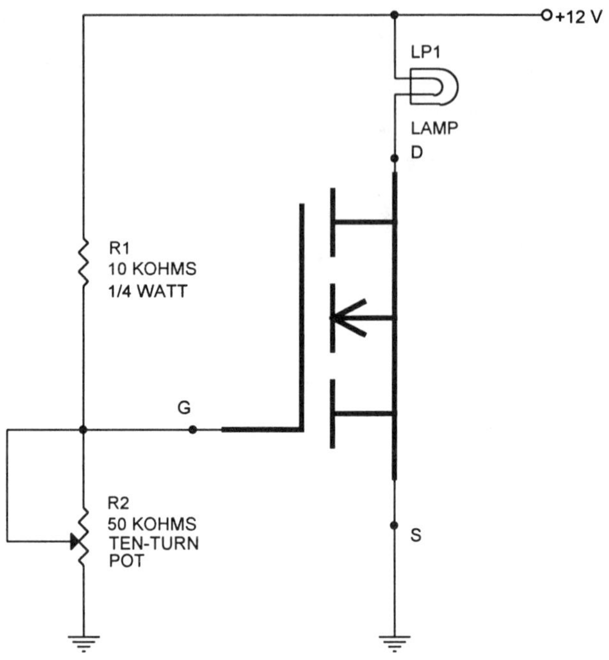

Figure 2.8: N-channel E-MOSFET switch circuit with voltage-divider bias.

3.2 Turn on your power supply. Use a digital voltmeter to measure V_{GS}. Slowly increase the potentiometer's resistance. Note the voltage at which the lamp just starts to turn on. Record this as your experimental value for $V_{GS(th)}$ in Table 2.5. Continue to increase the value of V_{GS} until the lamp reaches full brightness (V_{LAMP} approximately equals 11 volts). Record the corresponding value for V_{GS} in Table 2.5.
3.3 Turn off the power supply.
3.4 Consult the manufacturer's data manual for your E-MOSFET, and record the manufacturer's published value for $V_{GS(th)}$ in Table 2.5.

Table 2.5: Experimental and theoretical data for the circuit shown in Figure 2.8.

EXPERIMENTAL VALUE FOR $V_{GS(th)}$, VOLTS	
V_{GS} AT FULL LAMP BRIGHTNESS, VOLTS	
PUBLISHED VALUE FOR $V_{GS(th)}$, VOLTS	

PART 4: Circuit Design

4.1 Figure 2.9 shows an electronic switch using an NPN transistor to turn on a lamp <u>like the one used in this experiment</u> (i.e., the same current and voltage ratings). The base current is being provided by a +5 volt source. Calculate the value for R_B that will cause the transistor to go into saturation. If time permits, construct this circuit and demonstrate it to your instructor.

Calculations:

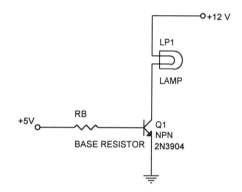

Figure 2.9: Transistor switch with base current provided by a 5 volt source.

4.2 Modify the circuit shown in Figure 2.7 so that Q1 becomes a 2N3905 transistor. Sketch your circuit in the space provided below. If time permits, construct this circuit and demonstrate it to your instructor.

PART 5: Questions

5.1 How is the operation of the circuit shown in Figure 2.4 different than that shown in Figure 2.7? What is the major difference?

5.2 Referring to the circuit shown in Figure 2.7, transistor Q1 was in cutoff when switch S1 was open and the lamp came on. Why was the value for V_{CE} for transistor Q1 not equal to 12 volts under those conditions?

5.3 If the transistor in Figure 2.4 failed as an open circuit, what voltage would you expect to measure for V_{CE} and V_{LAMP}? Explain the basis for your answer.

5.4 If switch S1 in Figure 2.5 is closed and R_B fails as an open circuit, what voltage would you expect to measure for V_{CE} and V_{LAMP}? Explain the basis for your answer.

5.5 What main symptom would result if the potentiometer in Figure 2.8 failed as a short circuit? Explain the basis for your answer.

5.6 To answer the following question, you will need to search the Motorola web site www.motorola.com. What is the maximum voltage you should select for V_{CC} when designing an electronic switch using a 2N3904 transistor? Hint: Find $V_{(BR)CEO}$.

5.7 To answer the following question, you will need to search the Motorola web site www.motorola.com. What is the maximum power dissipation, P_D, of a 2N3904 transistor at an ambient temperature of 25°C?

5.8 To answer the following question, you will need to search the Motorola web site www.motorola.com. What is the maximum power dissipation of a 2N3904 transistor at an ambient temperature of 40°C? Hint: Find the derating factor, mW/°C, and reduce the rated P_D by that amount for each degree above 25°C.

PART 6: Transistor Switch Formulas

$\beta_{DC} = I_C/I_B$ **2.1**

$I_{Csat} = V_{CC}/R_C$ **2.2**

$I_{Bsat} = I_{Csat}/\beta_{DC}$ **2.3**

$V_{BB} = V_{RB} + V_{BE}$ **2.4**

$V_{RB} = V_{BB} - V_{BE}$ **2.5**

$R_B = V_{RB}/I_{Bsat}$ **2.6**

Basic Control Relay Circuits: 3

INTRODUCTION:

A control relay is an electromagnetic switch. Its construction allows low-voltage power supplies to control or switch on and off high-voltage and/or high-current power supplies that in turn may be used to power electric motors, electric ovens, or other high-power consuming equipment. Modern automobiles may have as many as a dozen or more relays controlling various high-current loads such as headlights, windshield wiper motors, rear-window defrosters, etc. The switch (contacts) is opened or closed by energizing or de-energizing the control relay coil (electromagnet). Control relay coils may be energized by A.C. only, D.C. only, or may be energized by either power source depending on relay construction. The voltage source applied to the contacts, however, may be either D.C. or A.C. depending on the type of load that must be controlled. Therefore, when selecting a control relay for a particular application, the coil of the relay must be compatible with both the type (D.C. or A.C.) and the magnitude (voltage level) of the power supply. The relay contacts should be capable of handling the current and voltage required of the load being controlled. Sometimes relay contacts are rated according to the wattage or horsepower of the device that is to be operated. Control relay contacts may be normally open (N.O.), normally closed (N.C.), or both (SPDT). Some relays have several *sets* of contacts. In this particular instance, the word "set" includes a common connection, an open contact, and a closed contact. In this experiment, you will become familiar with the schematic symbols used for the coil of a relay and its contacts. You will also construct several different relay circuits, including the *holding* or *latching* relay circuit. Finally, you will observe the characteristic of *hysteresis* that relays and certain other electrical and electronic devices exhibit.

SAFETY NOTE:

Before starting this experiment, have your instructor review with you the identification of the connections to the control relay so that you do not mistakenly apply high voltage to the low-voltage coil or create a short by incorrectly wiring the contacts. As always, wear safety glasses, and remove all jewelry from your hands and fingers before the start of the experiment.

OBJECT:

Upon successful completion of this experiment and all reading assignments, the student should be able to:
- recognize the schematic symbols for a control relay and its contacts
- draw the schematic symbol for a control relay and its contacts, and properly label each
- given a relay, be able to identify the contacts' common connection, the normally-open and normally-closed contacts, and the relay coil inputs (connections)
- be able to construct a holding-relay circuit that will control a given load
- determine the effect that source polarity has on the operation of a control relay
- describe the phenomenon of hysteresis as it applies to a control relay

REFERENCES:

Chapter 11 of Maloney's Modern Industrial Electronics
National Electric Code® Handbook, 2000 Edition, copyrighted by the NFPA
www.midtex.com

MATERIALS:
- 2 - light bulbs, 28 volt rating
- 2 - light bulb sockets
- 1 - normally-open push button switch
- 1 - normally-closed push button switch
- 2 - single-pole, single-throw switches
- 1 - control relay socket
- 1 - 1 Kohms, 1/2 watt resistor
- 1 - 470 Ohms, 2 watt resistor
- 1 - 1N4003 rectifier diode or the equivalent
- miscellaneous lead wires and connectors

EQUIPMENT:
- 1 - adjustable D.C. power supply, 0 – 12 volts
- 1 - 24-volt D.C. power supply or 24-volt A.C. power supply
- 1 - digital multimeter
- 1 - 12 volt D.C. control relay, DPDT, with socket

PART 1: Identifying Relay Connections
Background:
Before wiring any piece of electrical or electronic equipment, it is a good idea to become familiar with all of the connections and the purpose of each. Figures 3.1 a), b), and c) show the schematic symbols for the coil and contacts of a control relay. Figure 3.1 a) shows one representation for a control relay coil. Figure 3.1 c) shows two alternative representations for a relay coil; however, the symbol on the right is most often used to represent a solenoid. Figure 3.1 b) shows the common connection and normally-open and normally-closed contacts for a relay with <u>two sets</u> of contacts. The symbol on the left in Figure 3.1 c) shows the alternate representation for a set of single-pole, double-throw contacts.

Figure 3.1 a): Schematic symbol for a control relay coil.

15.67 V

NC - Blue + White

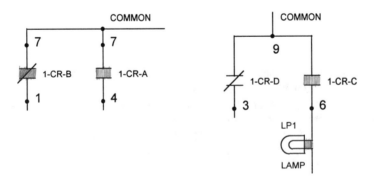

Figure 3.1 b): Schematic symbols for N.O. and N.C. relay contacts for a SPDT relay.

Figure 3.1 c): Alternate symbols for relay contacts and coil (electromagnet).

RELAY SPDT   COIL OR SOLENOID SYMBOL

Procedure:

1.1 Figure 3.1 b) shows contacts labeled with the actual pin numbers for a Midtex® 12 volt D.C. double-pole, double-throw general purpose relay. If you are using this particular relay, take time to confirm the pin numbers. If you are using a different relay, have your instructor assist you with the pin identification. Then label the schematic symbols in Figure 3.1 b) with the appropriate numbering or contact identification scheme for your relay. Use these pin numbers as a guide when wiring the circuits in this experiment.

1.2 Construct the circuit shown in Figure 3.2 a). If you are using two separate power supplies, then switches S1 and S2 can be the switches on your respective power supplies. If you are using one dual-outlet power supply, then you will need to obtain two single-pole, single-throw switches and wire each as shown in Figures 3.2 a) and 3.2 b). Do not yet wire the circuit shown in Figure 3.2 b).

Note to instructor: If your lab does not have both 12 VDC and 24 VDC supplies, you may replace the 24 VDC supply with a 24 VAC supply as shown in Figure 3.2 c).

1.3 Good technicians "build in stages and test in stages." Now test the circuit you just constructed by turning on power supply E1 and closing switch S1. Did you wire your circuit correctly? You should have heard a click or visually observed the relay's armature and contacts moving toward the coil. Turn off power supply E1.

1.4 Now wire the circuit shown in Figure 3.2 b). Have your instructor check your work.

1.5 Turn on power supply E2, and close switch S2. Did lamp LP1 turn on? If not, explain why.

Yes

Figure 3.2 a): Control relay coil circuit. Please note that a low voltage supply is used to power the coil.

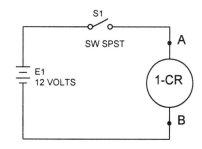

1.6 With switch S2 still closed, turn on power supply E1 and close switch S1. Describe what happened, and explain how it happened.

The light stayed on

25

1.7 Turn off all power. Reverse the connections to the relay coil. In other words, switch points A and B. Turn on the power supplies, and close switches S2 and S1. What happened? Was there any difference in the operation as a result of switching the coil connection?

_____ opposite _____

_____

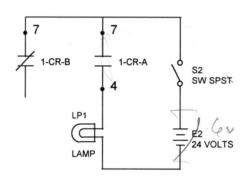

Figure 3.2 b): Lamp wired in series with normally-open contacts.

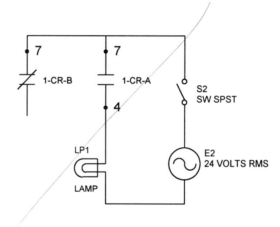

Figure 3.2 c): Alternative circuit for Figure 3.2 b) if you do not have a 24 VDC supply.

1.8 Turn off all power. Modify the circuit shown in Figure 3.2 b) so that it operates like the circuit shown in Figure 3.3.

1.9 Turn on power supply E2. Close switch S2. Describe what happened, and explain how it it happened.

_____

_____

_____

_____

Figure 3.3: Lamp wired in series with normally-closed contacts.

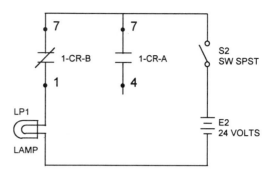

1.10 With switch S2 still closed, turn on power supply E1 and close switch S1. Describe what happened, and explain how it happened.

_____

_____

_____

_____

1.11 Turn off all power.

PART 2: The Holding-Relay Circuit
Background:
Take a moment to think about all of the electrical and electronic devices you have operated that turn on or start with the touch of a button. For example, many cars have rear-window defoggers or defrosters. By merely pressing and releasing the appropriate button, the window defogger will turn on and stay on for a certain length of time then turn off. The operation of this particular circuit involves several principles. In this part of the experiment, you will have the opportunity to study one of those principles—the holding relay circuit. If you have already studied flip-flops and latches in a digital electronics course, you may recall that these devices will change state (high to low or low to high) when supplied with a momentary pulse or clock input. The holding relay circuit that you will build exhibits a similar operational characteristic. In the case of the holding-relay circuit, you will press and release a push button switch, which will in turn cause your relay to be energized and stay energized or latched in place. This type of circuit is widely used in start-stop circuits that control industrial machinery such as A.C. motors. In many of these applications, *magnetic contactors* are used instead of control relays. Think of a magnetic contactor as the high-power big brother of the control relay.

Procedure:
2.1 With your power supplies turned off, construct the circuits shown in Figures 3.4 a) and b). Have your instructor check your wiring.

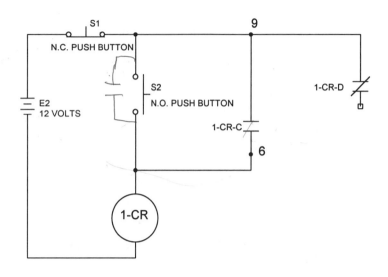

Figure 3.4 a): Latching or holding-relay circuit.

2.2 Turn on the power supplies. Press and release push button switch S2. This is the start switch. Describe what happened and explain how it happened. Explain why you didn't have to hold down S2 to keep the light on.

_____

_____

_____

_____

_____

_____

_____

2.3 Press and release push button switch S1. This is the stop switch. Describe what happened, and explain how it happened. Explain why you didn't have to hold down S1 to keep the light off.

_____

_____

_____

_____

_____

Figure 3.4 b): Load wired through the normally-open contacts being controlled by the holding-relay circuit and energized by a higher-voltage power supply.

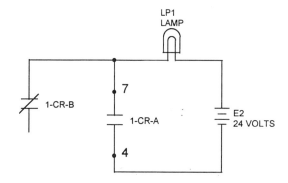

2.4 Turn off both power supplies. Modify the circuit shown in Figure 3.4 b) so that it operates like the circuit shown in Figure 3.5.

Figure 3.5: Load wired through the normally-closed contacts being controlled by the holding-relay circuit and energized by a higher voltage power supply.

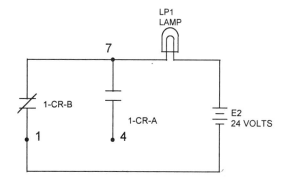

2.5 Turn on the power supplies. Press and release switch S2. Describe what happened, and explain how it happened.

_____

_____

_____

_____

_____

2.6 Press and release switch S1. Describe what happened, and explain how it happened.

_____

_____

_____

_____

_____

2.7 Turn off both power supplies. The coil in your relay is an inductor. As you may have already studied, inductors resist change in current. As a result, when the coil in any of the preceding circuits is turned off, there is an *inductive kick* associated with the collapsing magnetic field around the coil. To minimize damage or pitting of the switch contacts that are in series with the coil, manufacturers of D.C. powered relays will install a diode or a resistor (this is an ISO standard for automotive relays) in parallel with the coil. Construct the circuits shown in Figures 3.6 a) and b). Make sure that you correctly identify the anode and cathode of your diode when installing it. Demonstrate the circuit to your instructor. Explain its operation.

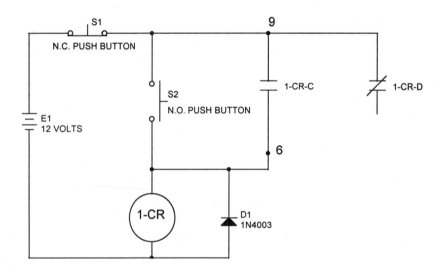

Figure 3.6a): Latching or holding-relay circuit with diode in parallel with the relay coil in order to minimize the effect of inductive kick.

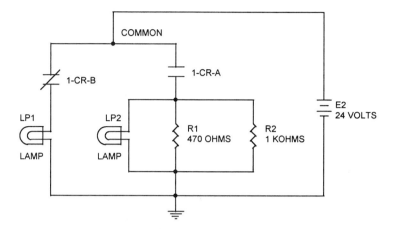

Students: Please note that while you can control more than one load with a single set of contacts, there are design rules that you must follow. As shown, your loads should be in parallel, not series. Also, the total current drawn by the loads should not be greater than the "current interrupting" capability of the contacts.

Figure 3.6 b): Practice circuit to demonstrate to instructor. This circuit demonstrates that more than one load can be controlled by a single relay contact.

PART 3: Hysteresis

Background:
Your control relay, like some other electronic devices and circuits you will study, exhibits the property of hysteresis. Devices that exhibit hysteresis turn on at one voltage level and turn off at another. The 74LS14 Schmitt-trigger inverter is a digital device that exhibits this property. In this part of the experiment, you will make current measurements demonstrating your relay's hysteresis characteristic.

Procedure:
3.1 With your power supply turned off and set for 0 volts, construct the circuit shown in Figure 3.7.

Figure 3.7: Circuit to determine control relay pull-in and drop-out currents.

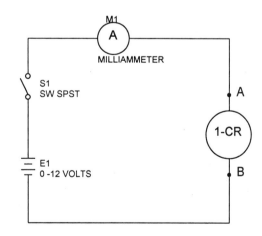

3.2 Turn on the power supply, and slowly increase the voltage until the relay just turns on (relay's armature comes into contact with the coil). Measure the current when the relay just turns on. This is referred to as the *pull-in* current. Record your measurement in Table 3.1.

3.3 Slowly decrease the power supply voltage until the relay just turns off. Measure the current when the relay just turns off. This is referred to as the *drop-out* current. Record your measurement in Table 3.1.

Table 3.1: Relay pull-in and drop-out currents.

PULL-IN CURRENT, mA	20
DROP-OUT CURRENT, mA	7.5

PART 4: Manufacturer's Ratings

4.1 Referring to your relay and/or a data manual for your relay, complete Table 3.2.

Table 3.2: Control relay specifications.

RELAY MANUFACTURER	
COIL VOLTAGE RATING	24v
CONTACTS RATING(S)	24v
RELAY TYPE (D.C. OR A.C.)	DC
NUMBER OF COMPLETE SETS OF CONTACTS	4
COIL RESISTANCE	64Ω
DUTY RATING (CONTINUOUS OR INTERMITTENT)	NA

4.2 If the rated coil resistance is not available in your data manual, you may measure the coil resistance by using an ohmmeter as shown in Figure 3.8.

Figure 3.8: Circuit to measure control relay coil resistance.

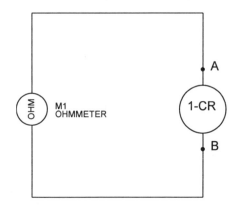

4.3 The *duty cycle* or *duty rating* of relays, solenoids, or magnetic contactors is often specified as either *intermittent* or *continuous*. A relay, solenoid, or any other magnetic device that is rated for continuous duty may be energized for long periods of time without overheating the coil and the subsequent damage that would otherwise result. Coils that have an intermittent rating, on the other hand, must be energized only for short periods of time, then turned off to allow for cooling. Coils with a continuous duty rating typically have more coils (turns or windings) of higher resistance wire. This results in less current draw and, in turn, less heat. The greater number of turns will still result in sufficient magnetomotive force to pull down the armature of a relay or pull in the plunger of a solenoid. If you are in doubt about a particular device's duty rating, contact the sales engineer for that particular manufacturer.

PART 5: Questions

5.1 What is the importance (value or use) of the information shown in Table 3.2 to an electronics technician? Explain at least two uses.

5.2 Why is the D.C. current (voltage) rating of a relay's contacts less than the A.C. current (voltage) rating of the contacts?

5.3 How does the construction of a relay whose coil is powered by A.C. differ from that of a relay whose coil is powered by D.C.? If necessary, make a sketch(es) to facilitate your explanation.

5.4 What is a shading coil (also known as a shading ring)?

5.5 Based on the results to step 1.7 in the procedure, is your relay a polarity-sensitive device? Explain your answer by referring to your experimental results.

5.6 Explain the difference between a switch or relay contact's "make current" and "break current" ratings. These two terms may also be referred to as "current at make" and "current at break".

PART 6: Troubleshooting

6.1 Assume that you are troubleshooting the circuits shown in Figures 3.2 a) and b). When you close switch S1, you hear the clicking sound associated with the closing of the contacts. However, when you close switch S2, the lamp does not turn on. What do you think is wrong? Describe how you would go about troubleshooting this circuit.

6.2 At the request of your supervisor you have just finished wiring the circuits shown in Figures 3.4 a) and b). You test your work by turning on both power supplies. Lamp LP1 is on, but it shouldn't be. Next you momentarily press then release the start button, S2. The light goes off and stays off. You press the stop button, S1, and the light comes back on. What do you think is wrong? Describe how you would go about troubleshooting this circuit.

6.3 At the request of your supervisor you have just finished wiring the circuits shown in Figures 3.4 a) and 3.5. You test your work by turning on both power supplies. Lamp LP1 is on. Next you momentarily press then release the start button, S2. The light momentarily goes out but then comes back on. What do you think is wrong? Describe how you would go about troubleshooting this circuit.

6.4 As part of a job assignment, you are to construct the circuits shown in Figures 3.6 a) and b). Having been well-schooled in the philosophy of build-in-stages, test-in-stages, you start by wiring push-button switches S1 and S2 in series with the coil. You turn on the power supply and press and hold S2. You hear the relay's armature move. You release the switch and observe the armature returning to its de-energized state. You turn off the power supply and wire contacts 1-CR-C in parallel with S2. You again turn on the power supply and press then release S2. The coil becomes energized and stays on. Successful so far, you turn off the power supply and install the diode in parallel with the coil. You again turn on the power supply and momentarily press then release S2. Nothing happens! So you again press S2, but this time you hold it down. You start to smell something, then your power supply's circuit breaker trips out. What's wrong? Comment on the ease or difficulty of troubleshooting this problem as compared to the preceding ones as a result of following the build-in-stages, test-in-stages philosophy.

PART 7: Practice Problems

7.1 Referring to your control relay data in Table 3.2, calculate the theoretical current that should flow in the coil after it is energized and reaches steady-state conditions.

7.2 Assuming the wire size in your control relay is American Wire Gage (AWG) 32, answer the following questions.

   a) What is the area of this wire in circular mils? Refer to the National Electric Code handbook or a set of wire tables.

   b) What is the diameter of the wire in mils?

c) What is the diameter of the wire in inches?

d) What is the resistance in Ohms per 1000 feet at 20°C for this wire?

e) Using the answer to part d) and the coil resistance from Table 3.2, calculate the length of the wire making up the coil in your control relay.

f) If the <u>average</u> length of each turn of wire in the coil of your control relay is 1.5 inches, how many <u>turns</u> are there in the coil?

g) The formula for the magnetomotive force (MMF) in a magnetic circuit is as follows:

$$MMF = (N) \cdot (I) \qquad 3.1$$

where N is the number of turns in the coil and I is the current through it. Calculate the MMF produced by your relay when operated at rated conditions.

PART 8: Circuit Design

8.1 Design a circuit that will meet the following design requirements: a) holding-relay circuit with push button operated start/stop control; b) when the start button is pressed, both a motor and a green light will be on; c) when the stop button is pressed, the motor and green light will go off and a red light will go on; d) the circuit will provide for *fail-safe* operation (ask your instructor what this means; the circuits shown in Figures 3.4, 3.5, and 3.6 provide fail-safe operation); e) the control relay is to be powered by a 12 volt D.C. source; f) the lamps and motor are to be powered by a standard 120 volts RMS sinusoidal source. Draw a schematic of this circuit showing all components in their normal or de-activated state. Label all components with proper identification, and attach your schematic to your lab report. If a computer-aided drafting software package is available at your institution, use it to produce your drawing. Print a copy and attach to your report. Finally, go to a manufacturer's web site such as www.midtex.com and find the part number of a control relay that will meet these specifications and handle at least 10 amps. Print a copy of the manufacturer's data sheet and attach to your report.

Transformer Basics: 4

INTRODUCTION:
The transformer is an electrical device based on the principles of electromagnetism. It is used to convert an alternating voltage waveform--typically sinusoidal--to an alternating waveform having either a larger or smaller amplitude. In applications such as the D.C. power supply at your lab station, a transformer is used to take the sinusoidal voltage of approximately 120 volts RMS from the wall outlet and reduce it down to a smaller sinusoidal voltage. A transformer wired to perform this operation is commonly referred to as a *step-down* transformer. This smaller A.C. voltage will then be rectified to a D.C. voltage. At your local electrical power-generating station, very large transformers are used to increase the voltage produced by steam-operated, turbine-driven, three-phase alternators to voltage levels of 100 kilovolts and higher. This voltage is then distributed to substations via overhead transmission lines. The transformers used to increase the alternator-produced voltage to these extremely high levels are referred to as *step-up* transformers. In this experiment, you will have the opportunity to investigate the voltage and current relationships in a center-tapped, step-down transformer.

SAFETY NOTE:
Before starting this experiment, have your instructor review with you the proper use of the oscilloscope to display the secondary voltages of a center-tapped transformer. It is important that you know how to use your oscilloscope to display the true voltage waveforms that appear across the secondary terminals of a center-tapped transformer. Incorrect use of the oscilloscope will result in either an incorrect waveform, such as displaying a floating voltage, or worse, causing a short circuit to ground through your oscilloscope. As always, wear safety glasses and remove all jewelry from your hands and fingers prior to the start of the experiment.

OBJECT:
Upon successful completion of this experiment and all reading assignments, the student should be able to:
- recognize and draw the schematic symbol for a center-tapped transformer
- identify the primary, secondary, and center-tap leads of a center-tapped transformer
- construct an operational circuit using a center-tapped transformer as a step-down transformer
- use an oscilloscope to display the secondary voltage waveforms in proper time phase for a center-tapped transformer
- experimentally determine the turns ratio for a given transformer
- determine the secondary voltage of a single-phase transformer given the input voltage and turns ratio
- determine the secondary current of a single-phase transformer for a known resistive load and known secondary voltage
- determine the primary current of a single-phase transformer given the secondary current and turns ratio

MATERIALS:
- 1 - 1 Kohms, 1/2 watt resistor
- 1 - 470 Ohms, 1/2 watt resistor
- 1 - 120:12.6 V_{RMS} center-tapped transformer
- 1 - fuse, 1/8 amp slow-blow
- 1 - in-line fuse holder
- 1 - solderless breadboard
- miscellaneous lead wires and connectors

EQUIPMENT:
1 - dual-trace oscilloscope
1 - handheld multimeter
2 - oscilloscope probes

PART 1: The Step-Down Transformer
Background:
The schematic symbol of a typical step-down transformer is shown in Figure 4.1. The left-hand side of the schematic is labeled as the *primary* or input side of the transformer. The right-hand side of the schematic is labeled as the *secondary* or output side of the transformer. The primary is constructed by wrapping many loops or *turns* of copper wire around a *core* of electrical steel. Electrical steel is an alloy of iron, carbon, and other elements that in combination produce desirable magnetic properties such as low *retentivity* and high *permeability*. The secondary is similarly connected with many loops of wire wrapped around a magnetically common core so that magnetic flux lines generated by the primary are linked to the secondary. The changing flux lines that flow from the primary to the secondary induce a voltage in the secondary.

Figure 4.1: Schematic symbol for a transformer showing primary and secondary windings.

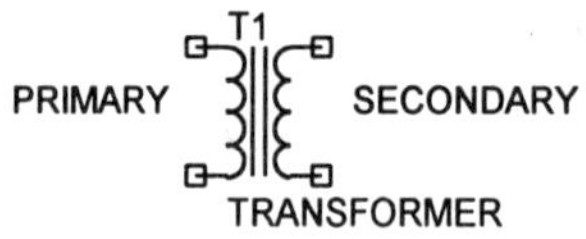

The magnitude of the secondary voltage depends mainly upon the voltage applied to the primary and the *turns ratio* of the transformer. We will refer to the number of loops or turns of wire wrapped around the primary as N_P and the number of turns comprising the secondary as N_S. The turns ratio of a transformer is then defined as the ratio of the number of turns on the primary to the number of turns on the secondary. Expressed mathematically, we have

$$\text{TURNS RATIO} = N_P/N_S \qquad \textbf{4.1}$$

For example, if a transformer is wrapped with 2000 turns of wire on the primary and 200 turns of wire on the secondary, the turns ratio is

$$2000/200 = 10/1$$

This ratio is typically expressed in the form 10:1. A transformer with more turns of wire on the primary than on the secondary will operate as a step-down transformer. Referring to Figure 4.2, note that a 120 volt RMS source is applied to a transformer having a turns ratio of 10:1. The resulting secondary voltage is 12 volts RMS. If you stop and think about it, you will discover that not only is the turns ratio 10:1 but the primary voltage and secondary voltage also have a 10:1 ratio. That leads us to another important transformer formula. The turns ratio of a transformer is also equal to the ratio of the primary voltage, V_P, to the secondary voltage, V_S.
Expressed mathematically, we have

$$\text{TURNS RATIO} = V_P/V_S \qquad \textbf{4.2}$$

Combining Formulas 4.1 and 4.2, we have

$$N_P/N_S = V_P/V_S \qquad 4.3$$

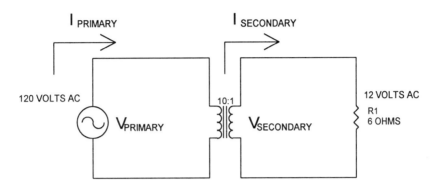

Figure 4.2: Step-down transformer circuit showing primary and secondary voltages and currents.

Thus far, the development of these formulas has been based on the assumption that a transformer is 100% efficient. In other words, we are assuming that the input or primary power, P_{IN}, equals the output or secondary power, P_{OUT}. While there are no man-made machines capable of 100% efficiency, a transformer, when operated at or near its rated load, is capable of very high efficiencies. For educational purposes, our assumption will produce very reasonable results. Recall that according to Watt's Law, electrical power is the product of voltage and current. Abbreviating the primary current as I_P and the secondary current as I_S, we have the following formulas for the input power and output power for a single-phase transformer:

$$P_{IN} = (V_P) \cdot (I_P) \qquad 4.4$$

and

$$P_{OUT} = (V_S) \cdot (I_S) \qquad 4.5$$

Combining Formulas 4.2, 4.4, and 4.5, we have the following ratio:

$$\text{TURNS RATIO} = I_S/I_P = V_P/V_S \qquad 4.6$$

Referring to Formula 4.6, notice the inverse relationship. In practical terms, this formula means that the <u>voltage</u> in a <u>step-down</u> transformer <u>is reduced</u> from the primary to the secondary by a factor equal to the turns ratio. However, the <u>current is increased</u> from the primary to the secondary by a factor equal to the turns ratio. Referring to our earlier example and Figure 4.2, we can apply Ohm's Law and determine the secondary current, which is 2 amps RMS.

Applying Formula 4.6, we can find the primary current as

$$I_P = I_S/(V_P/V_S)$$

$$= I_S/10$$

= 2 amps/10

= 0.2 amps

= 200 mA

A special type of step-down transformer is the center-tap transformer. A schematic of this type of transformer is shown in Figure 4.3.

Figure 4.3: Schematic symbol for a center-tapped transformer showing relationship of primary and secondary windings and center tap.

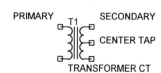

As you can see from the schematic, the center-tapped transformer has an additional lead or *tap* connected halfway between the first and last turn on the secondary. So, if we were to install a center tap on a transformer having 200 turns of wire wrapped on the secondary, we would place the center tap between turns 100 and 101. Even though we have added an additional lead, we can still use the preceding formulas to determine the currents and voltages in a center-tap transformer. Referring to Figure 4.4, we have modified the transformer in Figure 4.2 such that it now has a center tap. The overall turns ratio is still 10:1. In other words, there are still 2000 turns on the primary and 200 turns on the secondary. Therefore, we will still have 12 volts RMS across the <u>entire</u> secondary. Notice that across the top and bottom halves of the transformer secondary, we have exactly half of the total secondary voltage. This relationship can be explained in the following manner. First, we have 100 turns of wire in the top half of the secondary and 100 turns in the bottom half of the secondary. As a result, the ratio of primary turns to turns in the top half of the secondary becomes

TURNS RATIO = N_P/N_S TOP HALF OF SECONDARY

= 2000/100

= 20

The turns ratio for this part of the transformer is 20:1. In other words, the input voltage of 120 volts RMS is stepped down by a factor of 20. In other words,

SECONDARY VOLTAGE$_{TOP\ HALF}$ = $V_P/20$

= 120/20

= 6 volts RMS

The same line of thinking applies to the lower half of the transformer secondary. For *purely resistive* loads, the algebraic sum of the two secondary voltages will equal the voltage across the entire secondary. This type of transformer is similar in configuration to the transformer installed on the utility pole outside your home. That particular transformer steps the utility's line voltage down to two single-phase 120 volt

RMS sources and a combined 240 volt RMS single-phase source. Therefore, you have 120 volts available in your home for lights, radio, TV, etc., and 240 volts for electric ovens, hot water heaters, clothes dryers, etc.

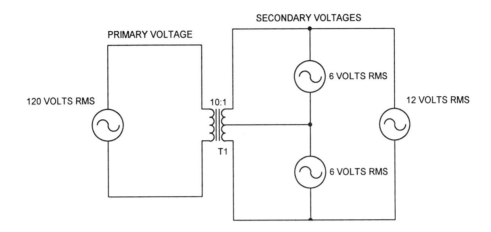

Figure 4.4: Center-tapped transformer circuit showing relationship of primary and secondary voltages.

Procedure:
1.1 Obtain all materials and equipment required to complete this experiment.
1.2 Use your handheld multimeter to measure the resistance of each fixed resistor and record in Table 4.1.

Table 4.1: Nominal and measured resistor values.

NOMINAL RESISTANCE	MEASURED RESISTANCE
470 OHMS	
1000 OHMS	

1.3 Take a moment to identify the primary, secondary, and center-tap leads of the transformer. Call your instructor if you need assistance or are in doubt.
1.4 Referring to Figure 4.5, use your handheld voltmeter to measure the voltages called for in Table 4.2 at the single-phase outlet at your work bench. Record your measurements in Table 4.2.

Figure 4.5: Schematic symbol for female three-wire plug.

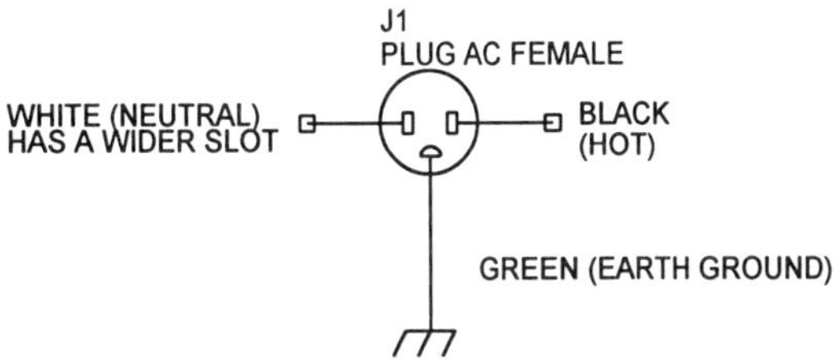

Table 4.2: Single-phase outlet voltages, RMS.

WALL OUTLET VOLTAGES	VOLTS RMS
HOT-TO-NEUTRAL	
HOT-TO-GROUND	
NEUTRAL-TO-GROUND	

1.5 Referring again to Figure 4.5, use your handheld voltmeter to measure the voltages called for in Table 4.3 at the single-phase outlet at your work bench. Record your measurements in Table 4.3.

FOR YOUR INFORMATION: The average or D.C. value of a sine wave (the form of electricity produced by most electric utilities) is 0 volts. As you probably expected, the D.C. value of the voltages measured at the single-phase outlet of your work bench were 0 volts. However, it is important for you to remember that in <u>some</u> circuits controlling industrial machinery, both A.C. and D.C. voltages may be combined (an A.C. waveform riding on top of a *D.C. offset*). So, when planning to work on any circuit, read the blueprint, schematic, or wiring diagram before making any measurements. Voltages such as these can be quite high and potentially hazardous! Special equipment such as a high-voltage probe may be required. Always follow established safety guidelines, and, if in doubt, ask for help from your supervisor or an experienced employee.

Table 4.3: Single-phase outlet voltages, D.C.

WALL OUTLET VOLTAGES	VOLTS D.C.
HOT-TO-NEUTRAL	
HOT-TO-GROUND	
NEUTRAL-TO-GROUND	

1.6 Before wiring your first circuit, be certain that the primary of your transformer has a polarized plug installed and that a 1/8 amp slow-blow fuse is installed in the hot lead of your transformer's primary circuit. Once your transformer is properly configured, you may proceed to the next step.

1.7 With the transformer unplugged, construct the circuit shown in Figure 4.6. Because you will not be using one of the secondary leads, insert it in an unused hole of your breadboard or locate it somewhere so that you will not inadvertently produce a short to one of the other leads. Have your instructor check your work.

1.8 Plug in the transformer. Use your handheld voltmeter to measure the A.C. voltage V_{CD}. This represents the voltage across load resistor R1. Record this voltage as $V_{SECONDARY}$ in Table 4.4. If there is no voltage present, check your meter to be sure that you have selected A.C. voltage and not D.C. voltage. Otherwise, unplug the transformer and check for a blown fuse. If the fuse is good, call your instructor for assistance. Have him or her explain an appropriate course of action for troubleshooting your circuit.

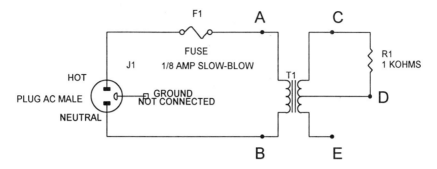

Figure 4.6: Center-tapped transformer circuit with load resistor connected in the upper half of the transformer secondary.

    1.9 Use the voltage measured from step 1.8 and the measured resistance value from Table 4.1 to calculate the current through R1, $I_{SECONDARY}$. Record in Table 4.4.

    Calculations:

    1.10 Use the primary voltage recorded in Table 4.2 and the measured secondary voltage from step 1.8 to calculate the turns ratio from the primary to the upper half of the secondary. Record in Table 4.4.

    Calculations:

    1.11 Use the value of the turns ratio just calculated and the secondary current recorded in Table 4.4 to calculate the current flowing in the primary windings, $I_{PRIMARY}$. Record in Table 4.4.

    Calculations:

Table 4.4: Experimental data for the circuit shown in Figure 4.6.

$V_{SECONDARY}$, VOLTS RMS	
$I_{SECONDARY}$, mA RMS	
Turns Ratio	
$I_{PRIMARY}$, mA RMS	

1.12 Unplug the transformer. Construct the circuit shown in Figure 4.7. Have your instructor check your work. Plug in the transformer, and use your hand-held voltmeter to measure the A.C. voltage V_{DE}. Record this voltage as $V_{SECONDARY}$ in Table 4.5.

1.13 Use the voltage measured in step 1.12 and the measured resistance value from Table 4.1 to calculate the current through R1, $I_{SECONDARY}$. Record in Table 4.5.

Calculations:

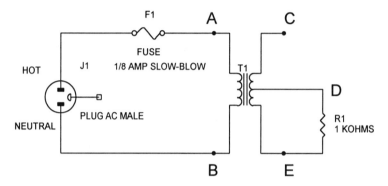

Figure 4.7: Center-tapped transformer circuit with load resistor connected in the lower half of the transformer secondary.

1.14 Use the primary voltage recorded in Table 4.2 and the measured secondary voltage from step 1.12 to calculate the turns ratio from the primary to the lower half of the secondary. Record in Table 4.5.

Calculations:

1.15 Use the value of the turns ratio just calculated and the secondary current recorded in Table 4.5 to calculate the current flowing in the primary windings, $I_{PRIMARY}$. Record in Table 4.5.

Calculations:

Table 4.5: Experimental data for the circuit shown in Figure 4.7.

V_{SECONDARY}, VOLTS RMS	
I_{SECONDARY}, mA RMS	
Turns Ratio	
I_{PRIMARY}, mA RMS	

1.16 Take a moment to compare the data in Table 4.4 to that in Table 4.5. If there is a significant difference in any of the corresponding sets of data, recheck your work!

1.17 Unplug the transformer. Construct the circuit shown in Figure 4.8. Have your instructor check your work. Plug in the transformer, and use your handheld voltmeter to measure the A.C. voltages V_{CE}, V_{CD}, and V_{DE}. Record these voltages in Table 4.6.

1.18 Use the value for V_{CE} recorded in Table 4.6 and the primary voltage recorded in Table 4.2 to calculate the turns ratio from the primary to the entire secondary. Record in Table 4.6.

Calculations:

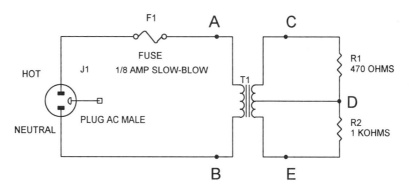

Figure 4.8: Center-tapped transformer circuit with load resistors connected in both halves of the transformer secondary.

Table 4.6: Experimental data for the circuit shown in Figure 4.8.

V_{CE}, VOLTS RMS	
V_{CD}, VOLTS RMS	
V_{DE}, VOLTS RMS	
Turns Ratio	

1.19 Unplug the transformer. Construct the circuit shown in Figure 4.9. Have your instructor check your work. Plug in the transformer, and use your handheld voltmeter to measure the A.C. voltages V_{CE}, V_{CD}, and V_{DE}. Record these voltages in Table 4.7.

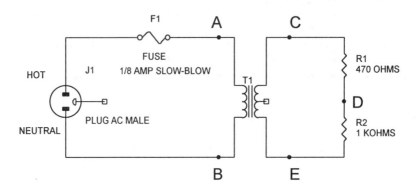

Figure 4.9: Center-tapped transformer circuit with load connected across the entire secondary and center tap disconnected.

Table 4.7: Experimental data for the circuit shown in Figure 4.9.

V_{CE}, VOLTS RMS	
V_{CD}, VOLTS RMS	
V_{DE}, VOLTS RMS	

1.20 Unplug the transformer. <u>Reconnect</u> your circuit as shown in Figure 4.8. Plug in the transformer. With the instructor's help, use the oscilloscope to display the voltage waveform across the entire secondary, V_{CE}, and the voltage across resistor R2, V_{DE}. Display both waveforms at the same time. Trigger on waveform V_{CE}. Display waveform V_{DE} in proper time phase with V_{CE}.
1.21 Draw both V_{CE} and V_{DE} in proper time phase with each other on Graph 4.1. Completely label the vertical and horizontal axes with voltage and time base values.
1.22 Now have your instructor show you the proper method to display the voltage across resistor R1, V_{CD}. This will involve using an isolation transformer and a two-to-three prong plug to isolate your transformer from the power line and ground. Or, you may use the subtract function on your oscilloscope to subtract V_{DE} from V_{CE} to get V_{CD}. On some oscilloscopes, this may involve the use of the *Add* and *Invert* functions. In any case, make sure you understand all safety precautions before you proceed. If in doubt, ask your instructor.

Graph 4.1: V_{CE} and V_{DE} voltage waveforms for the circuit shown in Figure 4.8.

V_{CE}: _____ volts/div _____ sec/div
V_{DE}: _____ volts/div _____ sec/div

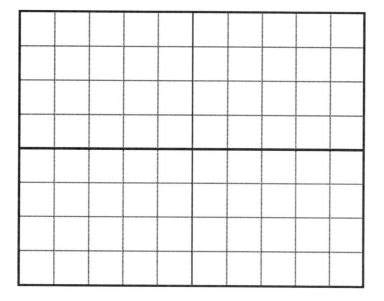

1.23 Unplug the transformer. Reconnect your circuit as shown in Figure 4.9, and plug in the transformer. With the instructor's help, use the oscilloscope to display the voltage waveform across the entire secondary, V_{CE}, and the voltage across resistor R2, V_{DE}. Display both waveforms at the same time. Trigger on waveform V_{CE}. Display waveform V_{DE} in proper time phase with V_{CE}.

1.24 Draw both V_{CE} and V_{DE} in proper time phase with each other on Graph 4.2. Completely label the vertical and horizontal axes with voltage and time base values.

1.25 Now have your instructor show you the proper method to display the voltage across resistor R1, V_{CD}. Draw the resulting waveform on Graph 4.2 in proper time phase with V_{CE} and V_{DE}.

1.26 Unplug the transformer.

1.27 Using the peak values from each of the three waveforms recorded on Graph 4.2, calculate the RMS value for each of these waveforms. Record your results in Table 4.8. Recall that the RMS value for a sine wave with no D.C. offset can be found using the following formula:

$$V_{RMS} = (0.707) \cdot (V_{PEAK}) \qquad \textbf{4.7}$$

Calculations:

Graph 4.2: V_{CE}, V_{CD}, and V_{DE} voltage waveforms for the circuit shown in Figure 4.9.

V_{CE}: _____ volts/div   _____ sec/div
V_{CD}: _____ volts/div   _____ sec/div
V_{DE}: _____ volts/div   _____ sec/div

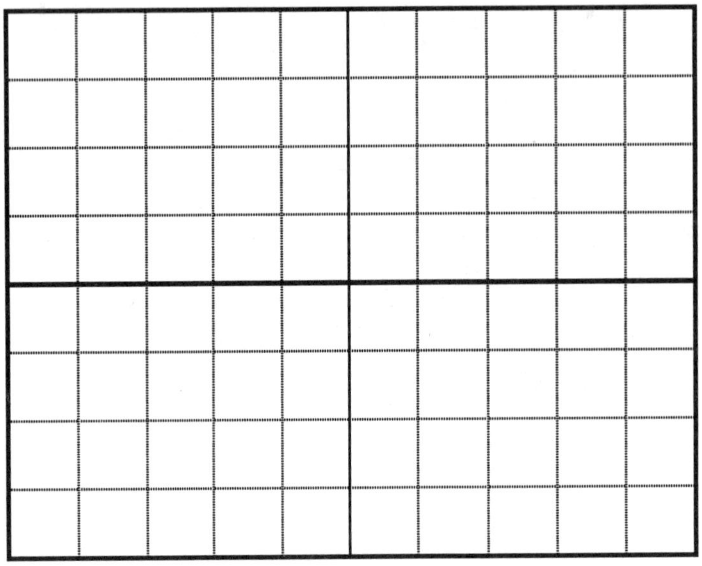

Table 4.8: Calculated RMS values for the circuit shown in Figure 4.9 using the peak voltages from Graph 4.2.

V_{CE}, VOLTS RMS	
V_{CD}, VOLTS RMS	
V_{DE}, VOLTS RMS	

PART 2: Manufacturer's Ratings

2.1 What is the rated primary voltage for your transformer?

2.2 What is the rated overall secondary voltage for your transformer?

2.3 What is the rated secondary current for your transformer?

2.4 How does your measured value for V_{CE} recorded in Table 4.6 compare with the manufacturer's rated secondary voltage? If they are not the same, explain any differences.

PART 3: Questions

3.1 What physical differences, if any, are there between the center tap and the other two secondary leads of your transformer? In other words, how did you determine which lead was the center tap?

3.2 Referring to Figure 4.8, what is the theoretical relationship between the voltages V_{CE} and V_{CD} for a center-tap transformer?

3.3 Do your experimental results recorded in Table 4.6 validate your answer to Question 3.2? Reconcile any differences.

3.4 Referring to Figure 4.8, what is the theoretical relationship between the voltages V_{CD} and V_{DE} for a center-tap transformer?

3.5 Do your experimental results recorded in Table 4.6 validate your answer to Question 3.4? Reconcile any differences.

3.6 Do any of the measured values from Table 4.7 differ from the calculated values in Table 4.8? Explain any differences.

3.7 Compare the values for V_{CD} and V_{DE} in Tables 4.6 and 4.7. Why are these voltages different from one table to the next?

PART 4: Practice Problems

4.1 Referring to the circuit shown in Figure 4.10, answer the following questions. Assume a step-down transformer with an applied voltage of 120 volts RMS and a turns ratio of 6:1. The load is a 10 Ohms resistor.

a) What is the secondary voltage?

b) What is the secondary current?

c) What is the primary current?

Figure 4.10: Transformer circuit for Practice Problem 4.1.

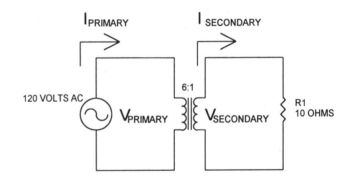

4.2 Referring to the circuit shown in Figure 4.11, answer the following questions. Assume a center-tapped, step-down transformer with an applied voltage of 120 volts RMS and an overall turns ratio from primary to secondary of 8:1. The load is a 1 Kohms resistor.

   a) What is the voltage across the load resistor, R1?

   b) What is the secondary current that flows through load resistor R1?

   c) What is the primary current?

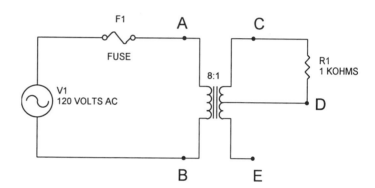

Figure 4.11: Transformer circuit for Practice Problem 4.2.

4.3 Referring to the circuit shown in Figure 4.12, answer the following questions. Assume a center-tapped, step-down transformer with an applied voltage of 120 volts RMS and an overall turns ratio from primary to secondary of 4:1. The load is a 220 Ohms resistor.

   a) What is the voltage across the load resistor R1?

   b) If you placed the leads of an A.C. voltmeter at points C and E in this circuit, what voltage would you expect to measure?

   c) If you placed the leads of an A.C. voltmeter at points C and D in this circuit, what voltage would you expect to measure?

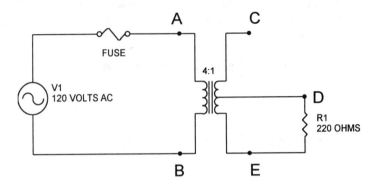

Figure 4.12: Transformer circuit for Practice Problem 4.3.

4.4 Referring to the circuit shown in Figure 4.13, answer the following questions. Assume a center-tapped, step-down transformer with an applied voltage of 2400 volts RMS and the voltage across the 24 Ohms load resistor R1 is 120 volts RMS.

a) What is the voltage across the load resistor R2?

b) If you placed the leads of an A.C. voltmeter at points C and E in this circuit, what voltage would you expect to measure?

c) If you placed the leads of an A.C. voltmeter at points D and E in this circuit, what voltage would you expect to measure?

d) What is the overall turns ratio from primary to secondary for this transformer?

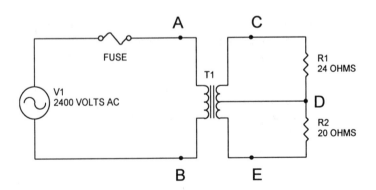

Figure 4.13: Transformer circuit for Practice Problem 4.4.

4.5 Referring to the circuit shown in Figure 4.14, answer the following questions. Assume a step-down transformer with an applied voltage of 480 volts RMS and a turns ratio from primary to secondary of 12:1.

a) If you placed the leads of an A.C. voltmeter at points C and E in this circuit, what voltage would you expect to measure?

b) If you placed the leads of an A.C. voltmeter at points C and D in this circuit, what voltage would you expect to measure?

c) What is the voltage across the load resistor R2?

Figure 4.14: Transformer circuit for Practice Problem 4.5.

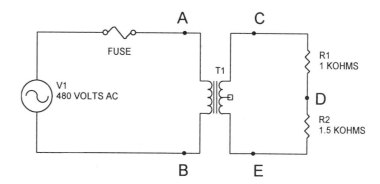

PART 5: Circuit Design

5.1 Design a circuit that will meet the following design requirements: a) holding-relay circuit with push button operated start/stop control; b) when the start button is pressed, both a motor and a green light will be on; c) when the stop button is pressed, the motor and green light will go off and a red light will go on; d) the circuit will provide for *fail-safe* operation (ask your instructor what this means; the circuits shown in Figures 3.4, 3.5, and 3.6 provide fail-safe operation); e) the control relay is to be powered by a 24 volt A.C. source; f) the lamps and motor are to be powered by a standard single-phase 120 volts RMS sinusoidal source; g) use a transformer to provide the 24 volts A.C. to the relay coil. Draw a schematic of this circuit showing all components in their normal or de-activated state. Label all components with proper identification. Include your schematic with your lab report. If a computer-aided drafting software package is available at your institution, use it to produce your drawing. Print a copy and attach it to your report.

PART 6: Transformer Formulas

TURNS RATIO = N_P/N_S **4.1**

TURNS RATIO = V_P/V_S **4.2**

$N_P/N_S = V_P/V_S$ **4.3**

$P_{IN} = (V_P) \cdot (I_P)$ **4.4**

$P_{OUT} = (V_S) \cdot (I_S)$ **4.5**

TURNS RATIO = $I_S/I_P = V_P/V_S$ **4.6**

$V_{RMS} = (0.707) \cdot (V_{PEAK})$ **4.7**

The Half-Wave Rectifier: 5

INTRODUCTION:

The rectifier diode is used in an unlimited number of applications. In industrial circuits, rectifier diodes are often used to convert a three-phase sinusoidal source into a high-power D.C. supply. The schematic symbol for a rectifier diode is shown in Figure 5.1 a). In an electronic circuit a diode operates much like a check valve does in a hydraulic circuit or in an automobile tire. A check valve, as shown in Figure 5.1 b), permits the flow of oil or air in one direction only. When putting air in one of your tires at the gasoline station, compressed air is pushed through the check valve into the inner walls of your tire. Inside the tire, back pressure seats the check valve seal in place, preventing the escape of air. Similarly, a diode permits the flow of electrons in one direction only. A resistor, on the other hand, is a bi-directional device permitting the flow of electrons in either direction. That is, a resistor is not polarity sensitive. Hence, we do not have to be concerned about inadvertently installing a resistor "backwards" in a circuit. However, because a diode is polarity sensitive, allowing electron flow only from cathode to anode, we must be careful to install or replace a diode in the orientation specified by the circuit designer. As shown in Figure 5.1 c), there is a small band or ring painted near the cathode lead of a diode. On high-power diodes, the schematic symbol of the diode may be imprinted on the side to indicate which end is the cathode and which end is the anode. The primary purpose of this lab is to observe the operational characteristics of a rectifier diode such as the 1N4003 in a half-wave rectifier circuit. Rectifier diodes such as the 1N4003 are primarily designed to operate in low-frequency circuits such as in 60 Hz single-phase or three-phase A.C. circuits. Diodes such as the 1N914, on the other hand, are designed to be used in high-frequency circuits because of their desirable high-speed switching characteristics. Before installing your rectifier diode in an A.C. circuit, you will have the opportunity to determine the D.C. operating characteristics of your diode.

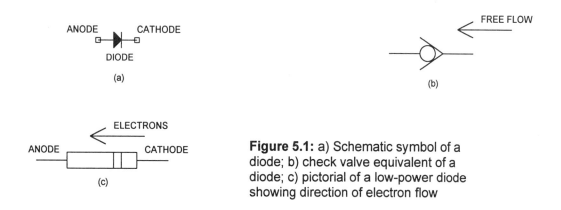

Figure 5.1: a) Schematic symbol of a diode; b) check valve equivalent of a diode; c) pictorial of a low-power diode showing direction of electron flow

SAFETY NOTE:

Before starting this experiment, have your instructor review with you the proper use of the oscilloscope to display the secondary voltages of a center-tapped transformer. It is important that you know how to use your oscilloscope to display the true voltage waveforms that appear across the secondary terminals of a center-tapped transformer. Incorrect use of the oscilloscope will result either in an incorrect waveform, such as displaying a floating voltage, or worse, causing a short circuit to ground through your oscilloscope. Also, if you are using a different transformer and/or diode than the one(s) specified in the materials list, have your instructor make sure that the diode you have selected will withstand the voltage and current it will be subjected to in all phases of your experiment. As always, wear safety glasses, and remove all jewelry from your hands and fingers before the start of the experiment.

OBJECT:
Upon successful completion of this experiment and all reading assignments, the student should be able to:
- recognize and draw the schematic symbol of a rectifier diode
- identify the anode and cathode of a rectifier diode
- experimentally determine the D.C. operating characteristics of a rectifier diode
- use a curve tracer (if available) to display the forward-biased I-V curves of a rectifier diode
- construct an operational half-wave rectifier circuit using a center-tapped transformer, load resistor, and rectifier diode
- use an oscilloscope to display the voltage across the load resistor and diode in a half-wave rectifier circuit in proper time phase with the secondary voltage of a step-down transformer
- calculate the D.C. value of the voltage across the load resistor in a half-wave rectifier circuit

REFERENCES:
www.motorola.com

MATERIALS:
- 1 - 10 Kohms ten, or twenty-turn trim pot
- 1 - 120:12.6 V_{RMS} center-tapped transformer
- 1 - fuse, 1/8 amp slow-blow
- miscellaneous lead wires and connectors
- 1 - rectifier diode, 1N4003 or an equivalent diode of equal or higher voltage and current ratings
- 1 - 1 Kohms, 1 watt resistor
- 1 - 220 Ohms, 1 watt resistor
- 1 - in-line fuse holder
- 1 - solderless breadboard

EQUIPMENT:
- 1 - analog or digital curve tracer (if available)
- 1 - dual-trace oscilloscope
- 1 - handheld multimeter
- 1 - digital ammeter
- 2 - oscilloscope probes

PART 1: D.C. Characteristics of the Rectifier Diode
Background:
As you may recall from your electronic devices class, a diode may be installed in a circuit to operate in one of two modes of operation—*forward*, or *reverse biased*. Figure 5.2 shows a circuit in which the diode is forward-biased. In this configuration, the diode will conduct. A small voltage drop—typically 0.7 volt for a silicon diode and 0.3 volt for a germanium diode—will be present across the diode with the remainder of the source voltage across the load resistor. Most rectifier diodes used in power supplies are silicon-based due to silicon's ability to withstand higher operating temperatures without breaking down. Please note, however, that it is a misconception to believe that all silicon diodes have a 0.7 volt drop when forward-biased. The actual forward-bias voltage drop, of course, depends on the amount of current flowing through the diode. Very high-power diodes typically have a forward voltage drop of more than 2 volts. However, the magnitude of this drop is small in comparison to the applied source voltage which may be several hundred volts. When operated in reverse-bias, a rectifier diode will act as an open, permitting at most a few microamperes of *reverse leakage current* through it. Figure 5.3 shows a circuit in which the diode is reverse-biased. In this phase of the experiment, you will construct the D.C. circuit shown in Figure 5.2 to determine the I-V characteristics of the 1N4003 diode when it is forward-biased. You will do this by varying the current through the diode and measuring the corresponding voltage drop across it.

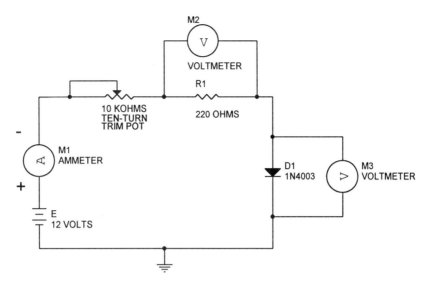

Figure 5.2: D.C. circuit to determine the forward-bias characteristics of a rectifier diode.

Procedure:

1.1 Obtain all materials and equipment required to complete this experiment.

1.2 Use your handheld multimeter to measure the resistance of each fixed resistor and record each in Table 5.1.

Table 5.1: Nominal and measured resistor values.

NOMINAL RESISTANCE	MEASURED RESISTANCE
220 OHMS	
1000 OHMS	

Figure 5.3: D.C. circuit to determine the reverse-bias characteristics of a diode.

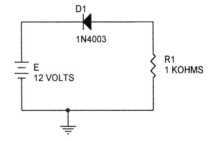

1.3 Construct the circuit shown in Figure 5.2. Notice that M2 and M3 are the <u>same</u> meter. Two voltmeters have been drawn to show the two circuit voltages that you will be measuring in this phase of the experiment. As a reminder, good voltage measurement techniques would prohibit using two voltmeters installed in a circuit simultaneously. Have your instructor review proper voltage measurement techniques using just <u>one</u> meter. Before turning on the power supply, have your instructor check your circuit.

NOTE TO INSTRUCTOR: If you do not have access to both an ammeter and a voltmeter, the students may use a voltmeter and Ohm's Law to obtain the current values shown in Table 5.2.

1.4 Vary the resistance of the 10 Kohms pot to obtain each of the current readings shown in Table 5.2. Measure and record the corresponding values for V_{R1} and V_D. Use Ohm's Law and the <u>measured</u> values for V_{R1} and R1 to <u>calculate</u> the series current and record it in Table 5.2. Food for thought: If the measured and calculated current values differ significantly, what should you do?

1.5 Once you have made all of the measurements for this circuit, turn off the power supply.

1.6 Using the measured current and V_D values from Table 5.2, plot the I-V curve for your diode in the space provided on Graph 5.1. Label both horizontal and vertical axes completely. Draw a smooth curve through the data points. As an alternative, your instructor may want you to use a computer software package to generate a plot of the data.

1.7 Construct the circuit shown in Figure 5.3. Use your handheld voltmeter to measure the voltage across the diode and the voltage across load resistor R1. Record your results in Table 5.3. Use Ohm's Law and the measured values for R1 and V_{R1} to calculate the circuit current. Record in Table 5.3.

1.8 Once you have made all of the measurements for this circuit, turn off the power supply.

Table 5.2: Forward bias I-V curve data for the diode shown in Figure 5.2.

E VOLTS	V_{R1} VOLTS	V_D VOLT	I_{DIODE} mA	$I_{CALCULATED}$ mA
12			2	
12			4	
12			6	
12			8	
12			10	
12			15	
12			20	
12			25	
12			30	
12			40	
12			50	

Graph 5.1: Plot of experimental data for the diode shown in Figure 5.2.

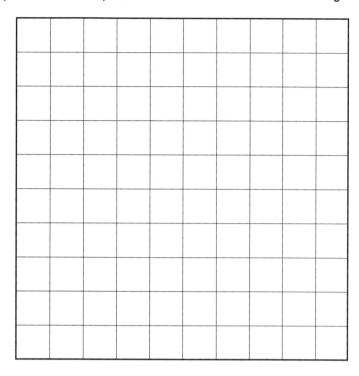

Table 5.3: Reverse bias I-V curve data for the diode shown in Figure 5.3.

V_{R1} VOLTS	V_D VOLTS	$I_{CALCULATED}$ mA

PART 2: The Curve Tracer

Background:
A curve tracer is an electronic instrument similar in many ways to an oscilloscope. While an oscilloscope is most often used to observe alternating voltage waveforms, a curve tracer is used to display the I-V characteristics of a wide range of electronic devices, such as the diode, transistor, SCR, etc. On an oscilloscope, you set the volts/div and sec/div amongst other controls to obtain a clear view of an alternating waveform. With the curve tracer, you have to choose appropriate volts/div on the horizontal axis and current/div on the vertical to produce a desirable I-V curve. If you are not already familiar with this procedure and the other controls on the curve tracer, have your instructor give you a demonstration.

Procedure:
2.1 Use a curve tracer to generate the forward-bias I-V curve for the diode that you used in Part 1 of this experiment. Select volts/div and current/div to produce a curve similar to the one you plotted in Graph 5.1. If your lab does not have a curve tracer, proceed to Part 3 of this experiment.

2.2 If you have a digital curve tracer and it is connected to a printer, make a copy of the diode's I-V curve and attach it to your lab report. If you do not have a printer or you are using an analog curve tracer, draw the I-V curve on Graph 5.2. Label both horizontal and vertical axes.

Graph 5.2: Forward-bias I-V curve data from curve tracer.

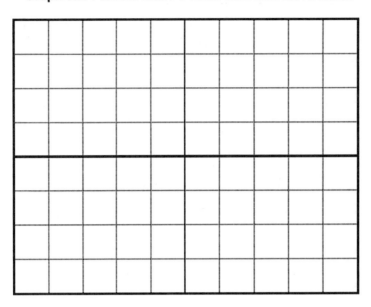

PART 3: The Half-Wave Rectifier

Background:

The forward, and reverse-bias characteristics of a rectifier diode (as discussed in Part 1 of this experiment) make it an excellent choice for converting A.C. sources to D.C. A typical circuit is shown in Figure 5.4. In a half-wave rectifier circuit, the diode will be forward-biased—dropping approximately 0.7 volt—during the positive alternation of the sine wave source and reverse-biased—acting as an open— during the negative alternation of the source. Current will then be permitted to flow through the load resistor only during the positive alternation of the source. This results in only a positive or D.C. voltage appearing across the resistor. The D.C. or average value of the voltage waveform that appears across the resistor may be found from the following formula:

$$V_{D.C.\ HALF-WAVE} = V_P/\pi \qquad \textbf{5.1}$$

V_P represents the peak of the voltage waveform across the load resistor. And, of course, $\pi = 3.14159$. The general formula for finding the average value of a waveform can be developed using calculus. In general terms, a waveform's average value can be found by dividing the total area under the curve by the total length of the curve. The average value for a half-wave rectified sine wave then is the area under the curve, which is $2V_P$ divided by the total length of the curve, 2π. This results in Formula 5.1.

Procedure:

3.1 If you have not already completed Experiment 4, Transformer Basics, have your instructor help you with the identification of the primary and secondary leads of the transformer. Once you are completely familiar with the transformer's construction and operation, proceed to step 3.2.

3.2 Have your instructor review with you the proper, hence safe, methods to display the voltages across the secondary of a transformer on your oscilloscope. Use D.C. coupling for all measurements made with your oscilloscope.

3.3 Construct the circuit shown in Figure 5.4. Once complete, have your instructor check your work.

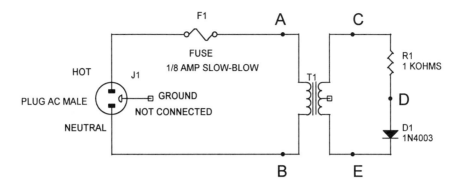

Figure 5.4: Center-tapped transformer with load resistor and rectifier diode arranged to produce a half-wave rectifier circuit.

    3.4 Plug in the transformer. With the instructor's help, use the oscilloscope to display the voltage waveform across the entire secondary, V_{CE}, and the voltage across the diode, V_{DE}. Display both waveforms at the same time. Trigger on waveform V_{CE}. Display waveform V_{DE} in proper time phase with V_{CE}. Draw waveform V_{CE} on Graph 5.3, and completely label the vertical and horizontal axes with voltage and time base values. Draw waveform V_{DE} on Graph 5.4 in proper time phase with V_{CE}, and completely label the vertical and horizontal axes with voltage and time base values.

    3.5 Now have your instructor show you how to display the voltage waveform across the resistor, V_{CD}. Draw waveform V_{CD} on Graph 5.5 in proper time phase with V_{CE}. Completely label the vertical and horizontal axes with voltage and time base values.

    3.6 Use your handheld voltmeter to measure the D.C. voltage across the load resistor. Record your measurement in Table 5.4.

    3.7 Unplug the transformer.

    3.8 Using the peak value of the load resistor waveform from Graph 5.5, calculate the average or D.C. value of that waveform. Record your result in Table 5.4.

    Calculations:

Graph 5.3: V_{CE} voltage waveform for the circuit shown in Figure 5.4.

V_{CE}: _____ volts/div _____ sec/div

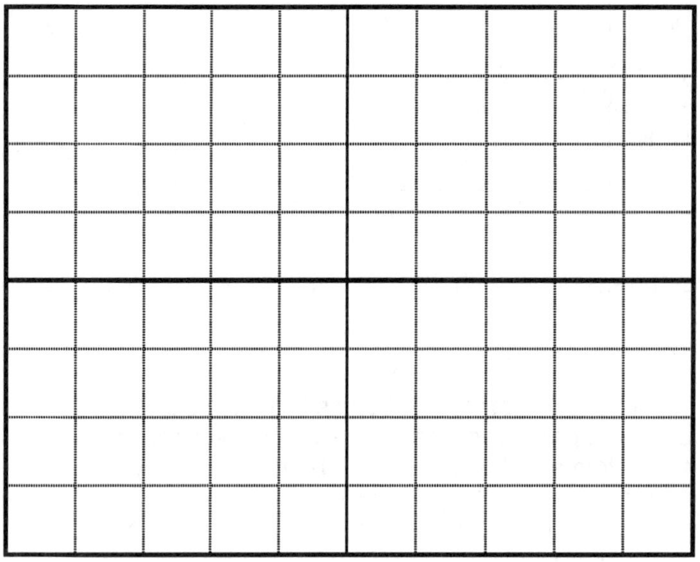

Graph 5.4: Diode voltage waveform, V_{DE}, for the circuit shown in Figure 5.4.

V_{DE}: _____ volts/div _____ sec/div

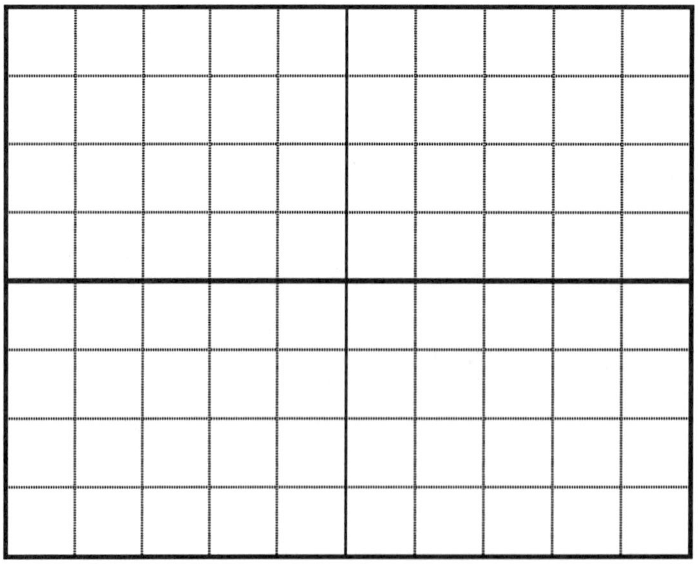

Graph 5.5: Load resistor voltage waveform, V_{CD}, for the circuit shown in Figure 5.4.

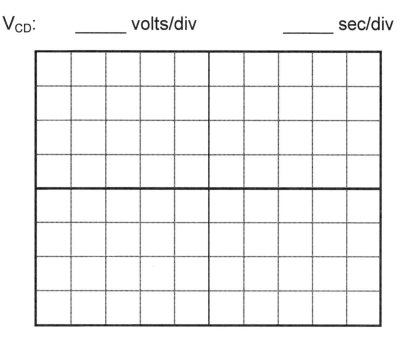

V_{CD}: _____ volts/div _____ sec/div

Table 5.4: Measured and calculated D.C. voltage values for the load resistor in the circuit shown in Figure 5.4.

V_{R1} MEASURED, VOLTS	
V_{R1} CALCULATED, VOLTS	

PART 4: Troubleshooting the Half-Wave Rectifier

Background:
While circuit failures may take many forms, such as opens, shorts, voltage levels drifting with time and/or temperature, intermittent failures, etc., most rectifier problems can be classified either as a short or an open. In this part of the experiment, you will simulate an open, then a short, in a rectifier diode. Remember that the symptoms exhibited by a short circuit include low voltage across, and high current through, the failed component. On the other hand, an open circuit will have high voltage across the open and no current through it.

Procedure:
4.1 Construct the circuit shown in Figure 5.5 by removing the diode from your circuit. This will simulate an open.

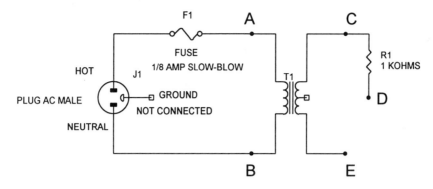

Figure 5.5: Half-wave rectifier circuit with an open diode.

4.2 Plug in the transformer. Use the oscilloscope to display the voltage waveform across the the open diode, V_{DE}. Trigger on the source waveform V_{CE}. Draw waveform V_{DE} on Graph 5.6 in proper time phase with V_{CE}, and completely label the vertical and horizontal axes with voltage and time base values.

4.3 Now display the voltage waveform across the resistor, V_{CD}. Draw waveform V_{CD} on Graph 5.7 in proper time phase with V_{CE}. Completely label the vertical and horizontal axes with voltage and time base values.

4.4 Unplug the transformer.

4.5 Construct the circuit shown in Figure 5.6 by replacing the diode with a jumper wire from point D to E. This will simulate a short.

4.6 Plug in the transformer. Use the oscilloscope to display the voltage waveform across the shorted diode, V_{DE}. Trigger on the source waveform V_{CE}. Draw waveform V_{DE} on Graph 5.8 in proper time phase with V_{CE}, and completely label the vertical and horizontal axes with voltage and time base values.

4.7 Now display the voltage waveform across the resistor, V_{CD}. Draw waveform V_{CD} on Graph 5.9 in proper time phase with V_{CE}. Completely label the vertical and horizontal axes with voltage and time base values.

4.8 Unplug the transformer.

Graph 5.6: V_{DE} voltage waveform for the circuit shown in Figure 5.5.

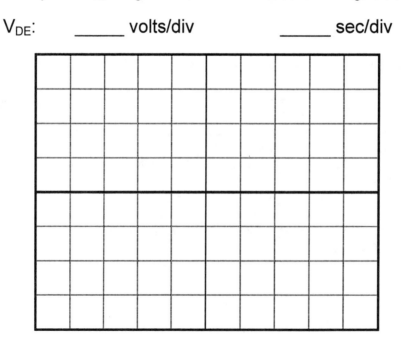

Graph 5.7: Load resistor voltage waveform, V_{CD}, for the circuit shown in Figure 5.5.

V_{CD}: _____ volts/div   _____ sec/div

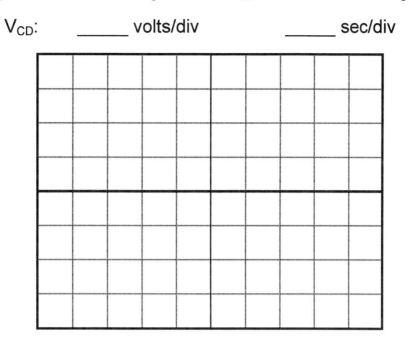

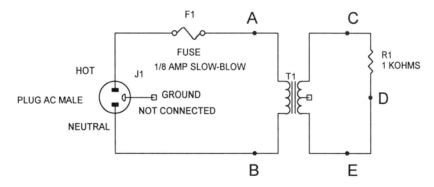

Figure 5.6: Half-wave rectifier circuit with shorted diode.

Graph 5.8: V_{DE} voltage waveform for the circuit shown in Figure 5.6.

V_{DE}: _____ volts/div   _____ sec/div

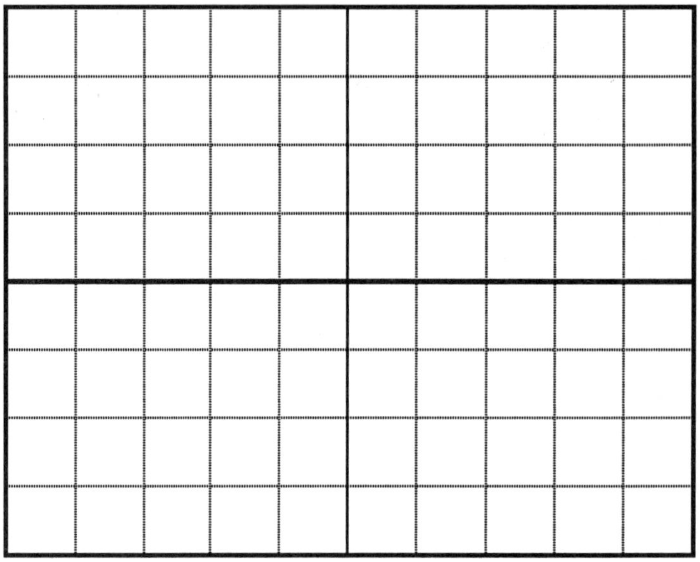

Graph 5.9: Load resistor voltage waveform, V_{CD}, for the circuit shown in Figure 5.6.

V_{CD}: _____ volts/div   _____ sec/div

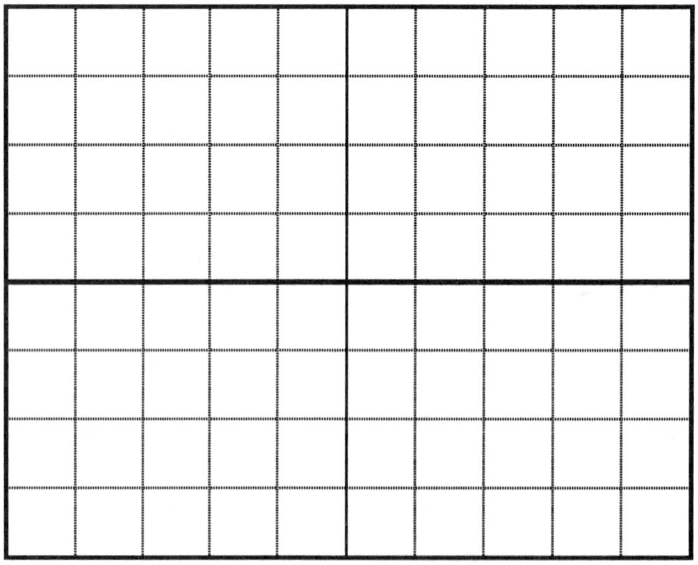

PART 5: Manufacturer's Ratings

5.1 Referring to the manufacturer's data sheet for the 1N4003 diode, complete Table 5.5.

Table 5.5: Manufacturer's data for the 1N4003 diode.

DIODE	V_{RRM} VOLTS	V_{RSM} VOLTS	I_O AMPS	I_{FSM} AMPS	T_J, T_{stg} deg. C
1N4003					

PART 6: Questions

6.1 Referring to your experimental data in Table 5.2, is your diode constructed from silicon or germanium? Explain your answer. Refer to your data as needed.

6.2 Referring to your experimental data in Table 5.3, do your results confirm what you know about a reverse-biased diode? Explain your answer. Refer to your data as needed.

6.3 How does your plot in Graph 5.1 of the experimental data you collected from the circuit in Figure 5.2 compare to the I-V curve that you produced on the curve tracer?

6.4 How do the measured and calculated D.C. values from Table 5.4 compare? Explain any differences.

6.5 Why was the voltage waveform for V_{CD} for the circuit shown in Figure 5.5 a flat line? Include Ohm's Law in your explanation.

6.6 What D.C. voltage would you expect to measure across the load resistor in Figure 5.5?

6.7 What D.C. voltage would you expect to measure across the open in Figure 5.5?

6.8 What is the rated secondary voltage of your transformer?

6.9 What A.C. voltage would you expect to measure across the load resistor in Figure 5.5?

6.10 What A.C. voltage would you expect to measure across the open in Figure 5.5?

6.11 Explain the resulting voltage waveform for V_{CD} for the circuit shown in Figure 5.6.

6.12 What D.C. voltage would you expect to measure across the load resistor in Figure 5.6?

6.13 What D.C. voltage would you expect to measure from point D to E in Figure 5.6?

6.14 What A.C. voltage would you expect to measure across the load resistor in Figure 5.6?

6.15 What A.C. voltage would you expect to measure across the short in Figure 5.6?

6.16 Describe what would happen to the circuit of Figure 5.4 if the load resistor R1 were to become shorted. Also discuss what circuit components you would most likely have to replace.

6.17 Describe how you would go about troubleshooting a circuit such as Figure 5.4 if you suspected that it had a shorted resistor. Discuss what symptoms you would look for and what voltages you would measure if your only measuring instrument was a handheld voltmeter capable of measuring both A.C. and D.C. voltages.

6.18 Describe how you would go about troubleshooting a circuit such as Figure 5.4 if you suspected that it had an open resistor. Discuss what symptoms you would look for and what voltages you would measure if your only measuring instrument was a handheld voltmeter capable of measuring both A.C. and D.C. voltages.

6.19 Define and explain what each of the following abbreviations from Table 5.5 means.

    a) V_{RRM}

    b) V_{RSM}

    c) I_O

    d) I_{FSM}

    e) T_J, T_{stg}

PART 7: Practice Problems

7.1 Referring to the circuit shown in Figure 5.7, answer the following questions. Assume a step-down transformer with an applied voltage of 120 volts RMS and a turns ratio of 6:1. The load is a 10 Ohms resistor.

a) What is the secondary voltage, V_{CE}? Express your answer in RMS units. If necessary, review the formulas at the end of Experiment 4.

b) What is the peak value of the secondary voltage V_{CE}?

c) Based on your answer to part b) and assuming that the diode is silicon with a nominal forward-bias voltage drop of 0.7 volt, what is the peak voltage across the load resistor?

d) Based on your answer to part c), what is the average value of the voltage across the load resistor?

e) Based on your answer to part d), what is the value of the current through the load resistor?

f) Based on your answer to part e), would a 1N4003 diode operate properly in this circuit or would it fail? Explain your answer while referring to the manufacturer's specifications.

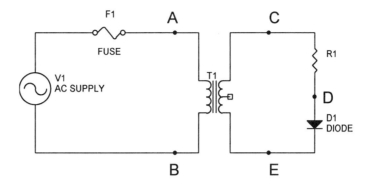

Figure 5.7: Half-wave rectifier circuit for practice problems.

7.2 Referring to the circuit shown in Figure 5.7, answer the following questions. Assume a step-down transformer with an applied voltage of 2400 volts RMS and a turns ratio of 10:1.

   a) Would a 1N4003 diode operate properly in this circuit, or would it fail? Explain your answer while referring to the manufacturer's specifications. Show all calculations.

   b) Assuming a factor of safety of 2, what diode would you select to be installed in this circuit if the load resistor is a 220 Ohms resistor? Your choices include the 1N4001, 1N4002, 1N4003, 1N4004, 1N4005, 1N4006, and the 1N4007. Explain your choice based on the manufacturer's specifications. Include all calculations necessary to solve this problem.

PART 8: Electronics Workbench®

8.1 Use Electronics Workbench® to create the circuit shown in Figure 5.4. Replace the male plug with a 120 volts RMS sinusoidal source. You do not have to include the fuse. Select a transformer turns ratio of 120:12.6. Select a silicon rectifier diode that has electrical properties as close as possible to the one used in your lab experiment. Use the oscilloscope function to display the voltage waveforms at points C and D. Print a copy of your schematic diagram and oscilloscope waveforms. How do these waveforms compare to your experimental results recorded in Graphs 5.3 and 5.4? Explain any differences, especially the secondary voltage.

PART 9: Half-Wave Formulas

$V_{D.C.\ HALF-WAVE} = V_P/\pi$ **5.1**

The Full-Wave Rectifier: 6

INTRODUCTION:

Experiment 5 involved constructing a functional single-phase, half-wave rectifier circuit. In this experiment, you will have the opportunity to construct two different full-wave rectifier circuits and compare their principles of operation, but first, let's consider the need for a full-wave rectifier. If you completed Experiment 5, you realize that the half-wave rectifier does not produce a smooth D.C. voltage. This is unacceptable for many applications. The full-wave rectifier circuits you are to construct in this experiment will help us approach the development of a D.C. power supply that will produce a smoother D.C. voltage with less *ripple*. Your first full-wave rectifier will be constructed using a step-down transformer connected to a bridge rectifier. The schematic symbol for a bridge rectifier is shown in Figure 6.1. Note that it is constructed from four diodes. Depending on the design requirements, the bridge rectifier may be constructed from four discrete diodes or the diodes may be integrated into one complete package. Referring to the pictorials of the bridge rectifier packages shown in Figure 6.2, note that there are four terminals. Two are for the single-phase A.C. inputs and the other two are for the (+) and (-) D.C. outputs. Always take your time to carefully identify the inputs and outputs of a bridge rectifier when installing or replacing one. Sometimes the only marking present is a small red dot near one corner, indicating the (+) output. The (-) output then is diagonally across from the (+) output. The other two pins then are the hot and neutral A.C. inputs.

The other full-wave rectifier circuit that you will construct uses a center-tap transformer with two diodes. A typical circuit is shown in Figure 6.8. At this stage, you might be saying to yourself, what is the benefit of using a center-tap transformer when a simple transformer with no centertap and a bridge rectifier can produce a full-wave output? There are several answers. One is that the center-tap transformer can provide us with more than one A.C. voltage level at the secondary, and hence, as with power supply transformers having many different taps on the secondary, more than one D.C. level can be produced. A second important use of the center-tap transformer is that it can be used to produce both (+) and (-) D.C. voltages from the same transformer. The primary purpose of this experiment is to study the operational characteristics of two common single-phase, full-wave rectifier circuits. As with Experiment 5, you will also have the chance to observe the effect that a failed diode has on the operation of these circuits.

Figure 6.1: Schematic symbol of a bridge rectifier.

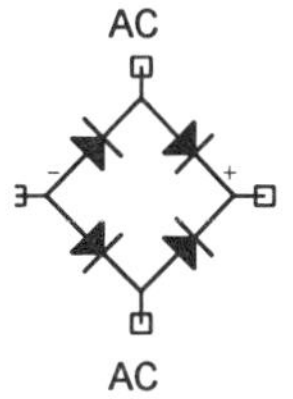

BRIDGE RECTIFIER

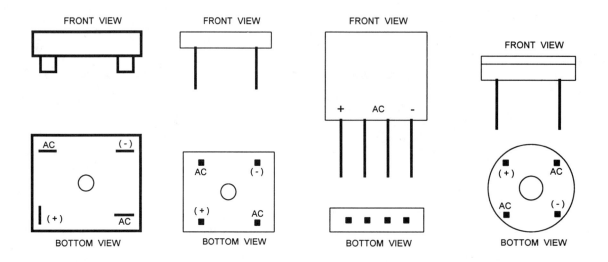

Figure 6.2: Pictorials of common bridge rectifier packages.

SAFETY NOTE:
Before starting this experiment have your instructor review with you the proper use of the oscilloscope to display the secondary voltages of a center-tapped transformer. It is important that you know how to use your oscilloscope to display the true voltage waveforms that appear across the secondary terminals of a center-tapped transformer. Incorrect use of the oscilloscope will result either in an incorrect waveform, such as displaying a floating voltage, or worse, causing a short circuit to ground through your oscilloscope. Also, if you are using a different transformer, diodes, or bridge rectifier than the one(s) specified in the Materials list, have your instructor make sure that the diodes and bridge rectifier that you have selected will withstand the voltage and current each will be subjected to in all phases of your experiment. As always, wear safety glasses, and remove all jewelry from your hands and fingers before the start of the experiment.

OBJECT:
Upon successful completion of this experiment and all reading assignments, the student should be able to:
- recognize and draw the schematic symbol of a bridge rectifier
- identify the inputs and outputs of a bridge rectifier
- construct an operational full-wave rectifier circuit using a step-down transformer, load resistor, and bridge rectifier package
- construct an operational full-wave rectifier circuit using a center-tapped transformer, load resistor, and two rectifier diodes
- use an oscilloscope to display the voltage across the load resistor and diodes in a full-wave rectifier circuit in proper time phase with the secondary voltage of a step-down transformer
- calculate the D.C. value of the voltage across the load resistor in a full-wave rectifier circuit
- troubleshoot and diagnose shorted and open components in full-wave rectifier circuits

MATERIALS:
- 1 - 1 Kohms, 1 watt resistor
- 4 - rectifier diodes, 1N4003 or an equivalent diode of equal or higher voltage and current ratings
- 1 - bridge rectifier, such as the SK3985 or SK5042 or an equivalent rectifier of equal or higher voltage and current ratings
- 1 - 120:12.6 V_{RMS} center-tapped transformer
- 1 - fuse, 1/8 amp slow-blow
- 1 - in-line fuse holder
- 1 - solderless breadboard

- miscellaneous lead wires and connectors

EQUIPMENT:
1 - dual-trace oscilloscope
1 - handheld multimeter
2 - oscilloscope probes

PART 1: The Full-Wave Bridge Rectifier
Background:
A full-wave bridge rectifier circuit is shown in Figure 6.3. The orientation of the diodes in the bridge is such that the flow of electrons will always be routed through the load resistor in the same direction regardless of the polarity of the A.C. source (transformer secondary in this case). The positive peak of the secondary's A.C. waveform will cause diodes D2 and D3 to be forward-biased, allowing current to flow through the load resistor. At the same time diodes D1 and D4 will be reverse-biased, acting as opens. The opposite will happen during the negative alternation of the secondary's A.C. waveform. Diodes D2 and D3 will be reverse-biased while diodes D1 and D4 are forward-biased. As already pointed out, current will flow through the resistor in the *same* direction during both the positive alternation and the negative alternation of the A.C. supply. The D.C. or average value of the voltage waveform that appears across the resistor may be found from the following formula:

$$V_{D.C.\ FULL-WAVE} = 2V_P/\pi \qquad \textbf{6.1}$$

V_P represents the peak of the voltage waveform across the load resistor. V_P will be approximately equal to the peak of the secondary voltage minus the voltage drop across the two forward-biased diodes (approximately 1.4 volts = 0.7 + 0.7). And, of course, π = 3.14159. The general formula for finding the average value of a waveform can be developed using calculus. In general terms, a waveform's average value can be found by dividing the total area under the curve by the total length of the curve. The average value for a full-wave rectified sine wave then is the area under the curve, which is $2V_P$ divided by the total length of the curve, π radians. This results in Formula 6.1. If you compare Formula 6.1 to Formula 5.1 (refer to Experiment 5), you should note that the full-wave voltage will be double that of the half-wave voltage for the same peak voltage. The average load current will then be

$$I_{AVG\ LOAD} = (2V_P/\pi)/R \qquad \textbf{6.2}$$

Because the diodes in the circuit are in series with the load but conduct only half of the time and are open the other half of the time, the average current flow through the diodes is

$$I_{AVG\ DIODE} = I_{AVG\ LOAD}/2 \qquad \textbf{6.3}$$

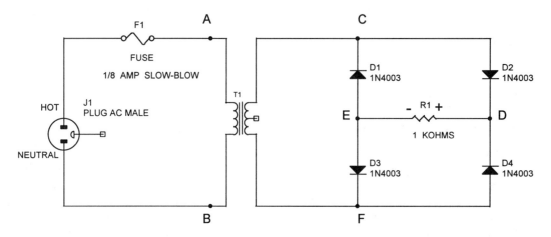

Figure 6.3: Full-wave bridge rectifier circuit constructed from discrete components and a step-down transformer.

The repetitive reverse voltage, V_{RRM}, that each diode will be exposed to is the peak of the full secondary waveform. You can confirm this by applying Kirchhoff's Voltage Law to the closed loop from point C to E to F in Figure 6.3. During the positive alternation of the secondary voltage, for example, diode D3 will be forward-biased while diode D1 is reverse-biased, having to withstand the entire secondary voltage while it is acting as an open. A similar analysis can be made for the outermost loop from point C to D to F. Hence, each diode's V_{RRM} rating must be <u>at least</u> equal to the peak of the secondary's A.C. waveform, V_P. As with any circuit design, an appropriate factor of safety should be chosen to prevent component failure due to unexpectedly high voltages or currents. If a circuit may be exposed to unexpectedly high voltages or currents (and almost all circuits that are ever turned on certainly will), then appropriate fuse, circuit breaker, and over-voltage (surge) protection devices must be installed under the supervision of a professional engineer.

Procedure:

1.1 Obtain all materials and equipment required to complete this experiment.
1.2 Use your handheld multimeter to measure the resistance of the 1 Kohms resistor and record in Table 6.1.

Table 6.1: Nominal and measured resistor values.

NOMINAL RESISTANCE	MEASURED RESISTANCE
1000 OHMS	

1.3 If you have not already completed either Experiment 4 or 5, have your instructor help you with the identification of the primary and secondary leads of the transformer. Once you are completely familiar with the transformer's construction and operation, proceed to step 1.4.
1.4 Have your instructor review with you the proper, hence safe, methods to display the voltages across the secondary of a transformer on your oscilloscope. Use D.C. coupling for all measurements made with your oscilloscope.
1.5 Construct the circuit shown in Figure 6.3. Once complete, have your instructor check your work.
1.6 Plug in the transformer. With the instructor's help, use the oscilloscope to display the voltage waveform across the entire secondary, V_{CF}. Draw waveform V_{CF} on Graph 6.1. Completely label the vertical and horizontal axes with voltage and time base values.

INSTRUCTOR NOTE: Please be certain that your students have mastered all proper safety procedures for using the oscilloscope (both portable and plug-in styles) when displaying single-phase and three-phase voltage sources at the primary and secondary of both low- and high-voltage transformers. Your presentation should include a discussion of proper grounding techniques, when to float an oscilloscope off-line, and the selection of high voltage oscilloscope probes. Students should also have mastered safe techniques for measuring high voltages with a handheld multimeter and high-voltage probe.

1.7 Now have your instructor show you how to display the voltage waveform across the resistor, V_{DE}. For safety reasons, you must follow the specific procedures outlined by your instructor when completing this step. This is an important learning event that will be very important in your future should you become employed in a position requiring you to use the oscilloscope to troubleshoot and/or install high-voltage equipment. Draw waveform V_{DE} on Graph 6.2 in proper time phase with V_{CF}, and completely label the vertical and horizontal axes with voltage and time base values. The resulting waveform should be a full-wave rectified waveform. If it is not, then you either have not followed proper oscilloscope measuring procedures and/or you have a faulty diode(s) in your bridge circuit. Should this unfortunate event happen, unplug your transformer and call your instructor for assistance.

1.8 Use your hand-held voltmeter to measure the D.C. voltage across the load resistor. Record your measurement in Table 6.2.

1.9 Unplug the transformer.

Graph 6.1: V_{CF} voltage waveform for the circuit shown in Figure 6.3.

V_{CF}: _____ volts/div   _____ sec/div

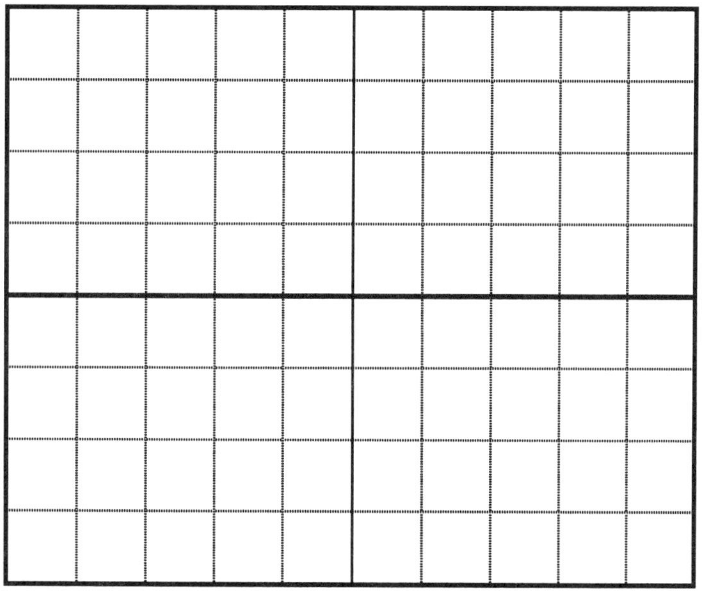

Graph 6.2: Load voltage waveform, V_{DE}, for the circuit shown in Figure 6.3.

V_{DE}: _____ volts/div _____ sec/div

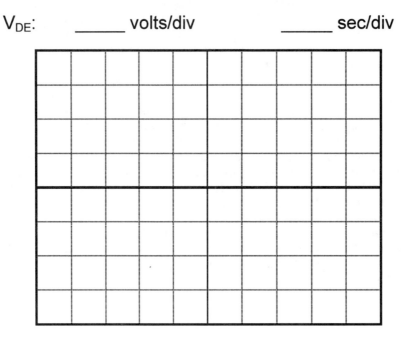

1.10 Using the peak value of the load resistor waveform from Graph 6.2, calculate the average or D.C. value of that waveform. Record your result in Table 6.2.

Calculations:

Table 6.2: Measured and calculated D.C. voltage values for the load resistor in the circuit shown in Figure 6.3.

V_{R1} MEASURED, VOLTS	
V_{R1} CALCULATED, VOLTS	

1.11 Construct the circuit shown in Figure 6.4 by removing diode D4 from your circuit. This will simulate an open.

1.12 Plug in the transformer, and use the oscilloscope to display the voltage waveform across the the resistor, V_{DE}. Draw waveform V_{DE} on Graph 6.3 in proper time phase with V_{CF}. Completely label the vertical and horizontal axes with voltage and time base values.

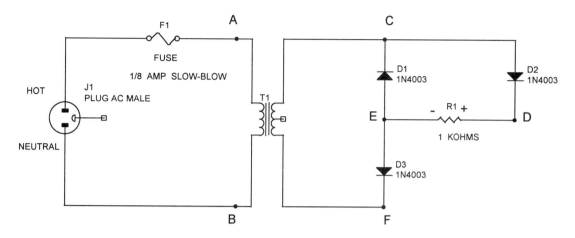

Figure 6.4: Full-wave bridge rectifier circuit with open diode in one arm of the bridge.

Graph 6.3: V_{DE} voltage waveform for the circuit shown in Figure 6.4.

V_{DE}: _____ volts/div _____ sec/div

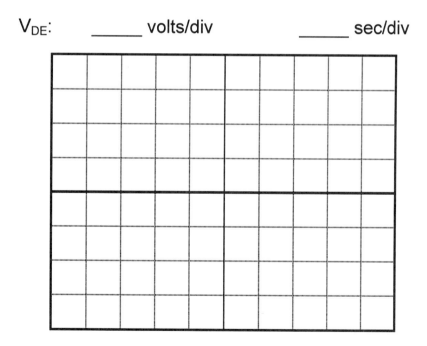

1.13 Use your handheld voltmeter to measure the D.C. voltage across the load resistor. Record your measurement in Table 6.3.
1.14 Unplug the transformer.
1.15 Using the peak value of the load resistor waveform from Graph 6.3, calculate the average or D.C. value of that waveform. Record your result in Table 6.3.

Calculations:

Table 6.3: Measured and calculated D.C. voltage values for the load resistor in the circuit shown in Figure 6.4.

V_{R1} MEASURED, VOLTS	
V_{R1} CALCULATED, VOLTS	

1.16 Replace your discrete diodes by a bridge rectifier package, and construct the circuit shown in Figure 6.5. Take every precaution to correctly identify the A.C. inputs and D.C. (+) and (-) outputs of your bridge rectifier before proceeding to wire the circuit.
1.17 Plug in the transformer, and use the oscilloscope to display the voltage waveform across the the resistor, V_{DE}. Draw waveform V_{DE} on Graph 6.4 in proper time phase with V_{CF}. Completely label the vertical and horizontal axes with voltage and time base values.

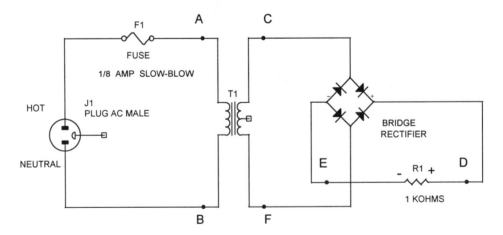

Figure 6.5: Full-wave rectifier circuit constructed from a bridge-rectifier package and a step-down transformer.

Graph 6.4: V_{DE} voltage waveform for the circuit shown in Figure 6.5.

V_{DE}: _____ volts/div   _____ sec/div

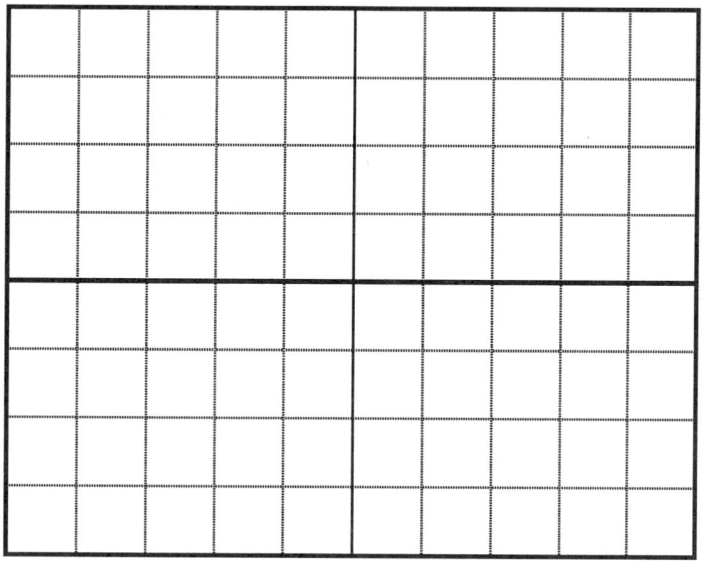

1.18 Use your handheld voltmeter to measure the D.C. voltage across the load resistor. Record your measurement in Table 6.4.
1.19 Unplug the transformer.
1.20 Using the peak value of the load resistor waveform from Graph 6.4, calculate the average or D.C. value of that waveform. Record your result in Table 6.4.

Calculations:

Table 6.4: Measured and calculated D.C. voltage values for the load resistor in the circuit shown in Figure 6.5.

V_{R1} MEASURED, VOLTS	
V_{R1} CALCULATED, VOLTS	

PART 2: Full-Wave Rectifier with Center-Tapped Transformer

Background:

The operation of the full-wave rectifier with center-tapped transformer can be explained by thinking of it as two half-wave rectifiers built into one. Your center-tapped transformer can be configured as a half-wave rectifier circuit as shown in Figure 6.6. The lower half of the transformer secondary can also be used as a half-wave rectifier as shown in Figure 6.7. When these two half-wave rectifiers are combined into one circuit as shown in Figure 6.8, the result, much like the application of the Superposition Principle from D.C./A.C. circuit analysis, is the whole full-wave rectifier circuit. As with the bridge rectifier, the orientation of the diodes in Figure 6.8 is such that the flow of electrons will always be routed through the load resistor in the same direction regardless of the polarity of the A.C. source (transformer secondary in this case). When point C becomes positive relative to point E, the center tap, diode D1 will be forward-biased. At the same moment, diode D2 will be reverse-biased acting as an open. Conventional current flow will then be from point C to E. Electron flow will, of course, be in the opposite direction. When point F becomes positive relative to point E, diode D2 will be forward-biased with conventional current flow from point F to point E; diode D1 will be reverse biased. As a result, the peak of the voltage waveform across the load resistor will be approximately the peak of the sinusoidal waveforms V_{CE} and V_{CF} less one forward-biased voltage drop of about 0.7 volt. In other words, the peak across the load is approximately one-half of the peak of the entire secondary voltage V_{CF}. The D.C. or average value of the voltage waveform that appears across the resistor may be found from the following formula:

$$V_{D.C.\ FULL\text{-}WAVE} = 2V_P/\pi \qquad \textbf{6.4}$$

You probably noticed right away that this is exactly the same formula as given earlier for the full-wave bridge rectifier. However, don't be fooled by the similarity! As with any formula, you must know how to apply it correctly or you will have worthless information. You will have a chance to apply this formula to the circuit shown in Figure 6.8 and compare your results to that for the bridge rectifier circuit. As with the formula for the bridge rectifier, V_P represents the peak of the voltage waveform across the load resistor. Realize, however, that this peak is produced by only half of the secondary voltage, not the entire secondary voltage. As for the bridge rectifier circuit, the average load current can be found using Ohm's Law as follows:

$$I_{AVG\ LOAD} = (2V_P/\pi)/R \qquad \textbf{6.5}$$

Again, this formula is the same as for the bridge rectifier. Like Formula 6.4, you must also properly apply this formula. Because the diodes in Figure 6.8 are in series with the load but conduct only half of the time and are open the other half, the average current flow through the diodes is then

$$I_{AVG\ DIODE} = I_{AVG\ LOAD}/2 \qquad \textbf{6.6}$$

The repetitive reverse voltage, V_{RRM}, that each diode will be exposed to is the peak of the <u>full</u> secondary waveform, not half of the secondary. This might seem to be a contradiction because the peak of the load waveform is roughly half of the secondary voltage. Applying Kirchhoff's Voltage Law to the closed loop from point C to D to F in Figure 6.8 will verify that when point C is positive, diode D1 will be forward-biased and diode D2 will be reverse-biased by the entire secondary voltage. Hence, each diode's V_{RRM} rating must be at least equal to V_P of the full secondary voltage waveform, V_{CF}.

Procedure:

2.1 Construct the circuit shown in Figure 6.6.

2.2 Plug in the transformer, and use the oscilloscope to display the voltage waveform across the entire secondary, V_{CF}. Select a volts/div setting such that the waveform will fill up as much of the oscilloscope screen as possible without clipping the waveform. Draw waveform V_{CF}

on Graph 6.5. Completely label the vertical and horizontal axes with voltage and time base values.

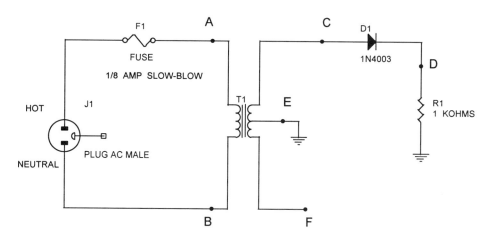

Figure 6.6: Half-wave rectifier circuit with the rectifier diode installed in the upper half of the transformer secondary.

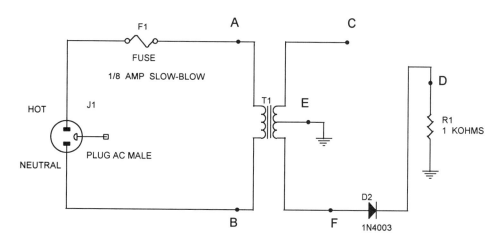

Figure 6.7: Half-wave rectifier circuit with the rectifier diode installed in the lower half of the transformer secondary.

2.3 Next display the voltage waveform V_{CE} in proper time phase with V_{CF}. Use the same volts/div as in step 2.2. Draw waveform V_{CE} on Graph 6.5 in proper time phase with V_{CF}.

2.4 Now display waveform V_{DE}. Display waveform V_{CE} at the same time, and trigger your oscilloscope on the channel that is displaying V_{CE}. Draw waveform V_{DE} on Graph 6.6 in proper time phase with V_{CE}. Completely label the vertical and horizontal axes with voltage and time base values.

2.5 Unplug the transformer.

2.6 Construct the circuit shown in Figure 6.7.

2.7 Plug in the transformer. Display waveform V_{DE} on the oscilloscope. Display waveform V_{CE} at the same time, and trigger your oscilloscope on the channel that is displaying V_{CE}. Draw waveform V_{DE} on Graph 6.7 in proper time phase with V_{CE}. Completely label the vertical and horizontal axes with voltage and time base values.

Graph 6.5: V_{CF} and V_{CE} voltage waveforms for the circuit shown in Figure 6.6.

V_{CF}: _____ volts/div   _____ sec/div
V_{CE}: _____ volts/div   _____ sec/div

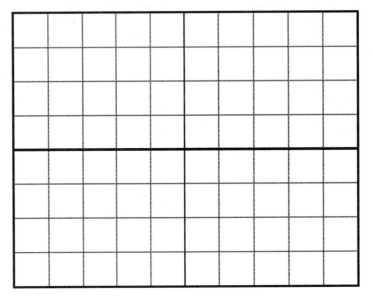

Graph 6.6: Load voltage waveform, V_{DE}, for the circuit shown in Figure 6.6.

V_{DE}: _____ volts/div   _____ sec/div

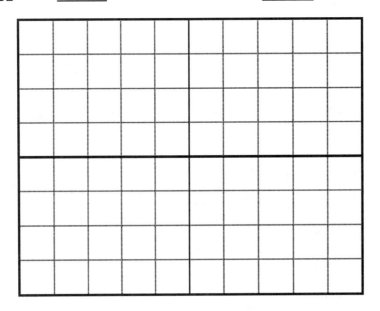

Graph 6.7: Load voltage waveform, V_{DE}, for the circuit shown in Figure 6.7.

V_{DE}: _____ volts/div   _____ sec/div

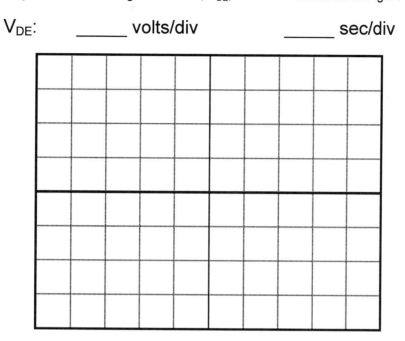

2.8 Unplug the transformer.
2.9 Construct the circuit shown in Figure 6.8.
2.10 Plug in the transformer. Display waveform V_{DE} on the oscilloscope. Display waveform V_{CE} at the same time, and trigger your oscilloscope on the channel that is displaying V_{CE}. Draw waveform V_{DE} on Graph 6.8 in proper time phase with V_{CE}. Completely label the vertical and horizontal axes with voltage and time base values.

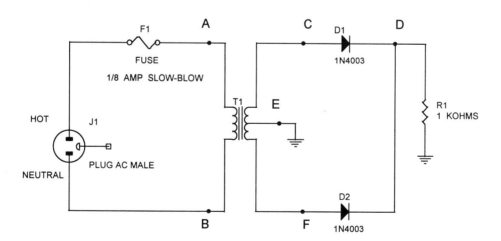

Figure 6.8: Full-wave rectifier circuit using a center-tap transformer and two rectifier diodes.

Graph 6.8: V_{DE} voltage waveform for the circuit shown in Figure 6.8.

V_{DE}: _____ volts/div   _____ sec/div

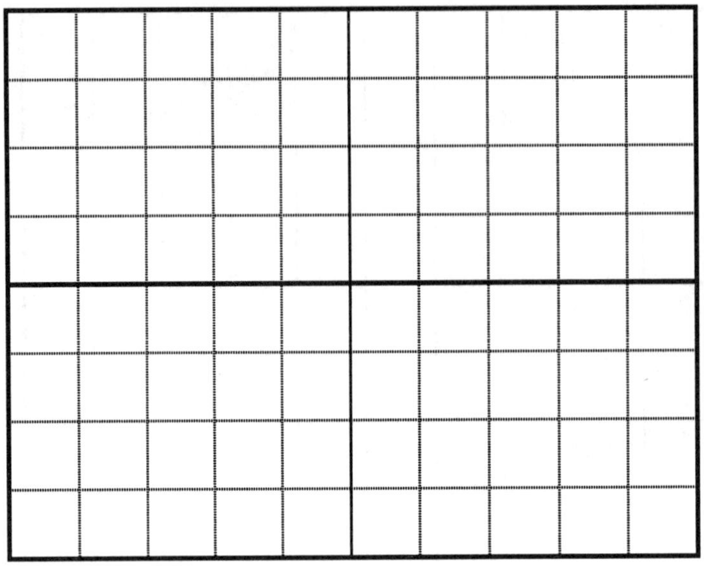

2.11 Use your handheld voltmeter to measure the D.C. voltage across the load resistor. Record your measurement in Table 6.5.
2.12 Unplug the transformer.
2.13 Using the peak value of the load resistor waveform from Graph 6.8, calculate the average or D.C. value of that waveform. Record your result in Table 6.5.

Calculations:

Table 6.5: Measured and calculated D.C. voltage values for the load resistor in the circuit shown in Figure 6.8.

V_{R1} MEASURED, VOLTS	
V_{R1} CALCULATED, VOLTS	

PART 3: Manufacturer's Ratings

3.1 Referring to the manufacturer's data sheets for your bridge rectifier, complete Table 6.6.

Table 6.6: Manufacturer's data for the bridge rectifier.

BRIDGE RECTIFIER	V_{RRM} VOLTS	V_{RSM} VOLTS	I_O AMPS	I_{FSM} AMPS	T_J, T_{stg} deg. C

PART 4: Questions

4.1 Referring to your experimental data for Figure 6.3, what is the absolute minimum value of V_{RRM} that should be specified for each diode? Explain how you arrived at your answer.

4.2 Referring to your experimental data for Figure 6.4, explain the shape (appearance) of the load resistor voltage waveform, V_{DE}.

4.3 Compare the measured and calculated voltages in Table 6.2. Explain any differences.

4.4 Compare the load voltages for Figures 6.3 and 6.5. Refer to Tables 6.2 and 6.4. Are they similar? Should they be? What might account for a <u>small</u> difference in your measured voltages for each circuit?

4.5 Referring to your experimental data for Figure 6.8, what is the absolute minimum value of V_{RRM} that should be specified for each diode? Explain how you arrived at your answer.

4.6 Referring to your experimental data for Figure 6.6, calculate the average value for the load voltage.

4.7 Referring to your experimental data for Figure 6.7, calculate the average value for the load voltage.

4.8 What is the relationship between the calculated values from Questions 4.6 and 4.7 and the voltages recorded in Table 6.5? What should the relationship be?

4.9 What is the relationship between voltages V_{CF} and V_{CE} in Figure 6.6?

4.10 Describe what would happen to the circuit of Figure 6.5 if the load resistor R1 were to become shorted. Also, discuss what circuit components you would most likely have to replace.

4.11 Describe how you would go about troubleshooting a circuit such as Figure 6.5 if you suspected that it had a shorted resistor. Discuss what symptoms you would look for and what voltages you would measure if your only measuring instrument was a handheld voltmeter capable of measuring both A.C. and D.C. voltages.

4.12 A fellow employee comes to you and says that a power supply similar to the one in Figure 6.8 is now producing only half as much voltage as it normally does. What would you expect is wrong? Explain your answer. What would you do to confirm your suspicions?

4.13 Explain what would happen if diode D4 in the full-wave bridge rectifier were to become shorted as shown in Figure 6.9.

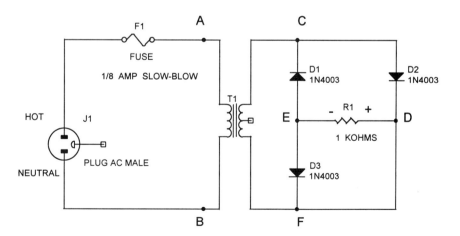

Figure 6.9: Full-wave bridge rectifier circuit with shorted diode in one arm of the bridge.

4.14 Describe how you would go about troubleshooting a circuit such as Figure 6.9 if you suspected diode D4 had shorted.

4.15 What is the relationship between the frequency of the sine wave in Graph 6.1 and the frequency of the load waveform in Graph 6.2?

PART 5: Practice Problems

5.1 Referring to the circuit shown in Figure 6.10, answer the following questions. Assume a step-down transformer with an applied voltage of 120 volts RMS and a turns ratio of 6:1. The load is a 10 Ohms resistor.

a) What is the peak of the secondary voltage waveform, V_{CF}?

b) Based on your answer to part a), and assuming that the diodes are silicon with a nominal forward-bias voltage drop of 0.7 volt, what is the peak voltage across the load resistor?

c) Based on your answer to part b), what is the average value of the voltage across the load resistor?

d) Based on your answer to part c), what is the value of the current through the load resistor?

e) What is the average current through the diodes?

f) Based on your answer to part a), what is the absolute minimum value of V_{RRM} that must be be specified for the diodes?

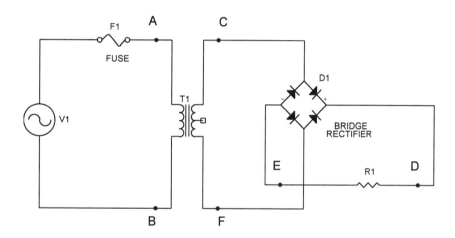

Figure 6.10: Full-wave rectifier circuit constructed from a bridge rectifier and a step-down transformer.

5.2 Assume that your engineering supervisor has directed you to design a full-wave rectifier circuit like that shown in Figure 6.11. She has told you that an average voltage of 54 volts D.C. must be provided to a 20 Ohms load resistor. The primary voltage to be applied to the transformer will be a 240 volt RMS single-phase source. Answer the following questions.

a) What is the average load current for this power supply?

b) What is the average current through each diode?

c) What is the peak voltage across the load resistor?

d) Assuming a voltage drop of 0.7 volt for each diode, what is the RMS value of voltage V_{CE}?

e) What is the RMS value of the voltage V_{CF}?

f) What turns ratio would you specify for this transformer?

g) Assuming a factor of safety of 3, what V_{RRM} would you specify for the diodes?

h) Again assuming a factor of safety of 3, what average forward current, I_O, would you specify for the diodes?

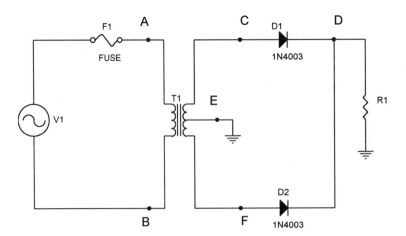

Figure 6.11: Full-wave rectifier circuit using a center-tap transformer and two rectifier diodes.

PART 6: Full-Wave Bridge Rectifier Formulas

$V_{D.C.\ FULL-WAVE} = 2V_P/\pi$ **6.1**

$I_{AVG\ LOAD} = (2V_P/\pi)/R$ **6.2**

$I_{AVG\ DIODE} = I_{AVG\ LOAD}/2$ **6.3**

PART 7: Full-Wave Rectifier w/C.T. Transformer Formulas

$V_{D.C.\ FULL-WAVE} = 2V_P/\pi$ **6.4**

$I_{AVG\ LOAD} = (2V_P/\pi)/R$ **6.5**

$I_{AVG\ DIODE} = I_{AVG\ LOAD}/2$ **6.6**

Power Supply Filtering: 7

INTRODUCTION:

Construction of a D.C. power supply involves a minimum of four steps: stepping up or stepping down a sinusoidal supply, converting the A.C. supply to a pulsating D.C. supply, filtering the pulsating D.C., and regulating the filtered output. In Experiment 4, you learned how the transformer can be used to take an input voltage and step it down to a smaller voltage at the transformer's secondary. In Experiments 5 and 6, you constructed circuits to convert a sinusoidal input to a pulsating D.C. output. In this experiment, you will have the opportunity to investigate voltage waveform filtering. Specifically, you will observe the effect that capacitors and inductors have on the shape of a full-wave rectified voltage waveform. Recall from your earlier studies that capacitors store charge and can release it into a load. Capacitors also resist change in voltage. Inductors, on the other hand, store energy in the form of a magnetic field and resist change in current. Inductors in high current D.C. power supplies can have a rating of several Henrys. Because of their size and cost, inductors--like those just described--are found less frequently in today's modern power supplies, having been replaced by capacitors, voltage regulators, and complex switching circuitry. Smaller inductors are still used in power supplies as noise filters (choke coils) and in some pulse-width modulated power supplies. The last phase of power supply design will be taken up in Experiments 8 and 9, in which you will investigate the principles of voltage regulation and work with two types of I-C voltage regulators.

SAFETY NOTE:

Before starting this experiment have your instructor review the operational steps and safety procedures to follow when using the oscilloscope to display the secondary voltages of a center-tapped transformer. It is important that you know how to use your oscilloscope to display the true voltage waveforms that appear across the secondary terminals of a center-tapped transformer. Incorrect use of the oscilloscope will result in either an incorrect waveform, such as displaying a floating voltage, or worse, causing a short circuit to ground through your oscilloscope. Also, if you are using a different transformer, diodes, or bridge rectifier than the one(s) specified in the Materials list, have your instructor make sure that the diodes and bridge rectifier that you have selected will withstand the voltage and current that each will be subjected to in all phases of your experiment. As always, wear safety glasses, and remove all jewelry from your hands and fingers before the start of the experiment.

OBJECT:

Upon successful completion of this experiment and all reading assignments, the student should be able to:
- construct an operational full-wave bridge rectifier circuit with purely resistive load and a single capacitor filter
- construct an operational full-wave bridge rectifier circuit with purely resistive load and filtering provided by a π-type filter
- experimentally determine the effect that varying load resistance has on the magnitude of the ripple content of a capacitor-filtered waveform
- compare the effectiveness of the single-capacitor filter to the π-type filter
- calculate the ripple factor and percent ripple for a given waveform

MATERIALS:
- 2 - 10 µF electrolytic capacitors, 150 WVDC
- 1 - choke coil in the range of 4 to 12 Henrys
- 1 - 120:12.6 V_{RMS} center-tapped transformer
  - miscellaneous lead wires and connectors
- 4 - rectifier diodes, 1N4003 or an equivalent diode of equal or higher voltage and current ratings
  or
- 1 - bridge rectifier, such as the SK3985 or SK5042, or an equivalent rectifier of equal or higher voltage and current ratings
- 1 - 22 Kohms, 1/2 watt resistor
- 2 - 1.5 Kohms, 1 watt resistors
- 1 - fuse, 1/8 amp slow-blow
- 1 - in-line fuse holder
- 1 - solderless breadboard

EQUIPMENT:
- 1 - dual-trace oscilloscope
- 2 - oscilloscope probes
- 1 - handheld multimeter

PART 1: Single-Capacitor Filter
Background:
A full-wave bridge rectifier circuit with load and capacitor filtering is shown in Figure 7.1. Recall from Experiment 6 that the voltage waveform across the load in an unfiltered full-wave bridge rectifier circuit would appear like the one shown in Figure 7.2. The D.C. or average value of this waveform can be found using the following formula:

$$V_{D.C.\ FULL-WAVE} = 2V_P/\pi \qquad\qquad 7.1$$

V_P represents the peak of the voltage waveform across the load resistor. V_P will be approximately equal to the peak of the transformer secondary voltage minus the voltage drop across the two forward-biased diodes (approximately 1.4 volts = 0.7 + 0.7). Note that the frequency of this waveform is twice the line frequency of 60 Hz, i.e., 120 Hz. This is important when determining a filter capacitor's capacitive reactance, X_C, and a choke coil's inductive reactance, X_L. For instance, when using a capacitor filter as shown in Figure 7.1, a good rule of thumb is that X_C should be less than or equal to $0.1(R_L)$. In this case, we have $X_C = 1/(2 \cdot \pi \cdot f \cdot C) = 1/[2\pi(120\ Hz)(10\mu F)] = 133$ Ohms $< 0.1(22\ Kohms) = 2.2$ Kohms. Filter capacitor C1 not only provides charge (current) to R_L, but also acts as a low resistance path to ground for the A.C. ripple.

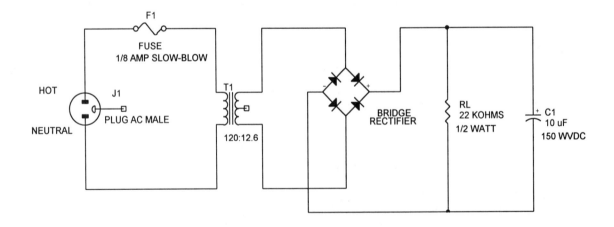

Figure 7.1: Full-wave bridge rectifier circuit with purely resistive load and single-capacitor filtering.

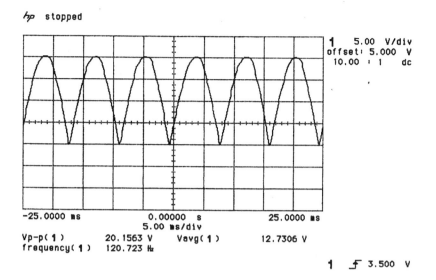

Figure 7.2: Full-wave rectified waveform without capacitor filtering.

The addition of the capacitor results in a waveform that appears much like the one shown in Figure 7.3. Notice that this waveform has more area under the curve than the one shown in Figure 7.2. As a result, the load waveform is now smoother and of greater magnitude (greater D.C. value). A figure of merit or quality of a power supply is the amount of variation or *ripple* exhibited by the output voltage waveform when the load draws the power supply's rated current. The ripple factor (R.F.) is calculated using the following formula:

$$R.F. = [(V_{P-P})/(2 \cdot \sqrt{3})]/V_{DC} \qquad 7.2$$

Sometimes the percent ripple is specified, which is simply the ripple factor times 100. V_{P-P} represents the peak-to-peak voltage of the ripple, or in other words, the voltage measured from the highest peak to the lowest. V_{P-P} is divided by $2 \cdot \sqrt{3}$ to convert the peak-to-peak voltage to RMS units. The $\sqrt{3}$ term is used rather than $\sqrt{2}$, because the ripple most nearly represents a sawtooth waveform rather than a sine wave. And, of course, V_{DC} is the D.C. or average value of this waveform. Substituting the values for V_{P-P} and V_{DC} from Figure 7.3 into Formula 7.2 we have:

$$R.F. = [(6.406 \text{ volts p-p})/(2 \cdot \sqrt{3})]/17.01 \text{ volts}$$

$$= 0.1087$$

$$\% \text{ ripple} = R.F. \times 100 \qquad 7.3$$

$$= 0.1087 \times 100$$

$$= 10.87\%$$

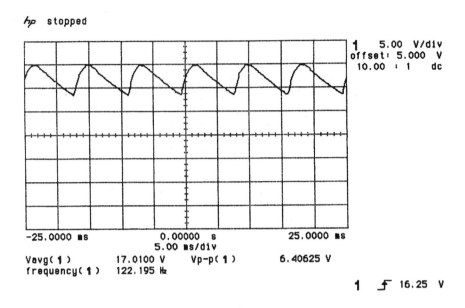

Figure 7.3: Full-wave rectified waveform with capacitor filtering.

While the waveform in Figure 7.3 is much smoother than the one shown in Figure 7.2, it is still not smooth enough for many applications. In Experiments 8 and 9, you will learn how voltage regulators can be used to eliminate nearly all variation or ripple present in the output voltage. Well, what are the drawbacks to using capacitors as filters, besides the added cost? In half-wave rectifiers, the peak reverse voltage (PRV) applied to the rectifier diode is doubled. For both full-wave and half-wave rectifiers, the peak current that flows through the diodes when the capacitor is charging is also much larger. Therefore, the PRV and current ratings of the diodes in capacitor-filtered rectifier circuits must be increased accordingly. In addition, the voltage regulation (V.R.) of circuits like the one shown in Figure 7.1 is poor for heavy (high-current drawing) loads. Like the ripple factor, voltage regulation is another figure of merit for power supplies. It is calculated using the following formula:

$$\% \text{ V.R.} = [(V_{NL} - V_{FL})/V_{FL}] \cdot 100 \qquad 7.4$$

V_{NL} is the voltage output from the power supply under no-load conditions. V_{FL} is the voltage output from the power supply under full-load conditions. Full-load is defined as the load condition under which rated current is drawn from the power supply.

INSTRUCTOR NOTE: Please be certain that your students have mastered all proper safety procedures for using the oscilloscope (both portable and plug-in styles) when displaying single-phase and three-phase voltage sources at the primary and secondary of both low- and high-voltage transformers. Your presentation should include a discussion of proper grounding techniques, when to float an oscilloscope off-line, and the selection of high-voltage oscilloscope probes. Students should also have mastered safe techniques for measuring high voltages with a handheld multimeter and high-voltage probe.

Procedure:
    1.1 Obtain all materials and equipment required to complete this experiment.
    1.2 Construct the circuit shown in Figure 7.1 but do <u>not</u> install the capacitor at this time.
    1.3 If you have not already completed Experiments 4, 5, or 6, have your instructor help you
        with the identification of the primary and secondary leads of the transformer. Once you are

completely familiar with the transformer's construction and operation, proceed to step 1.4.
1.4 Plug in the transformer. Use the oscilloscope to display the voltage waveform across the load resistor, and draw this waveform on Graph 7.1. Completely label the vertical and horizontal axes with voltage and time base values.
1.5 Use your handheld voltmeter to measure the D.C. voltage across the load resistor. Record your measurement in Table 7.1.
1.6 Unplug the transformer.

Graph 7.1: Load voltage waveform for the circuit shown in Figure 7.1 without capacitor filter.

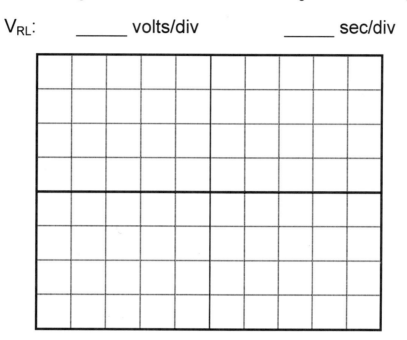

V_{RL}: _____ volts/div _____ sec/div

1.7 Using the peak value of the load resistor waveform from Graph 7.1, calculate the average or D.C. value of that waveform. Record your result in Table 7.1.

Calculations:

Table 7.1: Measured and calculated D.C. voltage values for the load resistor in the circuit shown in Figure 7.1 without capacitor filter.

V_{RL}, MEASURED, VOLTS D.C.	
V_{RL}, CALCULATED, VOLTS D.C.	

1.8 With the power off, install the filter capacitor in parallel with the load. Make sure that you properly identify the terminals of your capacitor, because electrolytic capacitors are polarity sensitive. If in doubt, ask your instructor for assistance. When connecting a capacitor across a load, the (+) terminal should be connected to the point of highest positive potential and the (-) terminal to the lowest potential (D.C. ground in this case). Before disconnecting a circuit with a capacitor, make sure that the capacitor is fully discharged. The capacitor should be allowed to discharge through the circuit load. When you are done, turn off the power supply and measure the voltage across your capacitor to make sure it is discharged.

1.9 Plug in the transformer. Use the oscilloscope to display the voltage waveform across the resistor. Use D.C. coupling, and estimate the D.C. value of the resulting waveform. Now switch to A.C. coupling and reduce the volts/div so that the ripple becomes more visible. Continue to reduce the attenuation until you can measure the peak-to-peak value of the ripple. Draw the resulting waveform on Graph 7.2, and completely label the vertical and horizontal axes with voltage and time base values. Be sure to label the magnitude of the D.C. offset. It may be necessary for you to use your oscilloscope's bandwidth limit function to eliminate any noise that may be riding on top of your ripple waveform.

Graph 7.2: Load voltage waveform for the circuit shown in Figure 7.1 with capacitor filter and a load resistance of 22 Kohms.

V_{RL}: _____ volts/div   _____ sec/div

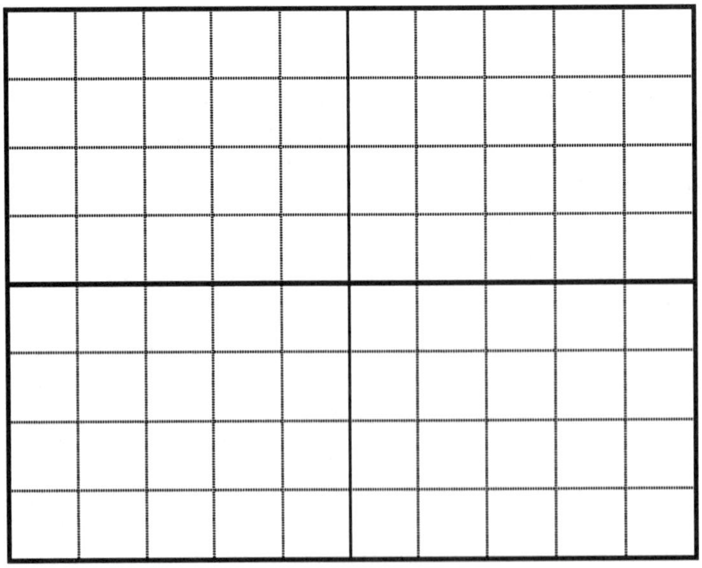

1.10 Use your handheld voltmeter to measure the D.C. voltage across the load resistor. Record your measurement in Table 7.2. How does this value compare to the value you estimated from the oscilloscope?
1.11 Unplug the transformer.
1.12 In Table 7.2 record the peak-to-peak value of the ripple shown in Graph 7.2. Use this value and your measured D.C. voltage recorded in Table 7.2 to calculate the ripple factor for your waveform. Also calculate the percent ripple. Record your results in Table 7.2.

Calculations:

Table 7.2: Experimental data for the circuit shown in Figure 7.1 with capacitor filter and 22 Kohms of load resistance.

V_{RL}, MEASURED, VOLTS D.C.	
$V_{PEAK-TO-PEAK}$ RIPPLE	
RIPPLE FACTOR	
% RIPPLE	

1.13 With the power off, replace the 22 Kohms load resistor with two 1.5 Kohms resistors connected in parallel to produce an effective load resistance of 750 Ohms.
1.14 Plug in the transformer. Use the oscilloscope to display the voltage waveform across the load. Use D.C. coupling, and estimate the D.C. value of the resulting waveform. Now switch to A.C. coupling and reduce the volts/div so that the ripple becomes more visible. Continue to reduce the attenuation until you can measure the peak-to-peak value of the ripple. Draw the resulting waveform on Graph 7.3. Completely label the vertical and horizontal axes with voltage and time base values. Be sure to label the magnitude of the D.C. offset.
1.15 Use your handheld voltmeter to measure the D.C. voltage across the load resistor. Record your measurement in Table 7.3. How does this value compare to the value you estimated from the oscilloscope?
1.16 Unplug the transformer.
1.17 Record in Table 7.3 the peak-to-peak value of the ripple shown in Graph 7.3. Use this value and your measured D.C. voltage recorded in Table 7.3 to calculate the ripple factor for your waveform. Also calculate the percent ripple. Record your results in Table 7.3.

Calculations:

Graph 7.3: Load voltage waveform for the circuit shown in Figure 7.1 with capacitor filter and a load resistance of 750 Ohms.

V_{RL}: _____ volts/div   _____ sec/div

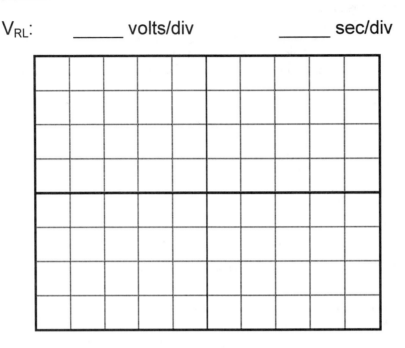

Table 7.3: Experimental data for the circuit shown in Figure 7.1 with capacitor filter and 750 Ohms of load resistance.

V_{RL}, MEASURED, VOLTS D.C.	
$V_{PEAK-TO-PEAK\ RIPPLE}$	
RIPPLE FACTOR	
% RIPPLE	

PART 2: Pi-Type Filter

Background:
Some of the advantages and disadvantages of the single-capacitor filter were discussed in the background information for Part 1 of this experiment. In this part of the experiment you will collect experimental data on the Pi-type filter shown in Figure 7.4. This name was chosen because the shape depicted by the connection of the two capacitors and the inductor resembles the Greek letter π. After you have collected your experimental data, you will have the opportunity to compare the performance of the two types of filters. Since the inductor is in series with the load, it should have as little resistance as possible. However, the inductive reactance, $X_L = 2 \cdot \pi \cdot f \cdot L$, should be $10(R_L)$ to filter the A.C. ripple.

Procedure:
2.1 Construct the circuit shown in Figure 7.4. You may use an inductor larger than 10 H if one is available. Check with your instructor.

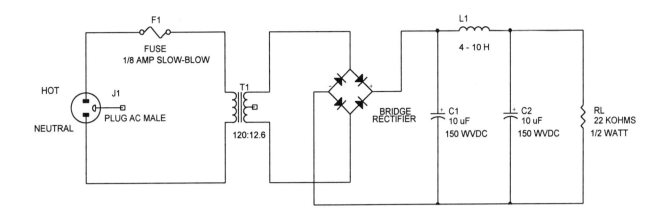

Figure 7.4: Full-wave bridge rectifier with Pi-type filter.

2.2 Plug in the transformer. Use the oscilloscope to display the voltage waveform across the resistor. Use D.C. coupling, and estimate the D.C. value of the resulting waveform. Now switch to A.C. coupling and reduce the volts/div so that the ripple becomes more visible. Continue to reduce the attenuation until you can measure the peak-to-peak value of the ripple. Draw the resulting waveform on Graph 7.4. Completely label the vertical and horizontal axes with voltage and time base values. Be sure to label the magnitude of the D.C. offset. It may be necessary for you to use your oscilloscope's bandwidth limit function to eliminate any noise that may be present riding on top of your ripple waveform.

2.3 Use your handheld voltmeter to measure the D.C. voltage across the load resistor. Record your measurement in Table 7.4. How does this value compare to the value you estimated from the oscilloscope?

2.4 Unplug the transformer.

2.5 Record in Table 7.4 the peak-to-peak value of the ripple shown in Graph 7.4. Use this value and your measured D.C. voltage recorded in Table 7.4 to calculate the ripple factor for your waveform. Also calculate the percent ripple. Record your results in Table 7.4.

Calculations:

Graph 7.4: Load voltage waveform for the circuit shown in Figure 7.4 with the Pi-type filter and a load resistance of 22 Kohms.

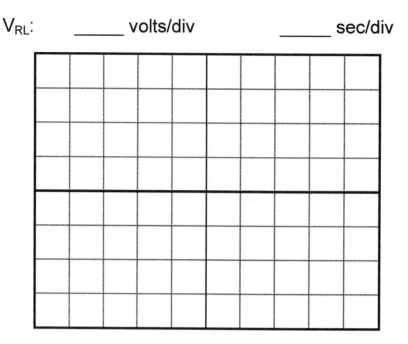

V_{RL}: _____ volts/div   _____ sec/div

Table 7.4: Experimental data for the circuit shown in Figure 7.4 with the Pi-type filter and 22 Kohms of load resistance.

V_{RL}, MEASURED, VOLTS D.C.	
$V_{PEAK-TO-PEAK\ RIPPLE}$	
RIPPLE FACTOR	
% RIPPLE	

2.6 With the power off, replace the 22 Kohms load resistor with two 1.5 Kohms resistors connected in parallel to produce an effective load resistance of 750 Ohms.

2.7 Plug in the transformer. Use the oscilloscope to display the voltage waveform across the load. Use D.C. coupling, and estimate the D.C. value of the resulting waveform. Now switch to A.C. coupling and reduce the volts/div so that the ripple becomes more visible. Continue to reduce the attenuation until you can measure the peak-to-peak value of the ripple and draw the resulting waveform on Graph 7.5. Completely label the vertical and horizontal axes with voltage and time base values. Be sure to label the magnitude of the D.C. offset.

Graph 7.5: Load voltage waveform for the circuit shown in Figure 7.4 with the Pi-type filter and a load resistance of 750 Ohms.

V_{RL}: _____ volts/div _____ sec/div

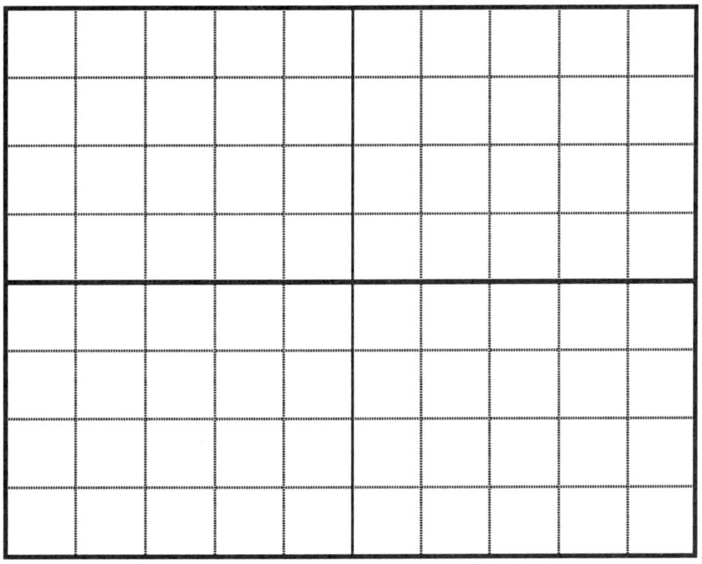

2.8 Use your handheld voltmeter to measure the D.C. voltage across the load resistor. Record your measurement in Table 7.5. How does this value compare to the value you estimated from the oscilloscope?

2.9 Unplug the transformer.

2.10 Record in Table 7.5 the peak-to-peak value of the ripple shown in Graph 7.5. Use this value and your measured D.C. voltage recorded in Table 7.5 to calculate the ripple factor for your waveform. Also calculate the percent ripple. Record your results in Table 7.5.

Calculations:

Table 7.5: Experimental data for the circuit shown in Figure 7.4 with the Pi-type filter and 750 Ohms of load resistance.

V_{RL}, MEASURED, VOLTS D.C.	
V_{PEAK-TO-PEAK RIPPLE}	
RIPPLE FACTOR	
% RIPPLE	

PART 3: Questions

3.1 Using the D.C. load voltage from Table 7.2 as the no-load voltage for the circuit shown in Figure 7.1 and the D.C. load voltage recorded in Table 7.3 as the full-load voltage, calculate the percent voltage regulation for that circuit.

3.2 Using the D.C. load voltage from Table 7.4 as the no-load voltage for the circuit shown in Figure 7.4 and the D.C. load voltage recorded in Table 7.5 as the full-load voltage, calculate the percent voltage regulation for that circuit.

3.3 <u>Compare</u> the two types of filters on the basis of the following criteria:

a) percent voltage regulation

b) amount of ripple under lightly loaded (22 Kohms) conditions

c) amount of ripple under heavily loaded (750 Ohms) conditions

d) magnitude of D.C. output voltage under lightly loaded (22 Kohms) conditions

e) magnitude of D.C. output voltage under heavily loaded (750 Ohms) conditions

3.4 The Pi-type filter had a better percent regulation with 750 Ohms applied but had a smaller D.C. output voltage compared to the single-capacitor filtered circuit under the same load conditions. Try to explain why the Pi-type filter had the lower output voltage. Hint: Measure the <u>resistance</u> of your inductor then analyze the circuit in Figure 7.4 by replacing the inductor with your measured resistance. Treat the capacitors as opens.

PART 4: Electronics Workbench®

4.1 Use Electronics Workbench® to create the circuit shown in Figure 7.1. Replace the male plug with a 120 volts RMS sinusoidal source. You do not have to include the fuse. Select a transformer turns ratio of 120:12.6. Select rectifier diodes or a bridge rectifier that have electrical properties as close as possible to the semiconductor(s) used in your lab experiment. Use the oscilloscope function to display the transformer secondary and load voltage waveforms. Print a copy of your schematic diagram and oscilloscope waveforms. Repeat these steps for a load resistance of 750 Ohms. How do these load waveforms compare to your experimental results recorded in Graphs 7.2 and 7.3? Explain any differences, especially the secondary voltage.

PART 5: Power Supply Filter Formulas

$V_{D.C.\ FULL-WAVE} = 2V_P/\pi$ **7.1**

$R.F. = [(V_{P-P})/(2 \cdot \sqrt{3})]/V_{DC}$ **7.2**

% ripple = R.F. X 100 **7.3**

$\% \ V.R. = [(V_{NL} - V_{FL})/V_{FL}] \cdot 100$ **7.4**

Voltage Regulator Principles: 8

INTRODUCTION:
In Experiment 4 you learned how transformers can be used to either lower or raise an A.C. voltage to a desired level. In Experiments 5 and 6 diodes were arranged to convert A.C. into pulsating D.C. Experiment 7 introduced you to the role that capacitors and inductors can play in filtering (smoothing) a pulsating D.C. source. If you had the opportunity to complete Experiment 7, you discovered the limitations of capacitors and inductors in filtering circuits. Capacitors in particular were not able to maintain a steady output (supply) voltage with widely fluctuating loads. One way to overcome this particular drawback is to install a *voltage regulator* in a power supply circuit to complement capacitor filtering. Voltage regulators are found in all modern power supplies. However, the methods employed to provide for voltage regulation vary widely. While it is possible that you may have never installed a voltage regulator in a circuit before today, you most certainly have relied on voltage regulators on numerous occasions. This not only includes the power supplies that you have been using during your electronics education but applications in your personal life as well. Just consider all of the electronic equipment that you already own. And, if that's not enough, as long as you have been driving a car you have relied on a voltage regulator under the hood of your car to maintain a nearly constant voltage to your car's electrical system. In this experiment, you will have the opportunity to investigate the operating principle of a number of different regulator circuits. This will include the two basic voltage regulator configurations, *series* and *shunt*. We will first review the zener diode, which will be followed by the use of discrete components to construct a series regulator. This will serve as a lead-in to Experiment 9, which covers I-C voltage regulators. You will use your bench-top D.C. power supply as the input to the voltage regulators. In a real-world application you would most likely take a single-phase source and drop the voltage down to a specified level, full-wave rectify it, filter it, and finally feed it into a voltage regulator circuit.

SAFETY NOTE:
Before starting this experiment, have your instructor review with you the manufacturer's ratings for the power transistor that you will be using in this experiment. As with any electronic device, it is important not to exceed the maximum power dissipation rating, $P_{D(MAX)}$. Also never exceed the device's maximum continuous current rating. Transistors can also fail if the differential voltage across any two of the three terminals becomes excessive. As always, wear safety glasses, and remove all jewelry from your hands and fingers before the start of the experiment.

OBJECT:
Upon successful completion of this experiment and all reading assignments, the student should be able to:
- design and construct a zener-diode based, shunt voltage regulator circuit to meet given specifications
- explain the principle of operation of series and shunt voltage regulators
- construct a series voltage regulator from a zener diode, a resistor, and a power transistor

REFERENCES:
www.motorola.com

MATERIALS:
- 1 - zener diode, 1N4742, 1 watt
- 1 - resistor, 220 Ohms, 1/2 watt
- 1 - resistor, 220 Ohms, 1 watt
- 1 - resistor, 470 Ohms, 1/2 watt
- miscellaneous lead wires and connectors
- 1 - 5 Kohms potentiometer
- 1 - NPN transistor, TIP29 or the equivalent
- 1 - 1N4003 rectifier diode or the equivalent
- 1 - solderless breadboard

EQUIPMENT:
- 1 - D.C. power supply
- 1 - digital multimeter

PART 1: The Zener Diode as a Shunt Regulator
Background:
Figure 8.1 shows the schematic symbols for a zener diode. Figure 8.2 shows the schematic symbol for a potentiometer. The potentiometer is the analog equivalent of a zener diode. As Figure 8.3 shows, a reverse-biased zener diode maintains a nearly constant voltage across it when operated in between the limits of the knee current, I_{ZK}, and the maximum current, I_{ZM}. The zener does this by automatically changing its internal resistance (impedance) as the amount of current through it changes. This is similar to manually adjusting a potentiometer to change its resistance.

Figure 8.1: Schematic symbols for a zener diode.

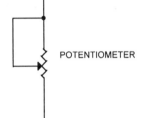

Figure 8.2: Schematic symbol for a potentiometer. A potentiometer is the analog equivalent of a zener diode.

While a zener diode maintains a nearly constant voltage over a range of currents, its nominal or rated voltage, V_{ZT}, occurs at a specific current, I_{ZT}. V_{ZT} and I_{ZT} are referred to as the zener test voltage and test current, respectively. These values are specified in the manufacturer's data manual. Figure 8.3 also shows the forward-biased I-V curve for a zener diode. As shown by this curve and in Figure 8.4, a forward-biased zener diode acts much like a forward-biased rectifier diode with a forward voltage drop of approximately 0.7 volt. Of course, we are more interested in the zener's reverse-bias operation as shown in Figure 8.5. In this figure, R_{SERIES} is a current-limiting resistor installed to prevent exceeding the zener's $P_{D(MAX)}$ or I_{ZM} when no load is attached. The potentiometer-based equivalent circuit for Figure 8.5 is shown in Figure 8.6. A circuit like that in Figure 8.6 can be used to maintain a constant voltage across the load resistor by manually adjusting the potentiometer's resistance.

Consider, for example, what would happen if a load resistance of 200 Ohms were installed in parallel with the potentiometer as shown in Figure 8.7. To maintain the desired output voltage of 10 volts, the potentiometer could be adjusted to a resistance of 200 Ohms. If the zener were used instead of the potentiometer, the zener's impedance would automatically change to 200 Ohms. If the load resistance were to decrease to 150 Ohms, as shown in Figure 8.8, the potentiometer could be adjusted to 300 Ohms to maintain V_{OUT} at 10 volts. In a similar manner, if the load resistance increased to 500 Ohms, an output voltage of 10 volts could still be obtained by changing the potentiometer's resistance to 125 Ohms, as in

Figure 8.9. This type of voltage regulator circuit is referred to as a shunt regulator because the regulating device is installed in parallel with the load.

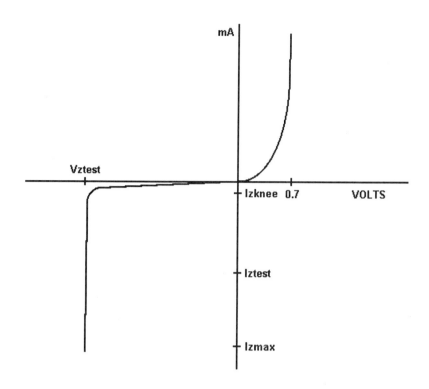

Figure 8.3: Forward and reverse-bias I-V curves for a zener diode.

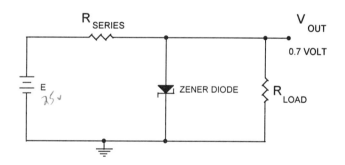

Figure 8.4: Forward-biased zener diode circuit. The output voltage, V_{OUT}, is limited to 0.7 volt.

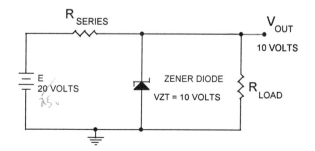

Figure 8.5: Reverse-biased zener diode acting as a shunt voltage regulator. Also shown are a load resistor and a series current-limiting resistor. The output voltage across the load is limited to the nominal or rated zener voltage, V_{ZT}.

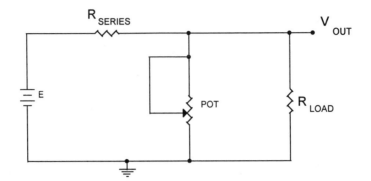

Figure 8.6: Equivalent circuit for Figure 8.5.

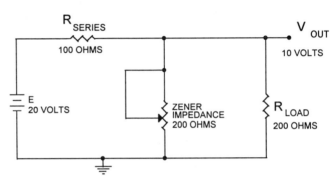

Figure 8.7: Equivalent circuit for Figure 8.5 with a load resistance of 200 Ohms and a zener impedance of 200 Ohms. The resulting output voltage is 10 volts.

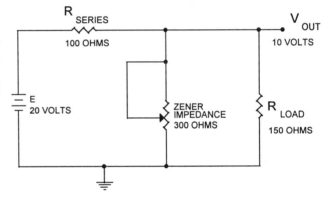

Figure 8.8: Equivalent circuit for Figure 8.5 with a load resistance of 150 Ohms and the zener impedance automatically adjusted to 300 Ohms. The output voltage is maintained at 10 volts.

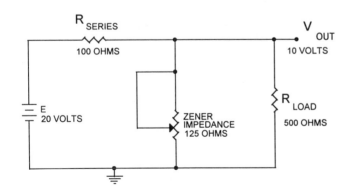

Figure 8.9: Equivalent circuit for Figure 8.5 with a load resistance of 500 Ohms and the zener impedance automatically adjusted to 125 Ohms. The output voltage is maintained at 10 volts.

Figure 8.10: Zener-diode based, shunt voltage regulator circuit designed to provide an output voltage of 12 volts.

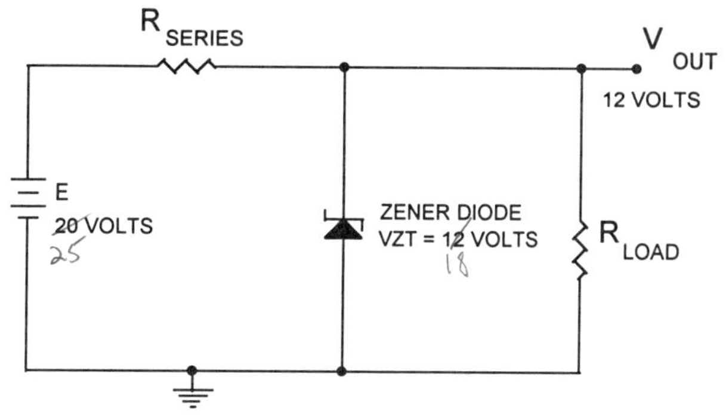

Let's now consider how we would go about designing a zener-diode based, shunt regulator to produce an output voltage of 12 volts. The basic circuit is shown in Figure 8.10. To <u>maintain</u> an output voltage of 12 volts it will be necessary to have a supply voltage greater than 12 volts. In this case, 20 volts has been chosen. To <u>obtain</u> an output voltage of 12 volts, we must select a zener diode having a V_{ZT} of 12 volts. The 1N4742 zener diode will satisfy this criteria. Assuming that the 1N4742 is rated for 1 watt, we must select a series resistor to limit the zener current to an amount that will not exceed the 1 watt rating when there is no load attached. This represents the worst case for the zener, because all of the current from the supply will flow through the zener. With a load attached the current will divide between the zener and the load. If it is not available in the manufacturer's data manual, the maximum allowable zener current can be approximated using the following formula:

$$I_{ZM} = P_{DMAX}/V_{ZT} \qquad \qquad 8.1$$

Solving for I_{ZM}, we would then have

$$I_{ZM} = 1 \text{ watt}/12 \text{ volts}$$

$$= 83.3 \text{ mA}$$

The minimum value of the series current-limiting resistor can then be found as follows:

$$R_{SERIES(MINIMUM)} = (E - V_{ZT})/I_{ZM} \qquad \qquad 8.2$$

Solving for R_{SERIES}, we would then have

$$R_{SERIES(MINIMUM)} = (20 \text{ V} - 12 \text{ V})/83.3 \text{ mA}$$

$$= 96 \text{ Ohms}$$

To prevent damage to the zener due to unexpected fluctuations in the supply voltage, we should choose an appropriate *factor of safety* (F.S.) and increase the series resistance by this factor. Choosing an appropriate factor of safety is largely a matter of considering the potential damage that could result if a component fails and the uncertainty of the conditions under which our product (voltage regulator in this case) might be used. For bridges and elevators, you want a very large factor of safety. In this case, a factor of safety of 2 would probably be more than adequate. We can now write a general formula for R_{SERIES} in terms of $R_{SERIES(MINIMUM)}$ and F.S. as follows:

$$R_{SERIES} = (R_{SERIES(MINIMUM)}) \cdot (F.S.) \qquad \qquad 8.3$$

Applying our factor of safety of 2, we can now solve for R_{SERIES} as follows:

$$R_{SERIES} = (96 \text{ Ohms}) \cdot (2)$$

$$= 192 \text{ Ohms}$$

Obviously this is not a standard resistor. Because our resistor is probably going to have a tolerance of 5 or 10%, we would be advised to choose a standard 220 Ohms resistor. Applying both Ohm's Law and Kirchhoff's Voltage Law, the current that will flow through this resistor when installed in the circuit shown in Figure 8.10 can be found from the following formula:

$$I_{SERIES} = (E - V_{ZT})/R_{SERIES} \quad\quad \textbf{8.4}$$

Solving for I_{SERIES}, we have

$$I_{SERIES} = (20 \text{ V} - 12 \text{ V})/220 \text{ Ohms}$$

$$= 36.36 \text{ mA}$$

Determining an appropriate wattage value for this resistor is left as an exercise for you. At this stage, we could go about constructing our circuit, but first, it might be nice to know over what range of load resistance our power supply will maintain 12 volts. In our previous design steps, we took care of the situation where no load would be connected, that is, an open circuit. Now we need to find out what the smallest load is that can be applied such that the output will still be approximately 12 volts. According to the Motorola® data manual, the 1N4742 zener diode has a knee current, I_{ZK}, of 0.25 mA. Now you might be saying to yourself, "Just what is the significance of this value?" Well, for the zener to maintain 12 volts, it must have no less than 0.25 mA through it. The maximum allowable current through the load resistance could then be found as follows:

$$I_{LOAD(MAX)} = I_{SERIES} - I_{ZK} \quad\quad \textbf{8.5}$$

Solving for $I_{LOAD(MAX)}$ for our circuit, we have

$$I_{LOAD(MAX)} = 36.36 - 0.25 \text{ mA}$$

$$= 36.11 \text{ mA}$$

Because the load resistor and the zener are in parallel, and assuming that the zener is maintaining its rated voltage, the smallest load resistor that could then be installed in our circuit could be found using the following formula:

$$R_{LOAD(MIN)} = V_{ZT}/I_{LOAD(MAX)} \quad\quad \textbf{8.6}$$

Substituting our values for V_{ZT} and $I_{LOAD(MAX)}$, we have

$$R_{LOAD(MIN)} = 12 \text{ V}/36.11 \text{ mA}$$

= 332 Ohms

Handwritten: B < E, 18v, 14mA

In the first part of this experiment, you will be constructing the shunt regulator circuit in Figure 8.11. With this circuit you will be able to determine the validity of our calculations.

Procedure:

1.1 Set your power supply for an output voltage of 20 volts. Turn off the power supply.
1.2 Construct the circuit shown in Figure 8.11. Adjust your potentiometer to approximately its middle position.
1.3 Turn on the power supply. Use your digital voltmeter to measure the output voltage V_{BC}. Does your voltmeter read approximately 12 volts? If not, turn off the power supply and check your circuit. If it does read close to 12 volts, go on to the next step. *18.47*
1.4 For each of the potentiometer settings shown in Table 8.1, measure and record the output voltage V_{BC}.

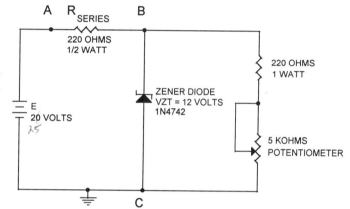

Figure 8.11: Test circuit to determine the voltage regulation for the zener-diode based, shunt voltage regulator.

Table 8.1: Shunt regulator output voltage for three different loads.

POTENTIOMETER SETTING	V_{BC}, VOLTS
MINIMUM RESISTANCE	12.25 v
MIDPOINT RESISTANCE	18.47 v
MAXIMUM RESISTANCE	18.58 v

PART 2: The Series Voltage Regulator
Background:
Voltage regulators can take the form of a shunt regulator as in Figure 8.11 or as series regulators. In a series voltage regulator circuit, the regulating element is in series with the load. A simple potentiometer-based series regulator is shown in Figure 8.12.

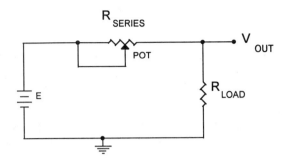

Figure 8.12: Potentiometer-based, series voltage regulator.

Let's again consider our 10 volt power supply that we discussed earlier. Again we will assume that we have a 20 volt input voltage. If our load resistance is 1000 Ohms, we can adjust our pot to 1000 Ohms to produce the desired 10 volts output as shown in Figure 8.13. Should the load resistance increase, we can increase the potentiometer resistance to compensate as in Figure 8.14. In a similar fashion, if the load resistance decreases, we can decrease the potentiometer resistance. As with the zener circuit, we will want to construct a circuit whose series resistance will change automatically. While the zener diode in Figure 8.11 was able to maintain an output of approximately 12 volts over a fairly wide range of resistance values, the voltage regulation provided by zener diodes is not adequate for regulating low-resistance, high-power loads. Rather, the zener finds its use primarily in serving as a reference voltage in lower-power circuits. In fact, we can use our 1N4742 zener as a reference voltage in a series regulator constructed from discrete components. Such a circuit might look like that shown in Figure 8.15. In this circuit the transistor is serving as the series regulator. The zener diode is being used to provide a reference voltage of approximately 12 volts at the base of the transistor. A value of 470 Ohms was chosen for R1 to provide the value of I_{ZT} specified by the manufacturer that would produce the rated zener voltage V_{ZT}. I_{ZT} for the 1N4742 is specified in the Motorola® data sheet to be 21 mA. The value for the series resistor R1 was found as follows:

$$R_{SERIES} = (E - V_{ZT})/I_{ZT} \qquad \textbf{8.7}$$

Substituting for E, V_{ZT}, and I_{ZT}, we have

$$R_{SERIES} = (20 - 12)/21 \text{ mA}$$

$$= 381 \text{ Ohms}$$

While there are 390 Ohm and 430 Ohm resistors, 470 Ohm resistors are more frequently found in most school laboratories. I therefore chose a 470 Ohm resistor for R1. The resulting zener current will still be well above the knee current. So just how does the circuit in Figure 8.13 work? First keep in mind that the zener diode is going to maintain a relatively constant 12 volts at the base of the transistor. Because the transistor is silicon-based, let's assume that the base-emitter junction has a forward-biased voltage drop of approximately 0.7 volt. This means that the voltage across the load is 12 V - 0.7 V = 11.3 volts. If the load changes or the input voltage changes such that the voltage across the load drops to say 11.2 volts, this means that the base-emitter voltage will increase to 12 V - 11.2 V = 0.8 volt. Because V_{BE} has just gotten larger, the transistor's base current, I_B, will also get larger. In turn the transistor's collector current, I_C, will increase. An increase in I_C means a corresponding decrease in the transistor's collector-to-emitter voltage, V_{CE}. A decrease in V_{CE} means an increase in the voltage across the load, returning the output voltage to its desired level. Conversely, if the load changes or the input voltage changes such that the voltage across the load increases to say 11.4 volts, this means that the base-emitter voltage will decrease to 12 V - 11.4 V = 0.6 volt. Because V_{BE} has just gotten smaller, the transistor's base current, I_B, will also get smaller causing the transistor's collector current, I_C, to decrease. A decrease in I_C causes a corresponding increase in the transistor's collector-to-emitter voltage, V_{CE}. An increase in V_{CE} produces a decrease in the voltage across the load, returning the output voltage to its desired level.

Figure 8.13: Potentiometer-based, series voltage regulator. Adjusting the pot to match the load resistance produces the desired output voltage of 10 volts.

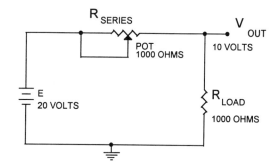

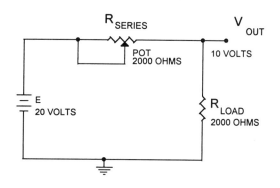

Figure 8.14: Potentiometer-based, series voltage regulator. The load resistance has increased to 2000 Ohms. Adjusting the pot to match the load resistance produces the desired output voltage of 10 volts.

Procedure:

2.1 Set your power supply for an output voltage of 20 volts. Turn off the power supply.

2.2 Construct the circuit shown in Figure 8.15. Adjust your potentiometer to approximately its middle position.

Figure 8.15: Series voltage regulator constructed from discrete components.

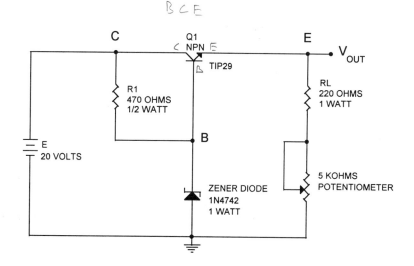

2.3 Turn on the power supply. Use your digital voltmeter to measure the output voltage, V_{OUT}, relative to ground. Does your voltmeter read approximately 11.3 volts? If not, turn off the power supply and check your circuit. If it does read close to 11.3 volts, go on to the next step. 17.44v

2.4 For each of the potentiometer settings shown in Table 8.2, measure and record the output voltage V_E.

Table 8.2: Series regulator output voltage for three different loads.

POTENTIOMETER SETTING	V_{OUT}, VOLTS
MINIMUM RESISTANCE	17.33 ✓
MIDPOINT RESISTANCE	17.44 ✓
MAXIMUM RESISTANCE	17.53 ✓

2.5 A measure of the quality of a power supply is its voltage regulation. The voltage regulation for a power supply is calculated as follows:

$$\% \text{ REGULATION} = [(V_{NL} - V_{FL})/V_{FL}] \cdot 100 \qquad 8.8$$

V_{NL} represents the no-load or open circuit output voltage. V_{FL} represents the output voltage at full load. Full-load voltage is also referred to as the rated load voltage, which produces the rated load output current. The smaller the % regulation is, the better. The perfect or ideal power supply would have a 0% regulation as the no-load voltage would equal the full-load regulation. This is one case where 100% is not good.

2.6 Assuming that the value for V_{OUT} at maximum resistance represents our regulator's no-load voltage and that V_{OUT} at the minimum resistance represents the full-load voltage, calculate the % regulation for the circuit shown in Figure 8.15. Record in Table 8.3 below.

Calculations:

$$\% = ((V_{NL} - V_{FL})/V_{FL}) \times 100$$

$$[(17.33 - 17.53)/17.53] \times 100$$

$$[(17.53 - 17.33)/17.33] \times 100$$

Table 8.3: Percent regulation for the circuit shown in Figure 8.15.

PERCENT REGULATION	1.15%

2.7 Now let's see how our regulator works when the load remains constant but the supply voltage changes. Adjust the potentiometer to its midpoint resistance.

2.8 Measure and record V_{OUT} for each of the supply voltages shown in Table 8.4. Supply voltages of 18 and 22 volts represent a fluctuation of 10% from the nominal voltage of 20 volts.

Table 8.4: Series regulator output voltage for three different supply voltages for the circuit shown in Figure 8.15.

SUPPLY VOLTAGE, E	V_{OUT}, VOLTS
22 18 VOLTS	17.2 v
25 20 VOLTS	17.44 v
28 22 VOLTS	17.77 v

2.9 In order to better understand how this regulator works to maintain a constant output voltage, measure the zener voltage, V_Z, the collector-to-emitter voltage, V_{CE}, the collector-to-base voltage, V_{CB}, and the base-to-emitter voltage, V_{BE}, for each of the three input voltages. Record your results in Table 8.5.

Table 8.5: Series regulator circuit data for three different supply voltages for the circuit shown in Figure 8.15.

SUPPLY VOLTAGE, E	V_Z, VOLTS	V_{CE}, VOLTS	V_{CB}, VOLTS	V_{BE}, VOLTS
22 18 VOLTS	17.77 v	4.88 v	4.31 v	.56 v
25 20 VOLTS	18.02 v	7.59 v	7.0 v	.56 v
28 22 VOLTS	18.35 v	10.3 v	9.72 v	.56 v

2.10 Modify the circuit in Figure 8.15 so that it appears as in Figure 8.16. Measure and record V_{OUT} for each of the supply voltages shown in Table 8.6. The potentiometer should still be set for its midpoint resistance.

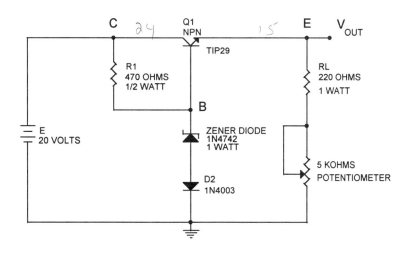

Figure 8.16: Series voltage regulator constructed from discrete components and a pull-up diode.

Table 8.6: Series regulator output voltage for three different supply voltages for the circuit shown in Figure 8.16.

SUPPLY VOLTAGE, E	V_{OUT}, VOLTS
22 18 VOLTS	17.81 v
25 20 VOLTS	18.15 v
28 22 VOLTS	18.47 v

PART 3: Manufacturer's Ratings

3.1 Referring to the manufacturer's data sheet for the 1N4742 zener diode, complete Table 8.7.

Table 8.7: 1N4742 zener diode specifications.

JEDEC Type No.	NOMINAL ZENER VOLTAGE, V_{ZT}, VOLTS	TEST CURRENT I_{ZT}, mA	Z_{ZT} OHMS	Z_{ZK} OHMS	KNEE CURRENT I_{ZK}, mA
1N4742 ~~4746~~ 4746	18 v	14 mA	20 Ω	N/A	N/A

PART 4: Questions

4.1 Referring to the data in Table 8.1, explain your measured value for V_{BC} when the potentiometer was set for minimum resistance. Include the concept of knee current in your explanation.

4.2 How did the addition of the 1N4003 rectifier diode to the series voltage regulator circuit shown in Figure 8.16 change its operation? Use Kirchhoff's Voltage Law in your explanation.

4.3 Referring to the circuit shown in Figure 8.16, why is it a good idea to include the 220 Ohms resistor in series with the potentiometer?

PART 5: Troubleshooting

5.1 Assume that you are troubleshooting the circuit shown in Figure 8.11. The circuit should provide an output voltage of 12 volts. However, when you measure the output voltage, V_{BC}, you find it to be 0.72 volt. What, if anything, is wrong with the circuit? What is the most likely problem?

5.2 Referring to the circuit shown in Figure 8.15, what output voltage, V_{OUT}, would you expect for each of the problems shown in Table 8.8? Record your answer in the table, then explain how you arrived at each of your answers.

Table 8.8: Troubleshooting data for the circuit shown in Figure 8.15.

COMPONENT FAILURE	PREDICTED OUTPUT VOLTAGE, VOLTS
SHORTED ZENER DIODE	low
OPEN ZENER DIODE	large
TRANSISTOR SHORTED FROM COLLECTOR TO EMITTER	low
TRANSISTOR OPEN FROM COLLECTOR TO EMITTER	large

a) Explanation for output voltage with shorted zener diode:

b) Explanation for output voltage with open zener diode:

c) Explanation for output voltage with shorted collector to emitter:

d) Explanation for output voltage with open collector to emitter:

PART 6: Practice Problems

6.1 The output of a power supply with no load applied is 44 volts. The output of the power supply when operating under rated load conditions is 40 volts. Calculate the percent regulation for this power supply.

6.2 A series regulator like that shown in Figure 8.16 has a supply voltage, E, of 24 volts. The output voltage is 15 volts. Answer the following questions.

   a) What is the voltage from the collector to the emitter of the transistor?

   b) What V_{ZT} value for the zener diode must be chosen to produce this output? Assume that your circuit is using the TIP29 transistor.

PART 7: Circuit Design

7.1 Design a zener-diode based, shunt regulator circuit that will meet the following design requirements: a) provide an output voltage of 13 volts with a supply voltage of 20 volts; b) use a 1N4743 zener diode rated for 1 watt; c) provide for a factor of safety of 3. In the space provided below, calculate the value required for the series current-limiting resistor and calculate the smallest load resistor $R_{L(MIN)}$ that can be installed in parallel with the zener diode such that the zener will stay in regulation. Draw a schematic of your final design, and label all components with proper identification. Attach your schematic to your lab report. As a guide to help you complete your design, refer to Formulas 8.1 through 8.6.

Calculations:

PART 8: Electronics Workbench®

8.1 Use Electronics Workbench® to create the circuit shown in Figure 8.15. Connect a D.C. voltmeter to measure the output voltage, V_{OUT}. Vary the potentiometer from minimum to maximum resistance. Print a copy of the schematic showing the results for each of these two conditions. Attach to your report. How do these results compare to those recorded in Table 8.2?

PART 9: Voltage Regulator Formulas

$I_{ZM} = P_{DMAX}/V_{ZT}$ **8.1**

$R_{SERIES(MINIMUM)} = (E - V_{ZT})/I_{ZM}$ **8.2**

$R_{SERIES} = (R_{SERIES(MINIMUM)}) \cdot (F.S.)$ **8.3**

$I_{SERIES} = (E - V_{ZT})/R_{SERIES}$ **8.4**

$I_{LOAD(MAX)} = I_{SERIES} - I_{ZK}$ **8.5**

$R_{LOAD(MIN)} = V_{ZT}/I_{LOAD(MAX)}$ **8.6**

$R_{SERIES} = (E - V_{ZT})/I_{ZT}$ **8.7**

% REGULATION $= [(V_{NL} - V_{FL})/V_{FL}] \cdot 100$ **8.8**

I.C. Voltage Regulators: 9

INTRODUCTION:

Experiment 8 provided you with an introduction to the principles of voltage regulation. As with many other electronic devices and circuits, manufacturers have developed I-Cs that perform tasks previously performed by discrete components. This includes voltage regulator circuits. In this experiment, you will have the opportunity to construct two power supply circuits using I-C voltage regulators. In the first part of this experiment, you will use a voltage regulator that produces a fixed output voltage. In the second part, you will use an adjustable voltage regulator. While these two devices can perform the same operation as several discrete components, these I-Cs are not a substitute for all power supply applications. In many instances, especially high-power applications, it is still necessary to use discrete components. Therefore, it is important to have a solid grasp on how zener diodes and power transistors operate in voltage regulator circuits.

SAFETY NOTE:

Before starting this experiment have your instructor review with you the manufacturer's ratings for the I-C voltage regulator you will be using in this experiment. As with any electronic device, it is important not to exceed the maximum power dissipation rating, $P_{D(MAX)}$. Also, never exceed the device's maximum continuous current rating. When working with voltage regulators, manufacturers often specify the maximum differential voltage that a regulator can withstand. Be absolutely certain to understand this concept before selecting and installing voltage regulators. As always, wear safety glasses, and remove all jewelry from your hands and fingers before the start of the experiment.

OBJECT:

Upon successful completion of this experiment and all reading assignments, the student should be able to:
- design and construct a regulated power supply using an I-C voltage regulator with fixed output
- design and construct a regulated power supply using an I-C voltage regulator with adjustable output

REFERENCES:

www.national.com

MATERIALS:

- 1 - 0.1 µF capacitor, ceramic disc square
- 1 - 0.33 µF capacitor blue
- 1 - 1 µF capacitor, aluminum or tantalum electrolytic yellow
  - miscellaneous lead wires and connectors
- 1 - fixed-output, I-C voltage regulator, LM7812CT, or SK3592, or equivalent 7812
- 1 - adjustable-output, I-C voltage regulator, LM317T, or SK9215, or the equivalent
  LM317
- 1 - 220 Ohms, 1/2 watt resistor
- 1 - 220 Ohms, 2 watt resistor
- 1 - 1 Kohms potentiometer
- 1 - 5 Kohms potentiometer
- 1 - solderless breadboard

EQUIPMENT:
1 - D.C. power supply
1 - digital multimeter

PART 1: The Fixed-Output I-C Voltage Regulator
Background:
The series voltage regulator introduced in Experiment 8 and redrawn in Figure 9.1 relied on three discrete components for maintaining a constant output voltage. The three components--enclosed by dashed lines in Figure 9.1--are the zener diode, a current limiting resistor to set up the zener's reference voltage, and the transistor.

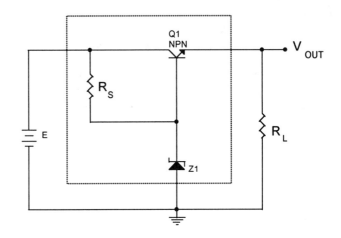

Figure 9.1: Series voltage regulator constructed from discrete components.

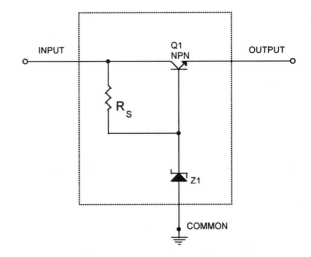

Figure 9.2: Simplified internal construction of a fixed-output voltage regulator.

In Figure 9.2, these same three components are shown without the supply voltage and the load resistance. In the configuration shown, these three electronic devices represent the basic components, or building blocks, for an I-C voltage regulator. In Figure 9.2, the three connections to our discrete-component regulator have been labeled as input, output, and common. Figure 9.3 shows the outline of a typical I-C voltage regulator. Note its three terminals. These three points correspond exactly to the three connections on the circuit shown in Figure 9.2. In reality, the internal circuitry of a voltage regulator such as the one in Figure 9.3 is much more complex. However, our circuit in Figure 9.2 serves as a simplified model to explain the operation and connections to an actual I-C voltage regulator. Figure 9.4 shows the complete circuit for our regulated power supply. C_{IN} is needed if the regulator is located more than a few inches from the power supply filter. Without this capacitor you might notice a slight fluctuation or drop in the output voltage when making a voltage measurement on your circuit. This will especially be true if you are using a long set of test leads to connect your bench-top power supply to the circuit shown in Figure

9.4. As with many electronic circuits, installing C_{OUT} will help to improve the transient response of the regulator by passing high-frequency noise to ground.

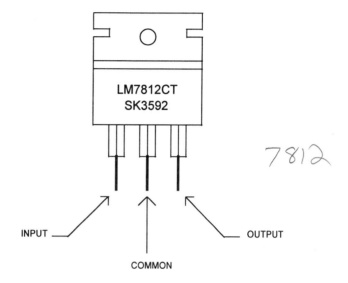

Figure 9.3: Package outline and pin identification of a fixed-output, I-C voltage regulator.

Procedure:

1.1 Set your power supply for an output voltage of 24 volts. Turn off the power supply.
1.2 Construct the circuit shown in Figure 9.4. Adjust your potentiometer to approximately its middle position.
1.3 Turn on the power supply. Use your digital voltmeter to measure the output voltage V_{BC}. Does your voltmeter read approximately 12 volts? If not, turn off the power supply and check your circuit. If it does read close to 12 volts, go on to the next step.
1.4 For each of the potentiometer settings shown in Table 9.1, measure and record the output voltage V_{OUT} (V_{BC}). After making your measurements, turn off your power supply.

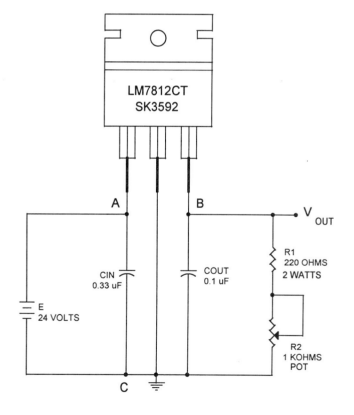

Figure 9.4: I-C voltage regulator installed in a power supply circuit to provide a constant output voltage with varying load and/or input voltage.

Table 9.1: I-C regulator output voltage for three different loads for the circuit shown in Figure 9.4.

R2 POTENTIOMETER SETTING	VOUT, VOLTS
MINIMUM RESISTANCE	11.9 v
MIDPOINT RESISTANCE	11.9 v
MAXIMUM RESISTANCE	11.9 v

1.5 A measure of the quality of a power supply is its voltage regulation. The voltage regulation for a power supply is calculated as follows:

$$\% \text{ REGULATION} = [(V_{NL} - V_{FL})/V_{FL}] \cdot (100) \quad 0\,\% \qquad 9.1$$

V_{NL} represents the no-load or open circuit output voltage. V_{FL} represents the output voltage at full load. Full-load voltage is also referred to as the rated load voltage which produces the rated load output current. Recall from Experiment 8 that a small % regulation is desirable. The perfect or ideal power supply would have a 0% regulation as the no-load voltage would equal the full-load voltage. This is one case where 100% is not good.

1.6 Assuming that the value for V_{OUT} at maximum resistance represents our regulator's no-load voltage and that V_{OUT} at the minimum resistance represents the full-load voltage, calculate the percent regulation for the circuit shown in Figure 9.4. Record in Table 9.2 below.

Calculations:

$$[(11.9v - 11.9v)/11.9v] \times 100$$

Table 9.2: Percent regulation for the circuit shown in Figure 9.4.

PERCENT REGULATION	0 %

1.7 Now let's see how our regulator works when the load remains constant but the supply voltage changes. Adjust the potentiometer to its midpoint resistance.

1.8 Measure and record V_{OUT} for each of the supply voltages shown in Table 9.3. Also measure and record the voltage V_{AB}. After making your measurements, turn off your power supply.

Table 9.3: I-C regulator circuit data for different supply voltages for the circuit shown in Figure 9.4.

SUPPLY VOLTAGE, E	V_{OUT}, VOLTS	V_{AB}, VOLTS
9 VOLTS	7.46 v	1.27 v
21 VOLTS	11.9 v	9.03 v
24 VOLTS	11.9 v	12.22 v
27 VOLTS	11.9 v	15.15 v

PART 2: The Adjustable Output I-C Voltage Regulator
Background:
Thus far in both Experiments 8 and 9, the voltage regulator circuits that you have built were all fixed-output voltage regulators. Now you will construct a variable-output power supply using an adjustable I-C voltage regulator. The package outline and pin identification for an adjustable voltage regulator are shown in Figure 9.5. Without going into the details of the internal operation of this device, once again consider our discrete component regulator circuit shown in Figure 9.2. Imagine how the operation of the circuit in Figure 9.1 would change if we could somehow vary the value of the reference voltage established by the zener diode. If there were such a thing as a zener diode with an adjustable V_{ZT}, we could install it in our circuit. However, keep in mind that there are other devices you have studied that could be installed in place of the zener to vary the reference voltage. For example, we could replace the zener with a high-gain transistor, Q2, as shown in Figure 9.6 and adjust the bias on the transistor's base to vary the collector voltage and, in turn, the reference voltage. Resistor R1 and potentiometer R2 establish an adjustable voltage-divider bias network to bias the base of Q2. Changing the setting of R2 will change the base current to Q2, in turn changing the collector current and the voltage from the collector to emitter. This in turn will change the value for the collector voltage of Q2, which acts as the reference voltage to the base of the series-regulating transistor, Q1.

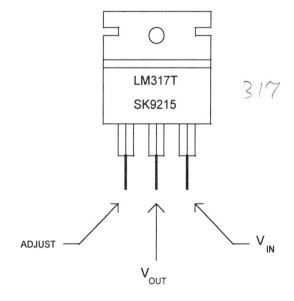

Figure 9.5: Package outline and pin identification of an adjustable-output, I-C voltage regulator.

Notice the similarities between the circuits shown in Figures 9.6 and 9.7. Each regulator has a connection for the input voltage and the output voltage to the load. Between the load and the output connection is a voltage-divider bias network. In Figure 9.6 the voltage across potentiometer R2 is applied to the base of Q2 while in Figure 9.7 the potentiometer's voltage is applied to the *adjust* terminal. While the internal construction of the adjustable I-C voltage regulator is much more complex than the circuit shown in Figure 9.6, the two are very similar in principle of operation. Although the manufacturer's data sheet calls for a 240 Ohms resistor for R1, a 220 Ohms resistor was selected because that is a standard value. The output of the adjustable voltage regulator can be approximated by the following formula:

$$V_{OUT} = (1.25\ V) \cdot (1 + R2/R1) \qquad 9.2$$

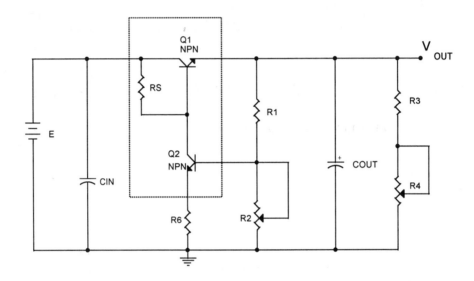

Figure 9.6: Adjustable-output, series voltage regulator constructed from discrete components.

Procedure:

2.1 Set your power supply for an output voltage of 24 volts. Turn off the power supply.

2.2 Construct the circuit shown in Figure 9.7. Adjust potentiometers R2 and R4 to approximately their middle position.

NOTE: While performing the next step of the procedure, you must be certain that potentiometer R4 is adjusted to provide 500 Ohms of resistance. This will prevent overheating resistor R3 due to excessive power dissipation. In all cases, do not exceed the power dissipation rating of the resistors nor the 15-watt rating (with heat sink) of the regulator.

2.3 Turn on the power supply. Use your digital voltmeter to monitor the output voltage, V_{OUT}, relative to ground. Vary the position of the adjustment pot R2 from minimum to maximum resistance. Does your voltmeter reading vary over a wide range of voltages? If not, turn off the power supply and check your circuit. If your voltmeter reading does vary over a wide range of voltage, go on to the next step.

2.4 With R4 set in the middle position (500 Ohms), adjust R2 so that the output voltage, V_{OUT}, is 14 volts. Do *not* allow V_{OUT} to exceed this voltage level for the remainder of the experiment.

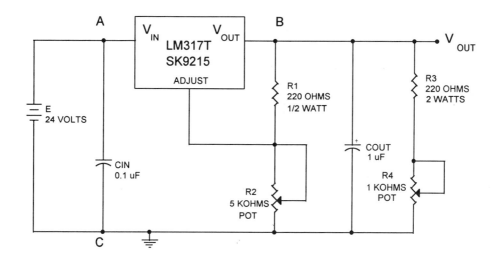

Figure 9.7: Adjustable-output, I-C voltage regulator installed in a power supply circuit to provide a range of output voltages. The output voltage can be set by adjusting the resistance of R2.

2.5 For each of the potentiometer settings for R4 shown in Table 9.4, measure and record the output voltage, V_{OUT}.

Table 9.4: I-C regulator output voltage for three different loads for the circuit shown in Figure 9.7.

R4 POTENTIOMETER SETTING	V_{OUT}, VOLTS
MINIMUM RESISTANCE	13.81 v
MIDPOINT RESISTANCE	14.00 v
MAXIMUM RESISTANCE	14.08 v

2.6 Assuming that the value for V_{OUT} at maximum resistance represents our regulator's no-load voltage and that V_{OUT} at the minimum resistance represents the full-load voltage, calculate the percent regulation for the circuit shown in Figure 9.7. Record your result in Table 9.5 below.

Calculations:

$$((13.81v - 14.08v)/14.08v) \times 100$$

Table 9.5: Percent regulation for the circuit shown in Figure 9.7.

PERCENT REGULATION	1.92%

2.7 Now let's see how our regulator works when the load remains constant but the supply voltage changes. Adjust potentiometer R4 to its midpoint resistance. R2 should still be adjusted to produce an output of 14 volts.

2.8 Measure and record V_{OUT} for each of the supply voltages shown in Table 9.6.

Table 9.6: I-C regulator circuit data for different supply voltages for the circuit shown in Figure 9.7.

SUPPLY VOLTAGE, E	V_{OUT}, VOLTS	V_{AB}, VOLTS
11 VOLTS	9.43v	1.54v
21 VOLTS	13.94v	7.09v
24 VOLTS	14.01v	10.0v
27 VOLTS	14.04v	12.95v

PART 3: Manufacturer's Ratings

3.1 Referring to the manufacturer's data manual for your fixed-output voltage regulator, complete Table 9.7.

Table 9.7: Fixed-output voltage regulator specifications.

I-C Part No.	OUTPUT TYPE	NOMINAL OUTPUT VOLTAGE, VOLTS	V_{IN}(Min) VOLTS	V_{IN}(Max) VOLTS	TOTAL POWER DISSIPATION, WATTS
LM7812	FIXED	12 v	12 v	12 v	12 W / 1A

3.2 Referring to the manufacturer's data manual for your adjustable-output voltage regulator, complete Table 9.8.

Table 9.8: Adjustable-output voltage regulator specifications.

I-C Part No.	OUTPUT TYPE	OUTPUT VOLTAGE RANGE, VOLTS	V_{IN}(Min) VOLTS	V_{IN}(Max) VOLTS	TOTAL POWER DISSIPATION, WATTS
317T	ADJUSTABLE	1.2v - 37v	1.2v	37v	55.5 W / 1.5A

PART 4: Questions

4.1 Is the voltage regulator shown in Figure 9.4 an example of a series regulator or a shunt regulator? Explain how you arrived at your answer.

Series, because its in series.

4.2 If you performed Experiment 8, how does the voltage regulation of your I-C regulator in Figure 9.4 compare to the regulation of the voltage regulator constructed from discrete components shown in Figure 8.15? Refer to your data in Tables 8.3 and 9.2.

The voltage regulator had a 1.15% regulation and the I-C regulator had a 0% regulation.

4.3 What is the relationship between the voltages V_{AB}, V_{BC}, and V_{AC} in Figure 9.4? Express your answer as a mathematical formula.

$V_{AB} - V_{BC} = V_{AC}$

One helps find the other

4.4 Referring to the circuit shown in Figure 9.4, why is it a good idea to include the 220 Ohms resistor in series with the potentiometer?

to cut down current & voltage to the pot. Its easier to adjust.

4.5 Referring to the data shown in Tables 9.3 and 9.6, what can you conclude about the operation of a voltage regulator? Specifically, discuss what is required to maintain a constant output of 12 volts in Figure 9.4 and 14 volts in Figure 9.7.

When they maintain their max voltage they will not go any higher. The amount of resistance in the potentiometer.

4.6 Referring to your data in Table 9.3, explain how the regulator shown in Figure 9.4 operates to maintain a constant output voltage. Refer to the voltages for E (V_{AC}), V_{AB}, and V_{OUT} (V_{BC}) in your explanation.

When it reaches its peak voltage it will not let any more go through the regulator. V_{AB} is constant but V_{BC} will still climb.

PART 5: Troubleshooting

5.1 Assume that you are troubleshooting the circuit shown in Figure 9.4. The circuit should provide an output voltage of 12 volts. You measure a supply voltage of 24 volts. However, when you measure the output voltage, V_{BC}, you find it to be 0.0 volt. You vary the resistance of R2, and there is no change. What is the most likely problem? How would you confirm your suspicions?

You have a bad regulator or perhaps have the leads wired up wrong.

5.2 Referring to the circuit shown in Figure 9.7, what output voltage, V_{OUT}, would you expect for each of the problems shown in Table 9.9? Record your answer in the table, then explain how you arrived at each of your answers.

Table 9.9: Troubleshooting data for the circuit shown in Figure 9.7.

COMPONENT FAILURE MODE	PREDICTED OUTPUT VOLTAGE, VOLTS
REGULATOR SHORTED	24 v
REGULATOR OPEN	0 v
R2 SHORTED	10 v
R2 OPEN	24 v

a) Explanation for output voltage with regulator shorted:

All voltage passes through the regulator for the output voltage.

b) Explanation for output voltage with regulator open:

No voltage can go through the regulator so there is no output voltage.

c) Explanation for output voltage with R2 shorted:

d) Explanation for output voltage with R2 open:

With the pot open

PART 6: Practice Problems

6.1 The output of a power supply with no load applied is 15.2 volts. The output of the same power supply when operating under rated load conditions is 15 volts. Calculate the percent regulation for this power supply.

6.2 A voltage regulator circuit like that shown in Figure 9.4 has a supply voltage, E, of 24 volts. The output voltage is 15 volts. What is the voltage from point A to point B, V_{AB}?

9v ✓

6.3 If the adjustment potentiometer R2 shown in Figure 9.7 is set for a resistance of 2860 Ohms, what should be the value for the output voltage across the load?

close to 17.2v

6.4 In Figure 9.7, what resistance setting would you have to select for potentiometer R2 to obtain an output voltage of 12.5 volts?

24v   17.2 at 286∩
24v   17.5 at 2.4k

PART 7: Circuit Design

7.1 Design a series regulator circuit that will meet the following design requirements: a) provide an output voltage of 15 volts with an input (supply) voltage of 24 volts; b) provide for input and output capacitor filtering as shown in Figure 9.4; and c) use a fixed-output I-C voltage regulator. Draw a schematic of your final design. Label all components with proper identification. Specify the manufacturer's part number for the voltage regulator that you choose and attach your schematic to your lab report. If a computer-aided drafting software package is available at your institution, use it to produce your drawing. Print a copy and attach it to your report.

7.2 Design a complete power supply that will meet the following design requirements: a) provide an output voltage of 15 volts; b) provide for input and output capacitor filtering as shown in Figure 9.7; c) use an adjustable-output I-C voltage regulator; d) the power supply is to be plugged into a 120 volt RMS conventional wall outlet; e) the 120 volts RMS is to be stepped down to an appropriate voltage to meet the remainder of your design criteria; f) the stepped-down A.C. voltage is to be full-wave rectified then filtered with a capacitor; and g) the filtered D.C. voltage is to then be fed into the regulator. Draw a schematic of your final design. Label all components with proper identification. Specify the manufacturer's part number for the voltage regulator that you choose and attach your schematic to your lab report. If a computer-aided drafting software package is available at your institution, use it to produce your drawing. Print a copy and attach it to your report.

PART 8: I.C. Voltage Regulator Formulas

% REGULATION = $[(V_{NL} - V_{FL})/V_{FL}] \cdot (100)$ **9.1**

$V_{OUT} = (1.25\ V) \cdot (1 + R2/R1)$ **9.2**

Current Sources and Regulators: 10

INTRODUCTION:

Experiments 8 and 9 provided you with the background necessary to enable you to construct constant voltage sources. Remember, an ideal voltage source will provide a constant voltage with varying load. If we then apply Ohm's Law and solve for the current ($I = V/R$), we realize that there is a corresponding change in the current for a change in load resistance when the supply voltage is held constant. In some applications, it is necessary to have a constant current source. A constant current power supply will provide a constant current with varying load. Applying Ohm's Law again, we can solve for the voltage in terms of the current ($V = I \cdot R$). From this formula, we can see that for a constant current with a varying load resistance, there must be a corresponding change in voltage. You might be thinking to yourself, "Why is he discussing the obvious? I mastered Ohm's Law a long time ago!" Well, as is usually the case, one can explain how most electronic devices work by referring to basic electronic principles such as Ohm's Law, Kirchhoff's Voltage Law, etc. Well, here's the bottom line, so to speak. To maintain a constant current to a varying load, the output of the power supply (the voltage across the load) must be automatically varied or adjusted as the load changes. To sense the change in load, there must be some feedback to the regulating part of the current supply. Recall the feedback connection to the I.C. voltage regulators and the transistor-based voltage regulators in the previous two experiments.

In this experiment, you will use two different electronic devices to construct constant current sources. The first device is the fixed-output I.C. voltage regulator used in Experiment 9. Yes, we can indeed construct a constant-current source from a voltage regulator. In the second circuit, you will use an I.C. designed specifically to act as an adjustable current source. Current sources can take many forms in addition to the two circuits that you will construct in this experiment. Circuits in which you may find constant current sources include differential amplifiers, display drivers, D/A converters, and instrumentation circuits, to name a few. An example of a high-power current source is the differentially-compounded generator used in mobile welding machines. The two common schematic symbols used to represent a current source are shown in Figure 10.1.

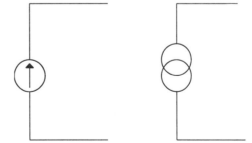

Figure 10.1: Schematic symbols for a current source.

SAFETY NOTE:

Before starting this experiment have your instructor review with you the manufacturer's ratings for the I-C voltage regulator and the adjustable current source you will be using in this experiment. As with any electronic device it is important not to exceed the maximum power dissipation rating, $P_{D(MAX)}$. Also, never exceed the device's maximum continuous current rating, I_{OUT}. When working with voltage regulators, manufacturers often specify the maximum differential voltage that a regulator can withstand. Be absolutely certain to understand this concept before selecting and installing voltage regulators. As always, wear safety glasses, and remove all jewelry from your hands and fingers before the start of the experiment.

OBJECT:
Upon successful completion of this experiment and all reading assignments, the student should be able to:
- design and construct a constant current source using an I-C voltage regulator with fixed output
- design and construct a constant current source using an I-C three-terminal adjustable current source

REFERENCES:
www.national.com
National Semiconductor's® Data Acquisition Databook

MATERIALS:
- 1 - LM334Z three-terminal adjustable current source
- 1 - solderless breadboard
  - miscellaneous lead wires and connectors
- 1 - fixed-output, I-C voltage regulator, LM7812CT, or SK3592, or equivalent
- 1 - 200 Ohms, ten-turn potentiometer
- 1 - 2 Kohms, ten-turn potentiometer
- 1 - 1 Kohms, 1 watt resistor
- 1 - 100 Ohms, 1/2 watt resistor

EQUIPMENT:
- 1 - D.C. power supply
- 1 - digital multimeter

PART 1: I-C Voltage Regulator-Based Constant Current Source

Background:
Before constructing the first circuit, let's take a moment to review the operation of the constant voltage circuit that you constructed from discrete components in Experiment 8. This circuit is shown in Figure 10.2.

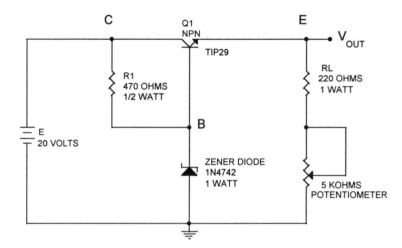

Figure 10.2: Series voltage regulator constructed from discrete components.

Recall that R1 and the zener diode set up a fixed reference voltage at the base of the transistor. If there is a change in the output voltage, there will be a corresponding change in the transistor's base current. An increase in the base current will cause the resistance from collector to emitter to decrease.

Conversely, a decrease in base current will cause the resistance from collector to emitter to increase. This automatic variation of the collector-to-emitter resistance will cause the collector-to-emitter voltage to vary, serving to keep the output voltage across the load at a constant value. Imagine that we now modify the constant-voltage circuit shown in Figure 10.2 so that it appears like that shown in Figure 10.3. In this circuit, a variable load is depicted by the potentiometer.

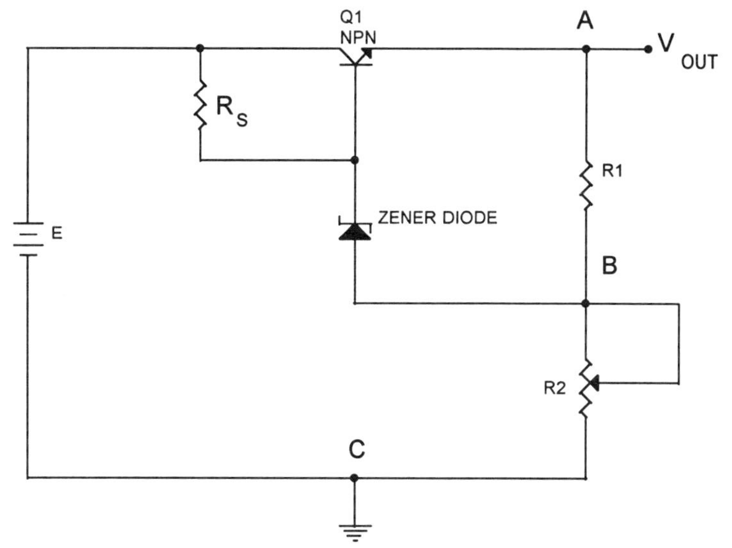

Figure 10.3: Series voltage regulator modified to act as a constant current source.

Recall from the introduction where I said there should be some form of feedback to our power supply control circuit to maintain a constant current. Note the major difference between the circuits shown in Figures 10.2 and 10.3. In Figure 10.3, the anode of the zener diode is no longer connected directly to ground but is connected instead to the load resistance. In Figure 10.2, a constant voltage was established at the base of the transistor. Now, in the circuit shown in Figure 10.3, if load resistance R2 varies, the reference voltage established at the base will also change. In turn, the collector-to-emitter voltage will change, which will cause the voltage V_{AC} to vary, keeping the output current through R1 and R2 constant. In this particular circuit, the voltage across R1, V_{AB}, will be constant because R1 and the current through it are constant. As explained in the introduction, the voltage across the load must change to maintain a constant current. Just how much the load resistance can change depends in large part on how much voltage the power supply can provide. This is an important concept! For example, if you are designing a current source to provide a constant 10 mA, the power supply will have to provide a voltage of 10 volts to a 1 Kohms load. A simple example of this is shown in Figure 10.4 a). But what if the load increases to 5 Kohms? That would mean that the power supply output voltage would have to increase to 50 volts to "push" 10 mA of current through 5 Kohms of resistance. Refer to Figure 10.4 b). You can see from this discussion that you would need a heavy-duty power supply to provide a constant current of more than a few milliamps to a circuit with a widely varying load. Fortunately, this will not be necessary for most electronic circuits requiring a constant current. There are some exceptions, however. Welding equipment is an example of electrical machinery that must provide a rather large constant current-- sometimes as much as 100 to 200 amperes or more. As experienced welders know, the exact amount of current depends largely on the size and type of material to be joined. Therefore, welding equipment must be capable of providing a wide range of currents depending on the work being performed. Once set, the welding equipment must be capable of maintaining a constant current even if the length of the welder's arc varies. As the length of the welder's arc varies so does the voltage drop from the weld rod to the metal being joined. Varying arc length is analogous to varying load resistance in an electronic circuit.

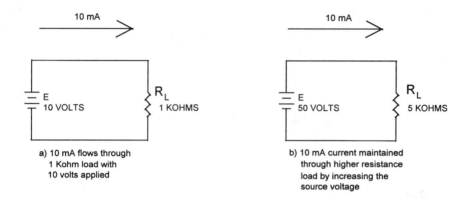

Figure 10.4: One example of how a constant current can be maintained through a varying load resistance.

Procedure:

1.1 Set your power supply for an output voltage of 30 volts. Turn off the power supply.
1.2 Construct the circuit shown in Figure 10.5. Adjust your potentiometer to provide a resistance of 200 Ohms.
1.3 Turn on the power supply. Use your digital voltmeter to measure the voltages V_{AC} and V_{AB}. Record your results in Table 10.1.

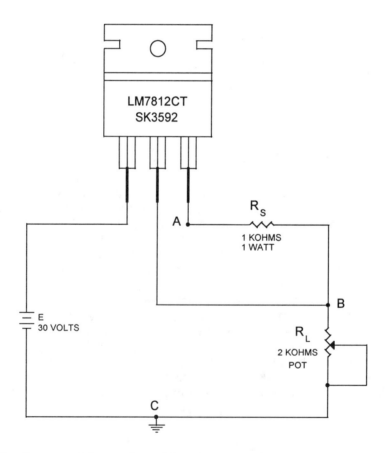

Figure 10.5: I-C voltage regulator configured to act as a constant current source with varying load.

1.4 For each of the potentiometer settings shown in Table 10.1, measure and record voltages V_{AC} and V_{AB}. Turn off the power supply and measure the resistance of R_S.

1.5 Now use Ohm's Law, the measured value for the resistance of R_S, and the values for V_{AB} to calculate the corresponding current through R_S, I_{RS}. Assuming that the current through R_S equals the current through the load resistance, R_L, then the value for I_{RS} is also the value for the load current, I_{RL}. Record the calculated values for I_{RL} in Table 10.1.

Calculations:

Table 10.1: Current regulator output voltages and load current for different loads for the circuit shown in Figure 10.5 with a supply voltage of 30 volts.

POTENTIOMETER SETTING	V_{AC}, VOLTS	V_{AB}, VOLTS	I_{RL}, mA
200 OHMS			
400 OHMS			
600 OHMS			
800 OHMS			
1000 OHMS			
1500 OHMS			

1.6 Turn on the power supply. Increase the power supply voltage to 35 volts. Repeat steps 1.4 and 1.5 of the procedure, and record your results in Table 10.2. Turn off the power supply.

Calculations:

Table 10.2: Current regulator output voltages and load current for different loads for the circuit shown in Figure 10.5 with a supply voltage of 35 volts.

POTENTIOMETER SETTING	V_{AC}, VOLTS	V_{AB}, VOLTS	I_{RL}, mA
200 OHMS			
400 OHMS			
600 OHMS			
800 OHMS			
1000 OHMS			
1500 OHMS			

PART 2: The Three-Terminal Adjustable I.C. Current Source
Background:
The package outline and pin identification for the LM334Z adjustable current source is shown in Figure 10.6. The circuit that you will build is shown in Figure 10.7. The potentiometer, R_{SET}, is used to adjust the current through the series-connected load and current regulator to the desired value. This particular I.C. current source can be operated from a voltage source in the range of 1 to 40 volts. The I.C. can be adjusted via R_{SET} to provide a constant current in the range of 1μA to 10 mA.

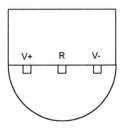

Figure 10.6: Bottom view and pin identification of the LM334Z three-terminal adjustable current source.

Procedure:
2.1 Set your power supply for an output voltage of 12 volts. Turn off the power supply.

2.2 Construct the circuit shown in Figure 10.7. Adjust potentiometer R2 so that the series resistance of R1 and R2 equals 1000 Ohms.

2.3 Turn on the power supply. Use your digital voltmeter to monitor the output voltage, V_{AB}. Vary the resistance of R_{SET} until V_{AB} is 5 volts. This corresponds to a current of 5 mA. Once adjusted, do not change the resistance of R_{SET} for the remainder of the experiment.

2.4 For each of the load resistance values (R1 + R2) shown in Table 10.3, measure and record the corresponding values for V_{AB} and V_{BC}. When you complete all measurements, turn off the power supply.

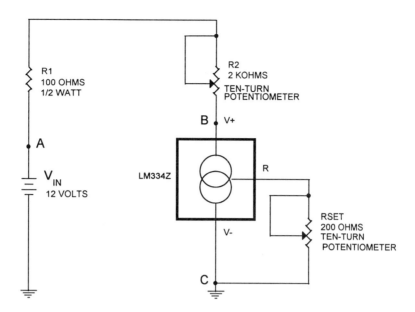

Figure 10.7: Current regulator constructed from three-terminal adjustable current source.

2.5 Using Ohm's Law, the values for the load resistance, and the corresponding values for V_{AB}, calculate the resulting load current, I_{LOAD}. Record your results in Table 10.3.

Calculations:

Table 10.3: Current regulator output voltages and load current for different loads for the circuit shown in Figure 10.7.

LOAD RESISTANCE	V_{AB}, VOLTS	V_{BC}, VOLTS	I_{LOAD}, mA
200 OHMS			
600 OHMS			
1000 OHMS			
1200 OHMS			
1600 OHMS			
2000 OHMS			

PART 3: Questions

3.1 Over what range of load resistance values did the circuit shown in Figure 10.5 maintain a nearly constant current?

3.2 Referring to the data shown in Table 10.1, explain how the voltage regulator operates to maintain a constant current. Refer to the voltages E, V_{AC}, and V_{AB} in your explanation.

3.3 In the original circuit shown in Figure 10.5, the current source started to go out of regulation (that is, the current started to drop off) with the application of the 1.5 Kohms load. Referring to the data in Table 10.1, explain why this happened.

3.4 When you increased the supply voltage to 35 volts, the current source went back into regulation (the current increased to the same value as for the other loads) with the application of the 1.5 Kohms load. Explain why (how) this happened.

3.5 What is the relationship between the voltages V_{AB}, V_{BC}, and V_{AC} in Figure 10.5? Express your answer as a mathematical formula.

3.6 Referring to the data shown in Table 10.3, comment on the I.C. current regulator's ability to maintain a constant current.

3.7 Referring to the data shown in Table 10.3, explain how the I.C. current regulator operates to maintain a constant current. Refer to the voltages V_{IN}, V_{AB}, and V_{BC} in your explanation.

Introduction to the Wattmeter: 11

INTRODUCTION:

In large manufacturing facilities, a major operating cost is the electric bill. Some manufacturing facilities have an electric utility bill that can average $500,000 per month. And you thought your utility bills were high! Plant engineers are not only responsible for keeping the equipment in their plant running, but also are concerned with the efficient operation of this equipment. Rotating machinery, in particular, consumes a large part of the electrical power load. Even a fraction of a percent improvement in the efficiency of the electric motors in a plant would likely save a large corporation several hundred, if not thousands, of dollars each month. This would obviously have a large economic impact. This is why plant engineers regularly monitor their operating costs and try to find areas for improvement. This requires information, which in turn requires tools. A set of tools for the electrical maintenance technician to determine if a given piece of electrical equipment is operating at an acceptable level of efficiency would include a voltmeter, clamp-on ammeter, and wattmeter. The results of the maintenance technician's work can then determine if a machine needs to be replaced with a more efficient model, or operated at higher loads (electrical machinery typically operates more efficiently at, or near, rated load), repaired (a motor's field windings may have some shorted turns due to overheating), have power-factor correction capacitors installed, or have some other form of corrective action taken. At some time or other, a maintenance technician may have to install a wattmeter in an instrument panel or replace a defective one. To prepare you for any of these scenarios in your future career, this experiment will provide you with the opportunity to install a wattmeter in an A.C. circuit with a purely resistive load and then with an inductive load. A wattmeter can also be used to measure the power consumed by a load in a D.C. circuit. The connection of the wattmeter to a D.C. load is the same as for the connection of a single-phase A.C. load. Refer to Part 1 of the experiment for a discussion of the wattmeter connections.

SAFETY NOTE:

Before starting this experiment, have your instructor help you to identify the connections to the wattmeter and review with you the proper procedure for connecting the wattmeter. You will also be operating the circuit at high voltage, so take extra precaution to prevent shorts. Make sure that there are no exposed leads or connectors. Also, the power resistors will become very hot. Do not touch them after energizing your circuit. Before circuit disassembly, allow 15 minutes for the resistors to cool down after turning off the power for the last time. Make sure that the resistors are placed on a nonconductive surface with a high melting point. Also have your instructor tell you what the maximum allowable power, current, and voltage ratings for your load resistors and inductor are so that you will not exceed any of these during the experiment. As always, wear safety glasses and remove all jewelry from your hands and fingers prior to the start of the experiment.

OBJECT:

Upon successful completion of this experiment and all reading assignments, the student should be able to:
- install a wattmeter in a single-phase circuit to measure the real or active power
- use a wattmeter to measure real power
- calculate the apparent power for a given circuit
- calculate the reactive power for a given circuit
- calculate the power factor for a given circuit

MATERIALS:
- 2 - 50 Ohms, 225 watt wire-wound power resistors
- miscellaneous lead wires and connectors

EQUIPMENT:
- 1 - single-phase or three-phase, variable A.C. power supply capable of 0 to 120 volts RMS
- 1 - hand-held multimeter
- 1 - in-line ammeter, 0 to 8 amps or hand-held clamp-on ammeter
- 1 - wattmeter, 0 to 500 Watts
- 1 - single-phase, 120 volt A.C. split-phase motor or equivalent inductive load

PART 1: The Wattmeter Circuit
Background:
An analog wattmeter has two coils--a current coil and a voltage coil. As with an ammeter, it will be necessary to connect the current coil in series with the power supply and the load. The voltage coil will be installed across the load. When determining the efficiency of an A.C. circuit, there are four important parameters to determine. They are *real power, apparent power, reactive power,* and *power factor*. Real power, also referred to as *true* or *active* power, is the useful electrical power that the machine can convert into mechanical power. Real power has the familiar units of watts, W. You will use the wattmeter to determine the real power in this experiment. The apparent power represents what the electric utility's industrial customers have to pay for. When reactive loads are present, the apparent power, S, is always greater than the real or useful power, P. If you are the customer, then you want the real power to be as close as possible to the apparent power. The apparent power, which has units of volt-amperes (abbreviated V-A), can be found by measuring the A.C. voltage across the load and multiplying it by the current through the load. The formula then is:

$$S = (V) \cdot (I) \qquad \qquad 11.1$$

The presence of reactive loads causes a *phase shift* between the load voltage and line current. We will refer to this phase shift or *phase angle* by the Greek symbol θ. The real or useful power may then be calculated as follows:

$$P = (V) \cdot (I)\cos(\theta) \qquad \qquad 11.2$$

In a purely resistive circuit, the phase angle is 0°. Substituting 0° for θ in Formula 11.2 results in Formula 11.1. In effect, what this tells us is that in a purely resistive circuit, the apparent power equals the real power and that the net or effective reactance present is zero. Therefore, in such circuits, the net or effective reactive power is also zero. Where there is a net or effective reactive component present, whether it be capacitance or inductance, the reactive power, Q, may be calculated as follows:

$$Q = (V) \cdot (I)\sin(\theta) \qquad \qquad 11.3$$

The unit of measure for reactive power is the volt-ampere-reactive, VAR. Using trigonometry, it can also be shown that apparent power, real power, and reactive power are related according to the Pythagorean Theorem as follows:

$$S^2 = P^2 + Q^2 \qquad \qquad 11.4$$

In Formula 11.2, the cosine of the phase angle is also referred to as the power factor, P.F. This is another measure or indicator of how resistive, or reactive depending on your viewpoint, a particular circuit is. Hence we have,

$$P.F. = \cos(\theta) \tag{11.5}$$

If we were to combine Formulas 11.1 and 11.2, then substitute into Formula 11.5, we would have another formula for the power factor as follows:

$$P.F. = P/S \tag{11.6}$$

Now that we have examined some of the mathematical relationships that are fundamental to the study of single-phase or three-phase circuits, you should be ready to proceed to the hands-on part of this experiment.

Procedure:
    1.1 Obtain all materials and equipment required to complete this experiment.
    1.2 Use your hand-held multimeter to measure the resistance of each power resistor, and record in Table 11.1. If either resistor is out of tolerance, notify your instructor immediately and obtain a replacement.

Table 11.1: Nominal and measured resistor values.

NOMINAL RESISTANCE	MEASURED RESISTANCE
R1 = 50 Ohms	
R2 = 50 Ohms	

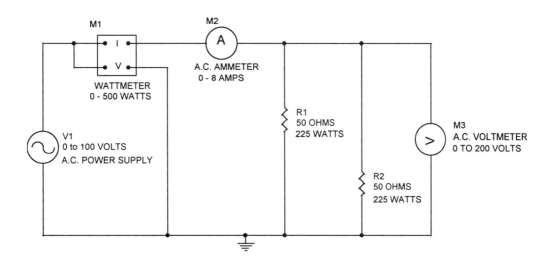

Figure 11.1: Wattmeter circuit with resistive load.

    1.3 With your power supply turned off and adjusted to produce zero volts when turned on, connect the circuit shown in Figure 11.1. You may use either an in-line ammeter to measure the load current or a clamp-on style ammeter. Be sure that the wattmeter's current coil is connected in series with the load and power supply and that the voltage coil is connected across the load. Have your instructor check your work.

1.4 If you are using any analog meters in this experiment, always check the polarity of your connections and be on the alert for any downscale deflection of the meter movement. It is possible for the meter movement of a wattmeter to deflect downscale. Should this happen, turn off the power supply immediately and call your instructor for assistance. Also be on the alert for excessive upscale deflection or pegging of the meter movement. Select an appropriate full-scale current value to prevent this from happening.

1.5 Turn on the power supply. <u>Slowly</u> increase the supply voltage until M3, the A.C. voltmeter, indicates 25 volts RMS. Record the corresponding ammeter and wattmeter readings in Table 11.2.

1.6 Repeat step 1.5 for a supply voltage of 50 V, 75 V, and 100 V RMS.

1.7 Reduce the supply voltage to 0 volts. Turn off the power supply.

1.8 Complete Table 11.2 by calculating the volt-amperes and power factor for each entry in the table.

Calculations:

Table 11.2: Wattmeter data for the circuit shown in Figure 11.1.

VOLTS	AMPERES	WATTS	VOLT-AMPERES	POWER FACTOR
25				
50				
75				
100				

1.9 If your results "appear" correct, disassemble your circuit and proceed to step 1.10. Otherwise, call your instructor, discuss your results, and come up with an appropriate plan of action.

1.10 With your power supply turned off and adjusted to produce zero volts when turned on, connect the circuit shown in Figure 11.2. The inductor represents the *main* or *running* winding of a single-phase, 120 volt A.C. split-phase motor or equivalent inductive load. If you have not previously worked with an A.C. motor, have your instructor help you identify the main winding connections. Do not connect the motor's starting windings. Even though the starting windings are <u>not</u> to be connected, the output shaft of the motor should still be covered with a protective shield should the motor shaft begin to turn.

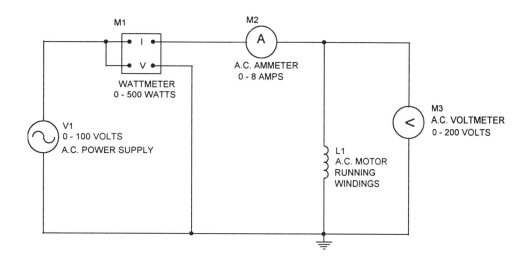

Figure 11.2: Wattmeter circuit with inductive load.

1.11 Turn on the power supply. <u>Slowly</u> increase the supply voltage until M3, the A.C. voltmeter, indicates 25 volts RMS. Record the corresponding ammeter and wattmeter readings in Table 11.3. Reduce your power supply to 0.0 volt.

1.12 Repeat step 1.11 for a supply voltage of 50 V, 75 V, and 100 V RMS. Return your power supply to 0.0 volt between each measurement and do not leave the power on once you complete your measurements. This will prevent overheating of the motor windings.

1.13 Reduce the supply voltage to 0 volts. Turn off the power supply.

1.14 Complete Table 11.3 by calculating the volt-amperes and power factor for each entry in the table.

Calculations:

Table 11.3: Wattmeter data for the circuit shown in Figure 11.2.

VOLTS	AMPERES	WATTS	VOLT-AMPERES	POWER FACTOR
25 VOLTS				
50 VOLTS				
75 VOLTS				
100 VOLTS				

1.15 If your results "appear" correct, disassemble your circuit. Otherwise, call your instructor, discuss your results, and come up with an appropriate plan of action.

PART 2: Questions

2.1 Are the results in Table 11.2 what you expected for a resistive load? Refer to your results, and compare them to what you would expect in theory.

2.2 Are the results in Table 11.3 what you expected for an inductive load? Refer to your results, and compare them to what you would expect in theory.

PART 3: Practice Problems

3.1 A single-phase motor is being powered by a 120 volt A.C. source. You measure the line current drawn by the motor to be 2 amperes. Using a wattmeter, you measure the electrical power to be 210 watts. Answer the following questions. Show units where required.

a) What is the apparent power?

b) What is the power factor?

c) What is the phase angle between the source voltage and line current?

3.2 An inductive load requires 100 volt-amperes of apparent power and 80 watts of real power. What is the reactive power?

3.3 An inductive load requires 200 volt-amperes of apparent power at a power factor of 0.90. Answer the following questions.

a) What is the real power?

b) What is the reactive power?

PART 4: Single-Phase Power Formulas

$S = (V) \cdot (I)$ **11.1**

$P = (V) \cdot (I)\cos(\theta)$ **11.2**

$Q = (V) \cdot (I)\sin(\theta)$ **11.3**

$S^2 = P^2 + Q^2$ **11.4**

$P.F. = \cos(\theta)$ **11.5**

$P.F. = P/S$ **11.6**

Delta and Wye Circuits: 12

INTRODUCTION:
The most common form of electrical power used to operate rotating machinery and many other types of equipment in industrial operations throughout the United States is three-phase power. In your home the most common form of electrical power is single-phase. This is nominally a 120 volt RMS supply that powers most of your home appliances and 240 volts RMS to electric clothes dryers, electric stoves, and some electric water heaters. The 240 volt source, as discussed in Experiment 4, can be thought of as two 120 volt RMS sources in series. A three-phase power supply produces three sinusoidal voltage waveforms of equal magnitude and frequency (typically 60 Hz). Each supply is phase shifted, or separated, from its closest neighbor by a phase angle of 120°. Refer to Figure 12.1. At this point you might be saying to yourself, "Why go to all that trouble to generate three sine waves that are each 120° apart? Why not just produce one large sinusoidal source?" Well, as it turns out, operating rotating machinery from a three-phase source results in smoother and more efficient operation than from a single-phase source of equal power. Your local electric utility uses large alternators to produce three-phase voltage waveforms, which are increased in magnitude by three-phase transformers, then passed via overhead transmission lines to substations where the voltages are stepped down to lower levels for distribution to local businesses. Finally, these three-phase voltages will be reduced further to a working level by a transformer or transformers owned and operated by individual businesses. The three A.C. sources produced at the electric utility may be thought of as three individual A.C. supplies, as shown in Figure 12.2. These three voltages are produced by three individual windings that are installed in the stator--stationary part--of the electric company's alternator and would be connected in a configuration similar to that shown in Figure 12.3. The three windings are physically arranged and electrically connected in a configuration such that the rotating magnetic field produced by the alternator's rotor will induce a voltage in each of the three windings. While it appears that there are just three voltages to analyze, there are actually six voltages. First, there are the three voltages measured from lines 1, 2, and 3 relative to their common, or *neutral*, connection, N. From now on we will refer to these voltages as V_{L1-N}, V_{L2-N}, and V_{L3-N}. These are the line-to-neutral voltages. Next we have the line-to-line voltages. These are the voltages measured from one line relative to another. Remember, we not only measure voltages in circuits relative to ground, but we also measure voltages at one point relative to another. In this case we are interested in the voltage at line 1 relative to line 2, line 2 relative to line 3, and line 1 relative to line 3. We will abbreviate these as V_{L1-L2}, V_{L2-L3}, and V_{L1-L3}. In this experiment you will have the opportunity to construct two common configurations of three-phase circuits, the *wye* connection and the *delta* connection. You will investigate the relationship between the voltages and currents in these two different three-phase connections.

SAFETY NOTE:
You will be operating each circuit at high voltage, so take extra precaution to prevent shorts. Make sure that there are no exposed leads or connectors. Also, the power resistors will become very hot. Do not touch them after energizing your circuit. Before circuit disassembly, allow 15 minutes for the resistors to cool down after turning off the power. Make sure that the resistors are placed on a nonconductive surface with a high melting point. Also have your instructor tell you what the maximum allowable power, current, and voltage ratings for your load resistors are so that you will not exceed any of these during the experiment. As always, wear safety glasses, and remove all jewelry from your hands and fingers before the start of the experiment.

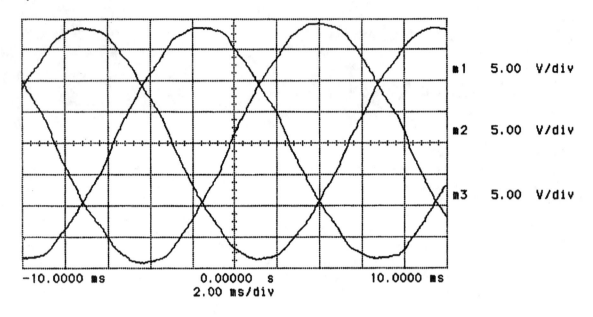

Figure 12.1: Sinusoidal voltage waveforms produced by a three-phase alternator displaced by 120°.

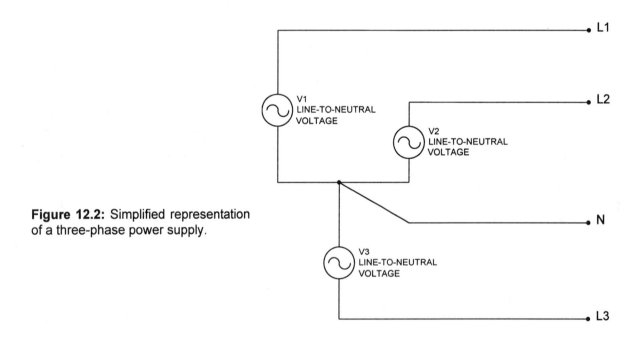

Figure 12.2: Simplified representation of a three-phase power supply.

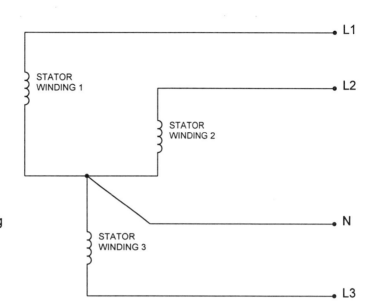

Figure 12.3: Simplified schematic of a three-phase alternator showing three stator windings or phases.

OBJECT:
Upon successful completion of this experiment and all reading assignments, the student should be able to:
- connect a three-phase purely resistive load in either wye or delta
- measure the line-to-line and line-to-neutral voltages in a three-phase circuit
- measure the line and phase currents in a three-phase circuit
- calculate all currents and voltages in a three-phase purely resistive circuit with balanced loads

MATERIALS:
    3 - 50 Ohms, 225 watt wire-wound power resistors
      - miscellaneous lead wires and connectors

EQUIPMENT:
    1 - three-phase, variable A.C. power supply capable of 0 to 120 volts RMS single-phase and
        0 to 208 volts line-to-line voltage
    1 - hand-held multimeter
    1 - in-line ammeter, 0 to 5 amps or hand-held clamp-on ammeter

PART 1: The Wye Connection
Background:
A three-phase source like that shown in Figures 12.2 and 12.3 may be used as three individual A.C. sources to provide 120 volts RMS to three individual loads. Or, it may be used to power equipment designed for three-phase operation. The three lines and neutral from the power supply could then be connected to a three-phase load such as that shown in Figure 12.4. This circuit is an example of a wye-connected load. Once connected, current will then flow in each line to the three loads. We will abbreviate the current in each line as I_L and the current in each load as I_P. The subscript P was chosen because we will refer to the current in each load as the phase current. In the wye connection shown, each line is in series with the load, so the line current will equal the phase current. This leads to our first formula for a three-phase, wye-connected circuit:

$$I_L = I_P \qquad \textbf{12.1}$$

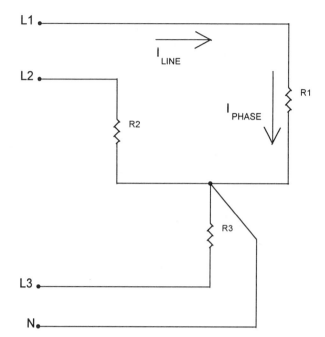

Figure 12.4: Three-phase wye connection with purely resistive loads.

Assuming each line-to-neutral voltage is 120 volts RMS, the resulting line-to-line voltages will be 208 volts RMS. It can be shown that the line-to-line voltage is related to the line-to-neutral voltage by a factor equal to the square root of three. Written as a mathematical formula, we then have for a wye-connected circuit

$$V_{L-L} = \sqrt{3}(V_{L-N}) \qquad 12.2$$

Take a moment to verify this relationship by finding the product of 120 volts times the square root of three. The relationship between this particular set of line-to-neutral and line-to-line voltages may be expressed as 120/208. This is not the only three-phase voltage source available in an industrial setting. Higher voltages such as 277/480 are also available. The exact voltages are determined by the turns ratio of the three-phase transformer providing the A.C. and the magnitude of the primary voltage applied to it. Because the line-to-neutral voltages are applied directly across their respective loads, the line-to-neutral voltage equals the load or phase voltage, V_P. This leads to our third mathematical relationship for a wye-connected load.

$$V_P = V_{L-N} \qquad 12.3$$

Assuming purely resistive loads, the current through a given load can then be calculated as follows:

$$I_P = V_P/R \qquad 12.4$$

As pointed out previously, the three line-to-neutral voltages are 120° out of phase. Because the loads are purely resistive, the three phase currents are also out of phase with each other by 120°. It can also be shown that the three line-to-line voltages are out of phase with each other by 120°, as are the three line currents. Because we are dealing with purely resistive loads, the current through each load and its corresponding line-to-neutral voltage are in phase. If the three-phase load were a motor, there would be a phase shift between each phase current and its corresponding line-to-neutral voltage. If the three loads are all equal in resistance or impedance, they are said to be *balanced*. For reasons of efficiency and safety, both the end user and the electric utility strive to achieve as nearly a balanced load condition as possible. If the loads are balanced, there will be no current in the neutral connection. If the loads are not balanced, there will be current flow in the neutral connection.

Procedure:

1.1 Obtain all materials and equipment required to complete this experiment.
1.2 Use your hand-held multimeter to measure the resistance of each power resistor and record your measurements in Table 12.1. If any of the resistors are out of tolerance, notify your instructor immediately and obtain a replacement.

Table 12.1: Nominal and measured resistor values.

NOMINAL RESISTANCE	MEASURED RESISTANCE
R1 = 50 Ohms	
R2 = 50 Ohms	
R3 = 50 Ohms	

1.3 With your power supply turned off and adjusted to produce zero volts when turned on, connect the circuit shown in Figure 12.5. You may use either an in-line ammeter to measure the load currents or a clamp-on style ammeter. It is not necessary to use four different ammeters as shown. You may use just one and move it from point to point to measure the different currents. Have your instructor check your work.
1.4 If you are using any analog meters in this experiment, be on the alert for excessive upscale deflection or pegging of the meter movement. Select an appropriate full-scale current value to prevent this from happening. If you have any doubt about what range to select, ask your instructor for assistance.

NOTE: Do not allow the line-to-neutral voltage to exceed 100 volts RMS nor the line-to-line voltage to exceed 173 volts RMS. If these voltages are exceeded, the power dissipation rating of the power resistors will be exceeded.

1.5 Turn on the power supply. <u>Slowly</u> increase the supply voltage until M6, your A.C. voltmeter, indicates 86.6 volts RMS. M6 indicates the voltage from line 2 to neutral, V_{L2-N}. You may use the same voltmeter to measure both the line-to-neutral voltages and line-to-line voltages. M5 is shown to indicate the connection to measure the voltage from line 2 to line 3, V_{L2-L3}.

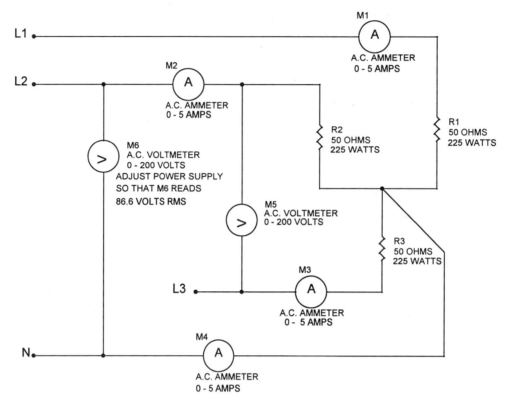

Figure 12.5: Three-phase wye circuit showing ammeter and voltmeter connections.

1.6 Measure and record all of the voltages and currents specified in Table 12.2. Reduce the power supply voltage to 0.0 volt. Turn off the power supply. Allow the resistors to cool before connecting the next circuit.

Table 12.2: Measured voltages and currents for the circuit shown in Figure 12.5.

LINE-TO-NEUTRAL VOLTAGES, VOLTS RMS		LINE-TO-LINE VOLTAGES, VOLTS RMS		LINE CURRENTS, AMPERES RMS	
V_{L1-N}		V_{L1-L2}		I_{L1}	
V_{L2-N}		V_{L2-L3}		I_{L2}	
V_{L3-N}		V_{L1-L3}		I_{L3}	

PART 2: The Delta Connection
Background:
Three-phase circuits with resistive loads may be configured like the circuit shown in Figure 12.6. This circuit is an example of a delta-connected load. As before, we will abbreviate the current in each line as I_L and the current in each load as I_P. Be aware that this connection results in a different relationship between the line and phase currents and the line and phase voltages. Notice how the incoming line current splits between R1 and R2. Because we are dealing with line currents that are 120° out of phase, the current in each resistor is not one-half of the line current (even if the resistors are equal in value).

Rather, it can be shown that the phase and line currents have the following relationship in a delta-connected circuit:

$$I_L = \sqrt{3}(I_P) \qquad 12.5$$

Notice also that now the line-to-line voltage is connected directly across the load. This leads to our next formula, which relates the line-to-line voltage to the phase or load voltage.

$$V_{L-L} = V_P \qquad 12.6$$

So, if the line-to-line voltage is 208 volts RMS, then the phase voltage or voltage across the load is also 208 volts RMS. Assuming purely resistive loads, the current through a given load can then be calculated as follows:

$$I_P = V_P/R \qquad 12.7$$

This is the same as Formula 12.4. However, V_P is now the line-to-line voltage; whereas, in Formula 12.4, V_P is the line-to-neutral voltage. By now you have probably noticed that there is no neutral connection in this particular configuration. Because there is no neutral connection for current to flow in when an unbalanced load condition occurs, it is important to try to maintain three equal loads when using the delta configuration. In the procedure that follows, you will have an opportunity to collect data from a delta circuit and compare the results to your wye circuit so you can observe the difference in operation of this configuration minus the neutral connection.

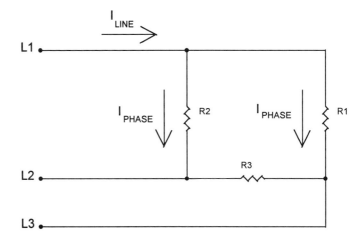

Figure 12.6: Three-phase delta connection with purely resistive loads.

Procedure:

2.1 With your power supply turned off and adjusted to produce zero volts when turned on, connect the circuit shown in Figure 12.7. You may use either an in-line ammeter to measure the load currents or use a clamp-on style ammeter. It is not necessary to use four different ammeters as shown. You may use just one and move it from point to point to measure the different currents. Have your instructor check your work.

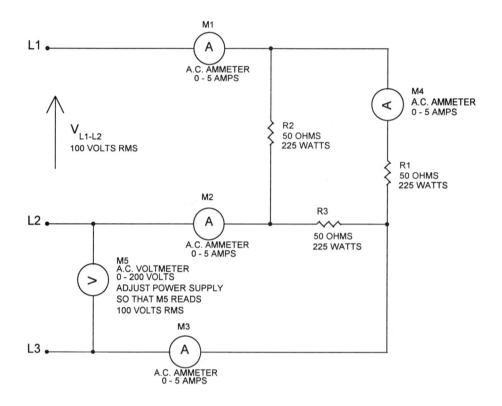

Figure 12.7: Three-phase delta circuit showing ammeter and voltmeter connections.

2.2 Turn on the power supply. <u>Slowly</u> increase the supply voltage until M5, your A.C. voltmeter, indicates 100 volts RMS. M5 indicates the voltage from line 2 to line 3, V_{L2-L3}. You may use the same voltmeter to measure all three line-to-line voltages. Measure and record all of the voltages and currents specified in Table 12.3.

2.3 Reduce the power supply voltage to 0.0 volt, and turn off the power supply. Allow the resistors to cool before disassembling.

Table 12.3: Measured voltages and currents for the circuit shown in Figure 12.7.

PHASE CURRENTS, AMPERES RMS		LINE-TO-LINE VOLTAGES, VOLTS RMS		LINE CURRENTS, AMPERES RMS	
I_{R1}		V_{L1-L2}		I_{L1}	
I_{R2}		V_{L2-L3}		I_{L2}	
I_{R3}		V_{L1-L3}		I_{L3}	

PART 3: Questions

3.1 Using the value you measured for V_{L1-N} from Table 12.2, calculate the theoretical line-to-line voltage.

3.2 How does your answer to Question 3.1 compare to your measured line-to-line voltages recorded in Table 12.2? In other words, did your experimental results confirm the theoretical value predicted by Formula 12.2?

3.3 Referring to your experimental results in Table 12.2, what is the current through R1 in the circuit shown in Figure 12.5?

3.4 Using the measured value for V_{L1-L2} from Table 12.3 and the measured value for R2 from Table 12.1, calculate the theoretical current that should flow through this particular resistor for the circuit shown in Figure 12.7.

3.5 How does your answer to Question 3.4 compare to your measured value for I_{R2} recorded in Table 12.3?

3.6 Using your answer to Question 3.4, calculate the theoretical line current for the circuit shown in Figure 12.7.

3.7 How does your answer to Question 3.6 compare to your measured line currents recorded in Table 12.3? In other words, did your experimental results confirm the theoretical value predicted by Formula 12.5?

PART 4: Practice Problems

4.1 A circuit like that shown in Figure 12.4 has a line-to-neutral voltage of 270 volts RMS and load resistors each having a resistance of 30 Ohms. Answer the following questions. Show units where required.

    a) What is the line-to-line voltage?

    b) What is the phase current?

    c) What is the line current?

4.2 A circuit like that shown in Figure 12.6 has a line-to-line voltage of 480 volts RMS and load resistors each having a resistance of 20 Ohms. Answer the following questions.

    a) What is the phase current?

    b) What is the line current?

PART 5: Three-Phase, Wye-Connection Power Formulas

$I_L = I_P$ **12.1**

$V_{L-L} = \sqrt{3}(V_{L-N})$ **12.2**

$V_P = V_{L-N}$ **12.3**

$I_P = V_P/R$ **12.4**

PART 6: Three-Phase, Delta-Connection Power Formulas

$I_L = \sqrt{3}(I_P)$ **12.5**

$V_{L-L} = V_P$ **12.6**

$I_P = V_P/R$ **12.7**

Measuring Three-Phase Power: 13

INTRODUCTION:
In Experiment 11, you were introduced to the wattmeter and single-phase power measurement. In Experiment 12, you were introduced to three-phase circuits and the voltage and current relationships in wye and delta circuits. This experiment will introduce you to three different methods for measuring three-phase power. Before performing each phase of the hands-on part of the experiment, you will be given an introduction to the mathematical formulas for calculating three-phase power for purely resistive loads and for circuits with reactive loads. An understanding of three-phase power is important to engineers and technicians working in industry and for those working for electric utilities. The ability of these professionals to develop and maintain equipment that will efficiently transmit and utilize three-phase power hinges on a thorough understanding of the principles presented in this lab.

SAFETY NOTE:
You will be operating each circuit in this experiment at high voltage, so take extra precaution to prevent shorts. Make sure that there are no exposed leads or connectors. Also, the power resistors will become very hot. Do not touch them while your circuit is energized. Allow several minutes for the resistors to cool down after turning off the power when making circuit modifications or when disassembling your circuit for the last time. Make sure that the resistors are placed on a nonconductive surface with a high melting point. As always, wear safety glasses and remove jewelry from your fingers and wrists. Also have your instructor tell you what the maximum allowable power, current, and voltage ratings for your load resistors are so that you will not exceed any of these ratings during the experiment.

OBJECT:
Upon successful completion of this experiment and all reading assignments, the student should be able to:
- ❑ measure three-phase power using the one-wattmeter method, the two-wattmeter method, and a three-phase wattmeter
- ❑ calculate three-phase apparent power, real power, and reactive power

MATERIALS:
      3 - 50 Ohms, 225 watt wire-wound power resistors
      - miscellaneous lead wires and connectors

EQUIPMENT:
      1 - three-phase, variable A.C. power supply capable of 0 to 120 volts RMS single-phase and
         0 to 208 volts line-to-line voltage
      1 - hand-held multimeter
      1 - in-line ammeter, 0 to 5 amps or handheld clamp-on ammeter
      2 - single-phase, analog wattmeters, 0 to 500 watts
      1 - three-phase, analog wattmeter, 0 to 600 watts

PART 1: The Single-Wattmeter Method
Background:
In Experiment 11, a single-phase wattmeter was used to measure A.C. power in a single-phase A.C. circuit. For analysis purposes, a three-phase circuit can be thought of as being composed of three individual single-phase circuits. In a similar fashion, the total power consumed by a three-phase circuit can be determined by summing the power consumed by each of the three loads or phases in a wye or delta circuit. As a formula, we have

$$P_T = P_{P1} + P_{P2} + P_{P3} \qquad \textbf{13.1}$$

Recall from Experiment 11 that the apparent power, S, for a single-phase circuit can be calculated as follows:

$$S = (V_P) \cdot (I_P) \qquad \textbf{13.2}$$

V_P represents the voltage across the load or phase, and I_P represents the current through the load. The total three-phase apparent power can then be found by calculating the apparent power for each of the three phases and finding the algebraic sum using Formula 13.1. Also recall from Experiment 11 that the real or useful power for a single-phase load may be found as follows:

$$P = (V_P) \cdot (I_P) \cos(\theta) \qquad \textbf{13.3}$$

The phase shift or phase angle between the phase voltage and phase current is represented by the Greek symbol θ. The total three-phase real power can then be found by calculating the real power for each of the three phases and finding the algebraic sum using Formula 13.1. If a technician needed to measure the total three-phase apparent power for a wye-connected load such as that shown in Figure 13.1, then he or she would measure each phase voltage, each phase current, and find the product of each phase voltage and current. The sum of these three products would equal the total three-phase apparent power. The total real power for the wye-connected circuit can be found using a single wattmeter as shown in Figure 13.1. For those circuits having a neutral connection, the wattmeter's voltage coil can be connected across one of the loads as shown. Note that one input to the voltage coil is one of the three lines, and the other input is the neutral wire. The connection shown will result in the wattmeter displaying the power consumed by resistor R3. To measure the power consumed by the other two loads would merely require moving the wattmeter's voltage coil connection from line 3 to line 1 to measure the power consumed by R1 then to line 2 to measure the power consumed by R2. The total three-phase power for this circuit could then be found by summing the three individual power readings for resistors R1, R2, and R3. The overall circuit power factor, P.F., for this circuit could then be found by dividing the total three-phase real power (P_T) by the total three-phase apparent power (S_T). As a formula, we have

$$P.F. = P_T / S_T \qquad \textbf{13.4}$$

Notice that this formula is in the same form as that used to calculate the power factor for a single-phase circuit (P.F. = P/S). The phase angle for any of the <u>individual</u> three-phase loads can be found by taking the inverse cosine of the power factor for the particular phase of interest as follows:

$$\theta = \cos^{-1}(P.F.) \qquad \textbf{13.5}$$

The reactive power, Q, for each phase can be found by applying the single phase formula

$$Q = (V) \cdot (I)\sin(\theta) \qquad \text{13.6}$$

The total three-phase reactive power can be found by calculating the reactive power for each individual phase and finding their algebraic sum. As you have already noticed, a solid understanding of single-phase power facilitates the study of three-phase power. Lastly then, we can relate total three-phase apparent power, real power, and reactive power with the Pythagorean Theorem as follows:

$$(S_T)^2 = (P_T)^2 + (Q_T)^2 \qquad \text{13.7}$$

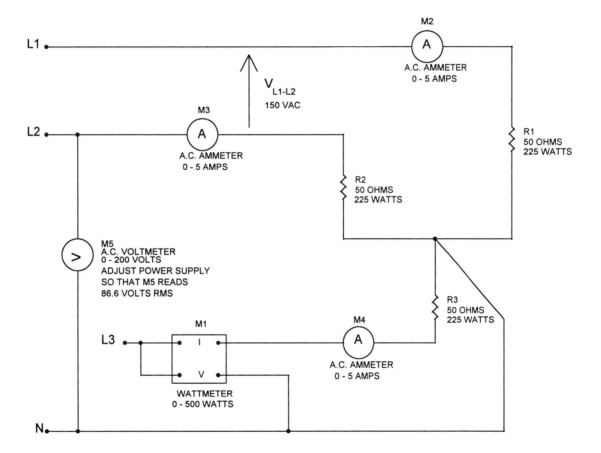

Figure 13.1: Three-phase, wye-connected circuit with a single wattmeter used to measure the real power consumed by resistor R3.

If you can be reasonably certain that all three loads in a three-phase circuit are balanced, that is, $P_{P1} = P_{P2} = P_{P3}$, then the total three-phase power can be found by measuring the power in just one of the three phases and multiplying it by three. As a formula, we would then have

$$P_T = 3(P_P) \qquad \text{13.8}$$

where P_P represents the power in any of the three phases in a three-phase circuit with balanced loads. Before starting the hands-on part of this experiment, I would like to point out that the single-wattmeter method can be expanded to three wattmeters as shown in Figure 13.2. The total circuit power is then found by summing the readings of the three meters. This has the obvious advantage of not having to disconnect the meter and reconnect it, as would be the case in the single-wattmeter method. Of course, the disadvantage is the cost of having two additional meters.

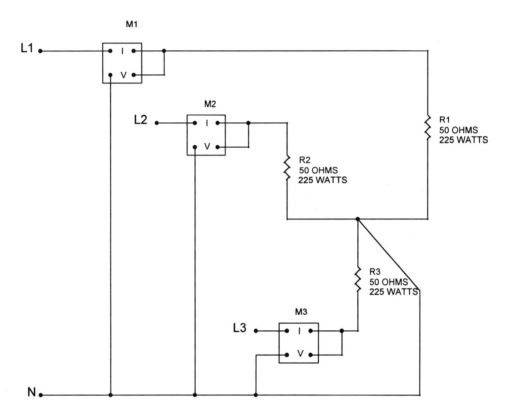

Figure 13.2: Three-phase, wye-connected circuit with three wattmeters used to measure the circuit's total real power. The sum of all three wattmeter readings equals the total three-phase real power.

Procedure:
1.1 Obtain all materials and equipment required to complete this experiment.
1.2 Use your hand-held multimeter to measure the resistance of each power resistor and record your measurements in Table 13.1. If any of the resistors are out of tolerance, notify your instructor immediately and obtain a replacement.

Table 13.1: Nominal and measured resistor values.

NOMINAL RESISTANCE	MEASURED RESISTANCE
R1 = 50 Ohms	
R2 = 50 Ohms	
R3 = 50 Ohms	

1.3 With your power supply turned off and adjusted to produce zero volts when turned on, connect the circuit shown in Figure 13.1. You may use either an in-line ammeter to measure the load currents or use a clamp-on style ammeter. It is not necessary to use three different ammeters as shown. You may use just one and move it from point to point to measure the different currents. Have your instructor check your work.

1.4 If you are using any analog meters in this experiment, be on the alert for excessive upscale deflection or pegging of the meter movement. Select an appropriate full-scale current value to prevent this from happening. If you have any doubt about what range to select, ask your instructor for assistance.

NOTE: Do not allow the line-to-neutral voltage to exceed 100 volts RMS nor the line-to-line voltage to exceed 173 volts RMS at any time during this experiment. If these voltages are exceeded, the power dissipation rating of the power resistors will be exceeded.

1.5 Turn on the power supply. Slowly increase the supply voltage until M5, your A.C. voltmeter, indicates 86.6 volts RMS. M5 indicates the voltage from line 2 to neutral, V_{L2-N}. You may use the same voltmeter to measure both the line-to-neutral voltages and line-to-line voltages. Measure and record all of the voltages and currents specified in Table 13.2.

1.6 Record in Table 13.3 the wattmeter reading for the power consumed by resistor R3.

1.7 Reduce the power supply voltage to 0.0 volt. Turn off the power supply.

1.8 Reconnect the wattmeter to measure the power for load resistor R2.

1.9 Turn on the power supply and increase the voltage until M5 once again reads 86.6 volts RMS.

1.10 Record in Table 13.3 the wattmeter reading for the power consumed by resistor R2.

1.11 Repeat steps 1.7 through 1.10 for resistor R1.

1.12 Reduce the power supply voltage to 0.0 volt, and turn off the power supply. Allow the resistors to cool before connecting the next circuit.

Table 13.2: Measured voltages and currents for the circuit shown in Figure 13.1.

LINE-TO-NEUTRAL VOLTAGES, VOLTS RMS		LINE-TO-LINE VOLTAGES, VOLTS RMS		LINE CURRENTS, AMPERES RMS	
V_{L1-N}		V_{L1-L2}		I_{L1}	
V_{L2-N}		V_{L2-L3}		I_{L2}	
V_{L3-N}		V_{L1-L3}		I_{L3}	

Table 13.3: Measured wattage values for the circuit shown in Figure 13.1.

LOAD RESISTOR	POWER CONSUMED, WATTS
R1	
R2	
R3	

PART 2: The Two-Wattmeter Method

Background:

On occasion you may have to measure the power in a wye circuit that does not have a neutral connection such as that shown in Figure 13.3, or you may have to measure the power in a delta-connected circuit such as that shown in Figure 13.4. As you examine Figure 13.3, note that the voltage coil inputs are connected to two line voltages, not between a line and neutral as in Figure 13.1. Well, you might ask, "What is the significance of this?" In Figure 13.1, the wattmeter was measuring the line-to-neutral voltage with its voltage coil and the line or phase current with its current coil. The product of these two values--which are indicated on the wattmeter's display--represents the power for that particular phase. And, as noted before, summing the power for the three phases will result in the total three-phase power. In the single-wattmeter method as shown in Figure 13.1, the line-to-neutral voltage and the load current being measured by the wattmeter are in phase. Now consider how the circuit in Figure 13.3 is different. First, each wattmeter is measuring a line-to-line voltage. As in Figure 13.1, each wattmeter is measuring the line or phase current. Remember, in the wye configuration, $I_L = I_P$. In the two-wattmeter method then, it might appear that each meter indicates the product of the line-to-line voltage and the phase current. As a formula, we would have

$$\text{WATTMETER READING} = (V_{L\text{-}L})(I_P) \qquad \textbf{13.9}$$

Compare this to Formula 13.2. Notice in this formula, the line-to-line voltage is used, not the phase voltage. There is a very subtle consequence of this difference. It can be shown using trigonometry and vectors that the line-to-neutral voltage and phase or line current are in phase with each other when the load is purely resistive. However, the line-to-line voltage and the phase or line current are out of phase with each other by 30°. Formula 13.9 is based on the assumption that the line-to-line voltage and the load current are in phase. But they are not! Formula 13.9 is <u>incorrect</u>. The wattmeter reading is actually found as follows:

$$\text{WATTMETER READING} = (V_{L\text{-}L})(I_P)\cos(30°) \qquad \textbf{13.10}$$

Do you remember the relationship between the line-to-line voltage and line-to-neutral voltage in a wye-connected circuit? I am sure that you recall it to be as follows:

$$V_{L\text{-}L} = \sqrt{3}\,(V_P) \qquad \textbf{13.11}$$

Substituting Formula 13.11 into Formula 13.10, we have

$$\text{WATTMETER READING} = \sqrt{3}\,(V_P)(I_P)\cos(30°) \qquad \textbf{13.12}$$

Carrying out the multiplication, we finally have

$$\text{WATTMETER READING} = 1.5(V_P)(I_P) \qquad \textbf{13.13}$$

Again, as a reminder, this represents the indication of <u>one</u> wattmeter in the <u>two</u>-wattmeter method. If we were to then sum the readings from the two wattmeters, we would have the following:

$$P_T = 1.5(V_P)(I_P) + 1.5(V_P)(I_P) = 3(V_P)(I_P) = 3(P_P) \qquad \textbf{13.14}$$

Note that we now have developed Formula 13.8. Therefore, the total three-phase power in the two-wattmeter method is the sum of the two meter readings.

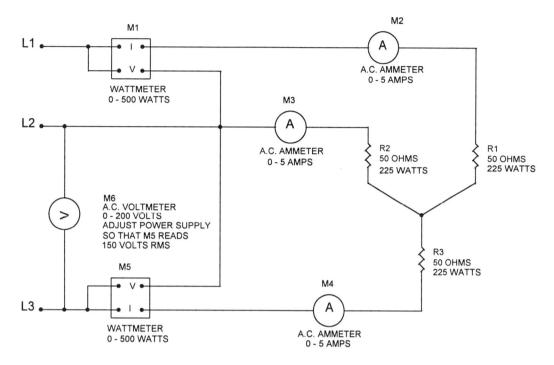

Figure 13.3: Three-phase, wye-connected circuit using the two-wattmeter method to measure the total three-phase real power. The sum of the two wattmeter readings equals the total three-phase real power.

Procedure:

2.1 With your power supply turned off and adjusted to produce zero volts when turned on, connect the circuit shown in Figure 13.3. It will not be necessary to measure all of the currents and voltages for this circuit, because they should be the same as for Figure 13.1. Have your instructor check your work.

2.2 Turn on the power supply. Slowly increase the supply voltage until M6, your A.C. voltmeter, indicates 150 volts RMS.

2.3 Record in Table 13.4 the wattmeter readings from M1 and M5.

2.4 Reduce the power supply voltage to 0 volts. Turn off the power supply. Allow the resistors to cool before you connect the next circuit.

Table 13.4: Measured wattage values for the circuit shown in Figure 13.3.

METER	READING IN WATTS
M1	
M5	

2.5 Connect the circuit shown in Figure 13.4.

175

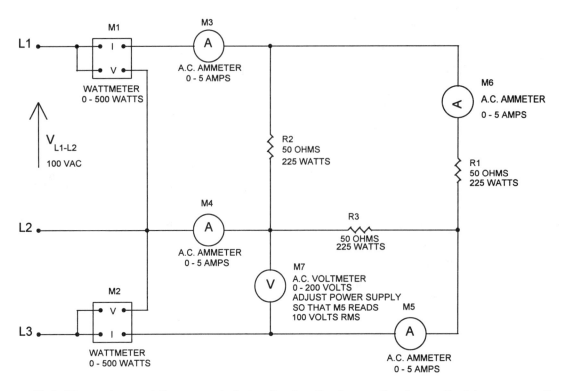

Figure 13.4: Three-phase, delta-connected circuit using the two-wattmeter method to measure the total three-phase real power. The sum of the two wattmeter readings equals the total three-phase real power.

2.6 Turn on the power supply. Slowly increase the supply voltage until M7, your A.C. voltmeter, indicates 100 volts RMS. M7 indicates the voltage from line 2 to line 3, V_{L2-L3}. Measure and record all of the voltages and currents specified in Table 13.5.
2.7 Record in Table 13.6 the wattmeter readings from M1 and M2.
2.8 Reduce the power supply voltage to 0 volts. Turn off the power supply.

Table 13.5: Measured voltages and currents for the circuit shown in Figure 13.4.

PHASE CURRENTS, AMPERES RMS		LINE-TO-LINE VOLTAGES, VOLTS RMS		LINE CURRENTS, AMPERES RMS	
I_{R1}		V_{L1-L2}		I_{L1}	
I_{R2}		V_{L2-L3}		I_{L2}	
I_{R3}		V_{L1-L3}		I_{L3}	

Table 13.6: Measured wattage values for the circuit shown in Figure 13.4.

METER	READING IN WATTS
M1	
M2	

PART 3: The Three-Phase Wattmeter

Background:
It is not necessary to measure three-phase power using only single-phase wattmeters. Meters are available to measure total three-phase power that have been integrated into one complete unit. If your lab has such a unit, have your instructor explain its principle of operation and the connections required to measure three-phase power in a wye- or delta-connected circuit. Some three-phase wattmeters have two meters built into one unit, and some have only one meter. For those units having two meters, the readings have to be added as with the two-wattmeter method.

Procedure:
3.1 With your power supply turned off and adjusted to produce zero volts when turned on, connect the circuit shown in Figure 13.5. It will not be necessary to measure all of the currents and voltages for this circuit, because they should be the same as for Figure 13.4. Have your instructor check your work.

3.2 Turn on the power supply. Slowly increase the supply voltage until M1, your A.C. voltmeter, indicates 100 volts RMS.

3.3 Record in Table 13.7 the total three-phase power indicated by your three-phase wattmeter.

3.4 Reduce the power supply voltage to 0 volts. Turn off the power supply.

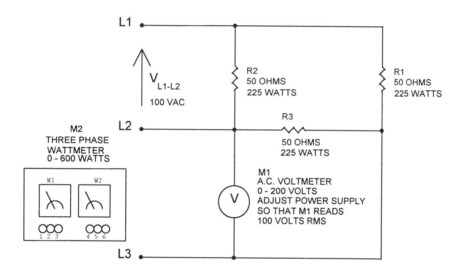

Figure 13.5: Three-phase, delta-connected circuit using a three-phase wattmeter to measure the total three-phase real power.

Table 13.7: Total three-phase power indicated by the three-phase wattmeter for the circuit shown in Figure 13.5.

METER	TOTAL POWER IN WATTS
M1	

PART 4: Questions

4.1 Using the data from Table 13.2, calculate the total three-phase apparent power for the circuit shown in Figure 13.1.

4.2 Using the data from Table 13.3, calculate the total three-phase real power for the circuit shown in Figure 13.1.

4.3 Considering the fact that power resistors were used in the circuit shown in Figure 13.1, how do the results from Questions 4.1 and 4.2 compare? If the resistors had absolutely no inductance, what should the relationship be between the apparent power and the real power for the circuit shown in Figure 13.1?

4.4 Using the data from Table 13.4, calculate the total three-phase real power for the circuit shown in Figure 13.3, which, of course, is the same as Figure 13.1.

4.5 How do your results from Question 4.4 compare to your results to Question 4.2? What should be the theoretical relationship between the results for Questions 4.4 and 4.2?

4.6 Using the data from Table 13.5, calculate the total three-phase apparent power for the circuit shown in Figure 13.4.

4.7 Using the data from Table 13.6, calculate the total three-phase real power for the circuit shown in Figure 13.4.

4.8 Considering the fact that power resistors were used in the circuit shown in Figure 13.4, how do the results from Questions 4.6 and 4.7 compare? If the resistors had absolutely no inductance, what should the relationship be between the apparent power and the real power for the circuit shown in Figure 13.4?

4.9 How do your results from Table 13.7 for the three-phase wattmeter compare to the results obtained for the delta circuit using the two-wattmeter method?

PART 5: Practice Problems

5.1 A circuit like that shown in Figure 13.1 has a line-to-neutral voltage of 270 volts RMS and load resistors each having a resistance of 30 Ohms. Answer the following questions. Show units where required.

   a) What is the phase current?

   b) Assuming each load is purely resistive, what is the total three-phase power?

5.2 A circuit like that shown in Figure 13.4 has a line-to-line voltage of 480 volts RMS, with each phase having a reactance of 20 Ohms. The loads are balanced having a power factor of 0.75. Answer the following questions?

   a) What is the current in each phase (load)?

b) What is the total three-phase apparent power?

c) What is the total three-phase real power?

d) What is the total three-phase reactive power?

e) What is the phase angle between the load voltage and the load current?

PART 6: The Single-Wattmeter Method Power Formulas

$P_T = P_{P1} + P_{P2} + P_{P3}$ **13.1**

$S = (V_P) \cdot (I_P)$ **13.2**

$P = (V_P) \cdot (I_P)\cos(\theta)$ **13.3**

$P.F. = P_T/S_T$ **13.4**

$\theta = \cos^{-1}(P.F.)$ **13.5**

$Q = (V) \cdot (I)\sin(\theta)$ **13.6**

$(S_T)^2 = (P_T)^2 + (Q_T)^2$ **13.7**

$P_T = 3(P_P)$ **13.8**

PART 7: The Two-Wattmeter Method Power Formulas

WATTMETER READING = $(V_{L-L})(I_P)$ **13.9**

WATTMETER READING = $(V_{L-L})(I_P)\cos(30°)$ **13.10**

$V_{L-L} = \sqrt{3}(V_P)$ **13.11**

WATTMETER READING = $\sqrt{3}(V_P)(I_P)\cos(30°)$ **13.12**

WATTMETER READING = $1.5(V_P)(I_P)$ **13.13**

$P_T = 1.5(V_P)(I_P) + 1.5(V_P)(I_P) = 3(V_P)(I_P) = 3(P_P)$ **13.14**

Three-Phase, Half-Wave Rectification: 14

INTRODUCTION:
Industrial electronics applications requiring high-voltage D.C. can be produced by rectifying the readily available three-phase A.C. supply found in most manufacturing settings. As with single-phase rectification studied earlier, three-phase A.C. power supplies can be converted to D.C. by either half-wave rectification or full-wave rectification. Other means of converting three-phase A.C. to D.C. are available and include the use of SCRs as adjustable rectifiers. In this lab, the focus will be on some of the principles involved in three-phase, half-wave rectification. Three-phase rectifier circuits are not only found in industrial applications but also include medical applications such as the power supply in radiology equipment.

SAFETY NOTE:
Prior to the start of this experiment, have your instructor review the proper way to use an oscilloscope to display three-phase voltages. You will also be operating each circuit at high voltage, so take extra precaution to prevent shorts. Make sure that there are no exposed leads or connectors. Also, the power resistor that is to be used as a load will become very hot. Do not touch it after energizing your circuit. Prior to circuit disassembly, allow this resistor to cool down after turning off the power for the last time. As always wear safety glasses and remove jewelry from your fingers and wrists. Finally, have your instructor tell you what the maximum allowable power, current, and voltage ratings for your rectifier diodes and power resistor are so that you will not exceed any of these ratings during the experiment.

OBJECT:
Upon successful completion of this experiment and all reading assignments, the student should be able to:
- construct an operational three-phase, half-wave rectifier circuit
- measure the line-to-line voltages and line-to-neutral voltage in a three-phase circuit
- use an oscilloscope to display the voltage across the load resistor in a three-phase, half-wave rectifier circuit
- given a three-phase, half-wave rectifier circuit and given the line-to-neutral voltage, calculate the peak load voltage, average (D.C.) load voltage, peak load current, average load current, peak diode current, and average diode current

MATERIALS:
- 3 - rectifier diodes, 1N5407, or SK9009, or the equivalent
- 1 - 1500 Ohms, 25 Watt power resistor
- 2 - switches, SPST
- miscellaneous lead wires and connectors

EQUIPMENT:
- 1 - three-phase, variable A.C. power supply capable of 0 - 120 volts RMS single-phase and 0 - 208 volts line-to-line voltage with circuit breaker
- 1 - hand-held multimeter
- 1 - isolation transformer
- 1 - oscilloscope

2 - oscilloscope probes
3 - D.C. ammeters

INSTRUCTOR'S NOTE: If your lab is not equipped with an adjustable three-phase power supply, you may still perform this lab if you have access to a fixed three-phase 120/208 A.C. supply with appropriate fuse and/or circuit breaker protection. You will still be able to use the 1500 Ohms, 25 watt power resistor, but be aware that it will be operating very near its maximum power dissipation rating. To provide an additional margin of safety, you could change the load resistor in Figure 14.7 to a 2500 Ohms power resistor with a wattage rating of 50 watts. SK9009 rectifier diodes or diodes with an equivalent or higher rating should be used. Also provide your students with all of the necessary safety precautions and safety training required for working with three-phase voltages of this magnitude.

PART 1: The Three-Phase, Half-Wave Rectifier
Background:
Before discussing the three-phase, half-wave rectifier, let's review some fundamental three-phase circuit fundamentals. Recall that a three-phase A.C. power supply can be described as three single-phase A.C. sources with each sinusoidal voltage waveform separated from the other two by a phase shift of 120° as shown in Figure 14.1.

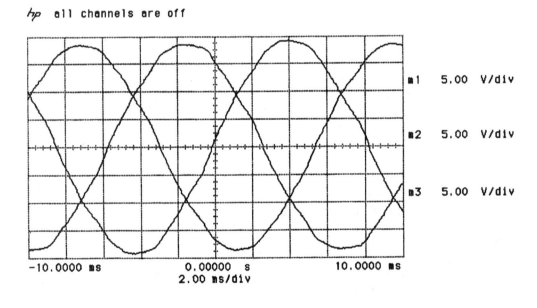

Figure 14.1: Sinusoidal voltage waveforms produced by a three-phase alternator displaced by 120 degrees.

Recall also that the mathematical relationship between the peak voltage, V_P, of a single-phase sinusoidal A.C. waveform and it RMS value, V_{RMS} (V_{AC}), is as follows:

$$V_{RMS} = (0.707) \cdot (V_P) \qquad \textbf{14.1}$$

This formula also applies to three-phase line-to-neutral and three-phase line-to-line voltages. Figure 14.2 shows the relationship between the three line-to-neutral voltages, V_{L1-N}, V_{L2-N}, V_{L3-N} and the three line-to-line voltages V_{L1-L2}, V_{L2-L3}, V_{L1-L3}. For example, if the peak of the line-to-neutral voltage of a particular three-phase A.C. supply is 392 volts, the RMS value would be

$$V_{RMS} = (0.707) \cdot (392)$$

$$= 277 \text{ volts RMS}$$

If the RMS value of the line-to-line voltage of a particular three-phase supply is 480 volts RMS, then the peak of this waveform would be

$$V_P = (V_{RMS})/(0.707)$$

$$= 480/(0.707)$$

$$= 679 \text{ volts peak}$$

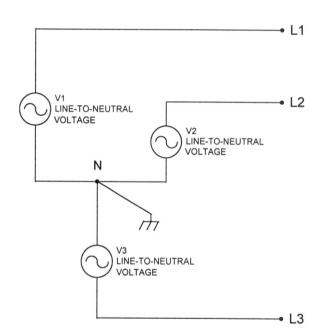

Figure 14.2: Simplified representation of a three-phase power supply.

The formula relating the line-to-neutral voltage to the line-to-line voltage of a sinusoidal three-phase A.C. supply is as follows:

$$V_{L-L} = \sqrt{3}(V_{L-N}) \qquad \textbf{14.2}$$

For example, if the RMS value of the line-to-neutral voltage of a particular A.C. supply is 277 volts RMS, then the RMS value of the line-to-line voltage is

$$V_{L-L} = \sqrt{3}(277 \text{ volts RMS})$$

= 480 volts RMS

When referring to the current that flows in the line connections from the power supply to the load, abbreviate it as I_L. When referring to the current that flows in the windings of a three-phase alternator, three-phase motor, or other three-phase load (whether wye or delta-connected), abbreviate it as I_P. Now let's go about explaining how a three-phase, half-wave rectifier works. Let's assume that the three-phase A.C. source shown in Figure 14.2 is connected to the circuit shown in Figure 14.3. Figure 14.3 shows a three-phase transformer that is being used to step-down the supply voltage. A three-phase transformer works on the same principle as a single-phase transformer, but there are three sets of windings instead of one. Each set of windings steps down the voltage applied to the primary by a factor equal to the turns ratio for that particular set of windings. The primary and secondary of a three-phase transformer may be either wye or delta connected. In this example both the primary and the secondary are wye-connected. The abbreviations H2 and H1 designate the primary connections and X2 and X1 designate the secondary connections. Each line coming out of the secondary of the transformer is connected to the anode of a rectifier diode. The three cathodes are tied together at a common point and are in turn connected to a purely resistive load. The low potential side of the resistor is connected to the neutral of the transformer's secondary. In effect, what we have are three line-to-neutral voltages individually being applied to a series-connected, diode-resistor network. In other words, we have three single-phase, half-wave rectifiers like the one shown in Figure 14.4. The voltage waveform across the load for the circuit shown in Figure 14.4 is shown in Figure 14.5. Now, imagine if you will, what would happen if we superimposed three of these half-wave rectified waveforms each separated from the other by a phase shift of 120°. The resulting load waveform would look like that shown in Figure 14.5. Note how much smoother this waveform is compared to the single-phase, half-wave rectified waveform. Note also how much more area under the curve there is now, especially since the waveform does not drop to zero at any point along the x-axis. Only one diode conducts at a time. At any given moment, the diode that conducts is the one that "sees" the most positive voltage from the incoming three-phase supply. The other two diodes are off, not conducting, as they are reverse biased. Each diode conducts for one-third of the entire cycle. This is an important point. Keep it in mind! In this configuration the peak of the voltage waveform across the load is equal to the peak of the applied line-to-neutral voltage minus the forward-bias voltage drop of the conducting diode. Since we will be using rather high power diodes for circuits such as this, the forward-bias drop will likely be more than 0.7 volt and may be as high as 2.0 volts or more due to the heavy doping of silicon-based power diodes and due to the high currents that will flow through each diode. However, this voltage drop is still relatively small compared to the peak of the supply voltage (typically greater than 100 volts). The average value of the output voltage across the load resistor may be found by integrating the area under the voltage curve. If we assume that each of the peaks is sinusoidal with each pulse having a width of 120° ($4\pi/6$), the area under the first pulse could be found by integrating from 30° to 150°. If this area is then divided by the length of that curve segment ($4\pi/6$), the result is the following formula for the D.C. or average value of the voltage waveform across the resistor:

$$V_{DC} = (0.827) \cdot (V_P) \qquad \textbf{14.3}$$

In this formula, V_P is the peak of the voltage waveform across the resistor. The average current then that flows through the resistor can be found using Ohms' Law:

$$I_{DC(RESISTOR)} = V_{DC(RESISTOR)}/R \qquad \textbf{14.4}$$

I know that all of you have a firm grasp on Ohm's Law by now, so it was not my intent to insult anyone's intelligence by including it amongst our list of formulas. Rather, I wanted to include it as an appropriate reminder of how I went about determining the current rating for each of the diodes. Since each of the diodes conducts for one-third of the time, the average current through each diode in Figure 14.3 can be found as follows:

$$I_{AVERAGE(DIODE)} = I_{DC(RESISTOR)}/3 \qquad \textbf{14.5}$$

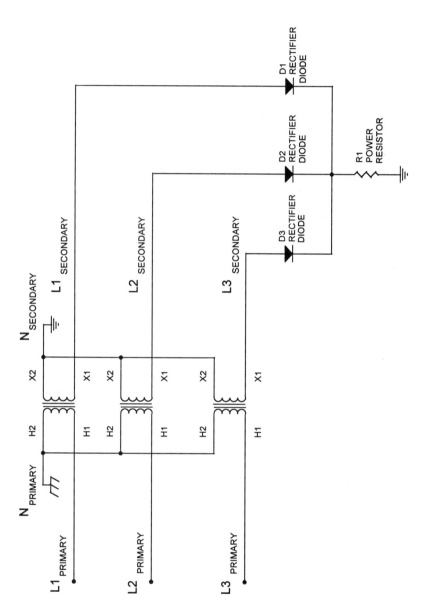

Figure 14.3: Schematic diagram of a wye-wye connected three-phase, step-down transformer. The three-phase secondary is connected to three diodes and a load resistor resulting in a three-phase, half-wave rectifier circuit.

Figure 14.4: Half-wave rectifier circuit.

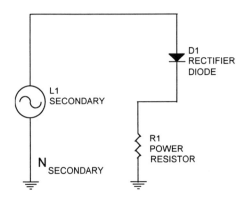

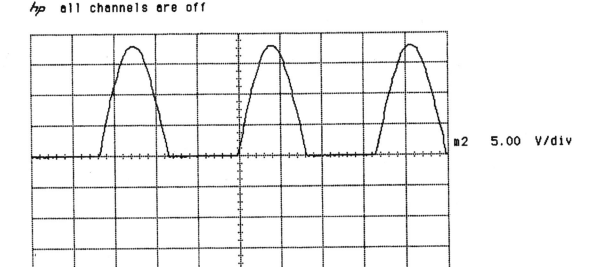

Figure 14.5: Single-phase, half-wave rectified load waveform for the half-wave rectifier circuit shown in Figure 14.4.

Another important parameter to specify when installing diodes in rectifier circuits is the peak repetitive voltage, V_{RRM}. If you study the three-phase waveform shown in Figure 14.1 closely, you will notice that when the first phase is at its most positive point, the other two waveforms are negative. The differential voltage from the peak of the positive waveform to where the other two negative waveforms cross is the maximum reverse-bias voltage any one of the diodes in Figure 14.3 will be subjected to. This will make more sense if you realize that at the same time the positive peak voltage of the first waveform (less one diode's forward-bias voltage drop) is appearing across the load and reverse biasing diodes D2 and D3, the other waveforms--both negative at this same point--are also reverse biasing diodes D2 and D3. The total reverse bias voltage then is the sum of the absolute value of these two instantaneous voltages. This is similar to what happens in a voltage doubler circuit. This differential voltage is approximately equal to one and one-half times the peak of the line-to-neutral voltage. In other words, the maximum peak reverse voltage, V_{PRV}, for any one of the diodes is

$$V_{PRV} = (1.5) \cdot (V_{PEAK(LINE-TO-NEUTRAL)}) \quad\quad 14.6$$

As always, choose an appropriate factor of safety when selecting rectifier diodes or any other electronic component for that matter. Never use a factor of safety of 1. <u>At least</u> double the expected load voltage, current, or power when establishing electronic device specifications. Always consult with a professional engineer before making a final selection.

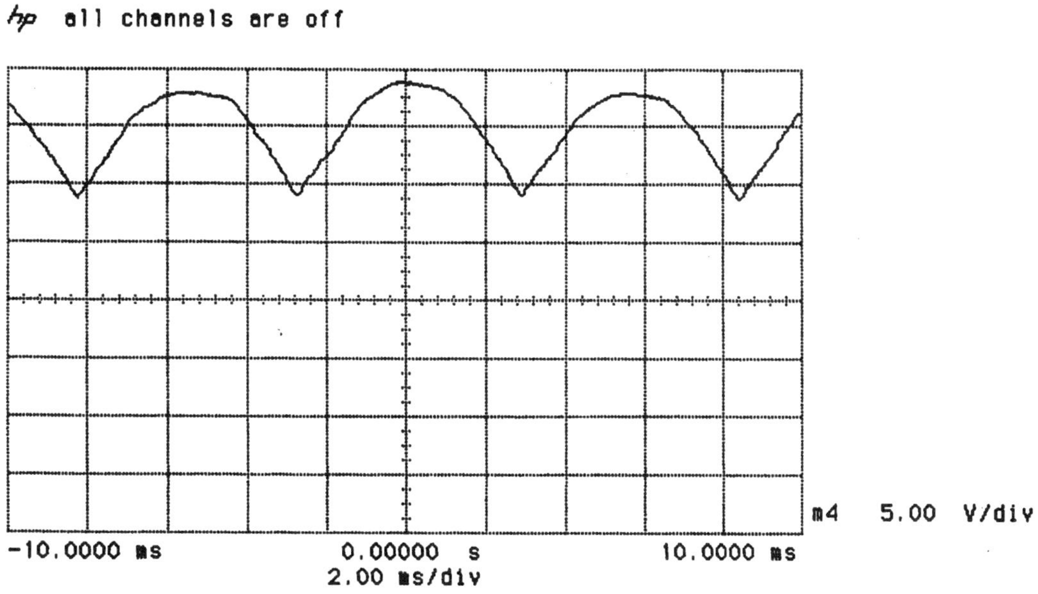

Figure 14.6: Three-phase, half-wave rectified load waveform for the half-wave rectifier shown in Figure 14.3.

Procedure:
1.1 Obtain all materials and equipment listed above.
1.2 Use your hand-held multimeter to measure the resistance of the power resistor and record your measurement in Table 14.1. If the resistor is out of tolerance, notify your instructor immediately and obtain a replacement.

Table 14.1: Nominal and measured load resistor values.

R1, NOMINAL RESISTANCE	R1, MEASURED RESISTANCE

1.3 With your power supply turned off, and adjusted to produce zero volts when turned on, connect the circuit shown in Figure 14.7. It is not a requirement to use three different ammeters as shown. You may use just one and move it from point to point to measure the different currents. Have your instructor check your work.

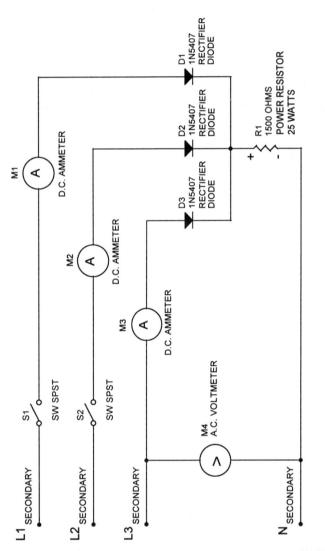

Figure 14.7: Three-phase, half-wave rectifier circuit.

1.4 If your are using any analog meters in this experiment, be on the alert for excessive upscale deflection or pegging of the meter movement. Select an appropriate full-scale current value to prevent this from happening. If you have any doubt about what range to select, ask your instructor for assistance.

1.5 Turn on the power supply. <u>Slowly</u> increase the supply voltage until M4, your hand-held A.C. voltmeter, indicates 80 volts RMS. M4 indicates the voltage from line 3 to neutral, V_{L3-N}. Close both switches. Measure and record all of the voltages and currents specified in Table 14.2.

Table 14.2: Measured voltages and currents for the circuit shown in Figure 14.7.

LINE-TO-NEUTRAL VOLTAGES, VOLTS RMS		LINE-TO-LINE VOLTAGES, VOLTS RMS		LINE CURRENTS, AMPERES D.C.	
V_{L1-N}		V_{L1-L2}		I_{L1}	
V_{L2-N}		V_{L2-L3}		I_{L2}	
V_{L3-N}		V_{L1-L3}		I_{L3}	

1.6 Using your hand-held multimeter as a D.C. voltmeter, measure the voltage across the load resistor. Be sure to observe proper polarity when making your measurement, especially if using an analog voltmeter. Record your measurement in Table 14.3.

Table 14.3: Output voltage for different switch positions for the circuit shown in Figure 14.7.

SWITCH SETTINGS	OUTPUT VOLTAGE, D.C. VOLTS
BOTH SWITCHES CLOSED	
S1 OPEN; S2 CLOSED	
S1 AND S2 OPEN	

1.7 Use your oscilloscope to display the voltage waveform across the load resistor. Draw the resulting waveform in the space provided on Graph 14.1. Trigger your oscilloscope on line voltage, L1.

1.8 Open switch S1. Using your hand-held multimeter as a D.C. voltmeter, measure the voltage across the load resistor. Record your measurement in Table 14.3.

1.9 Use your oscilloscope to display the voltage waveform across the load resistor. Draw the resulting waveform in the space provided on Graph 14.2. Trigger your oscilloscope on line voltage, L1.

1.10 Open switch S2. Switch S1 should still be open. Using your hand-held multimeter as a D.C. voltmeter, measure the voltage across the load resistor. Record your measurement in Table 14.3.

1.11 Use your oscilloscope to display the voltage waveform across the load resistor. Draw the resulting waveform in the space provided on Graph 14.3 below. Trigger your oscilloscope on line voltage, L1.

1.12 Reduce the power supply voltage to 0 volts. Turn off the power supply.

Graph 14.1: V_{R1} voltage waveform for the circuit shown in Figure 14.7.

V_{R1}: _____ volts/div _____ sec/div

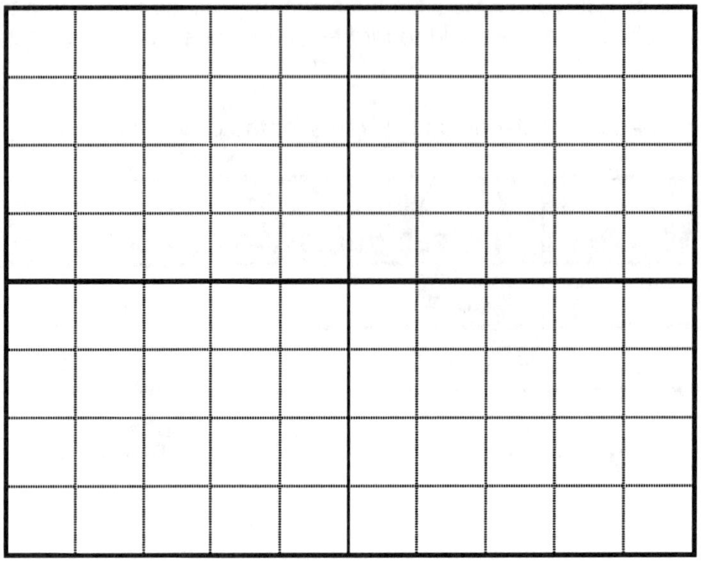

Graph 14.2: V_{R1} voltage waveform for the circuit shown in Figure 14.7 with switch S1 open.

V_{R1}: _____ volts/div _____ sec/div

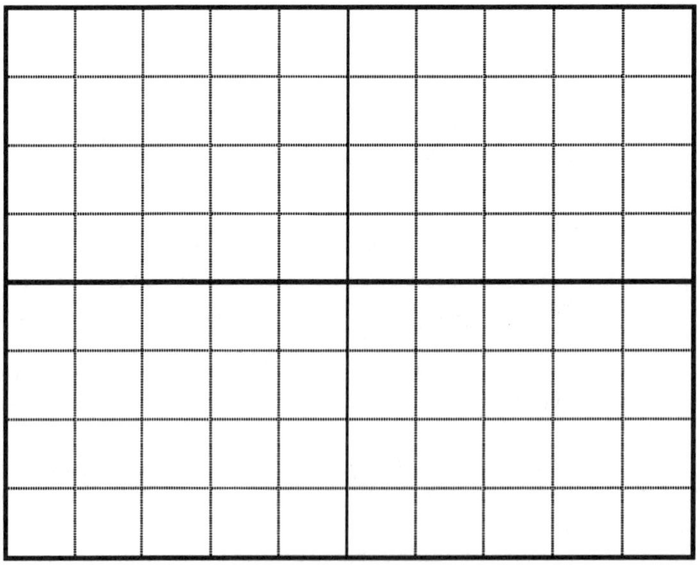

Graph 14.3: V_{R1} voltage waveform for the circuit shown in Figure 14.7 with switches S1 and S2 open.

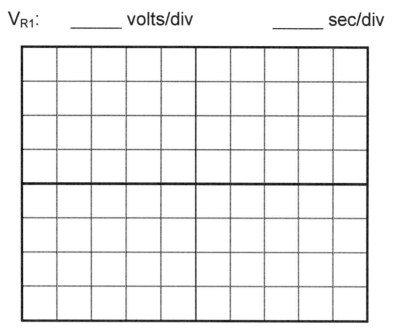

V_{R1}: _____ volts/div   _____ sec/div

PART 2: Manufacturer's Ratings

2.1 Referring to the manufacturer's data manual or web site for your rectifier diodes, complete Table 14.4.

Table 14.4: Rectifier diode specifications.

Part Number	V_{RRM}, VOLTS	V_{RSM}, VOLTS	I_O, AMPS	V_F, VOLTS	I_R, uAmps

PART 3: Questions

3.1 What is the relationship between the line currents in Table 14.2 and the currents through the diodes for the circuit shown in Figure 14.7?

3.2 Use the RMS value you measured for V_{L1-LN} from Table 14.2 to calculate the theoretical line-to-line voltage.

3.3 How does your answer to Question 3.2 compare to your measured line-to-line voltages recorded in Table 14.2? In other words, do your experimental results confirm the theoretical value predicted by Formula 14.2?

3.4 Referring to your experimental results in Table 14.3 and the measured resistance for the load resistor, calculate the current through R1 in the circuit shown in Figure 14.7 when all of the switches are closed.

3.5 Using the value for the load current calculated in Question 3.4, calculate the average value for the current through each diode in Figure 14.7.

3.6 How does the average diode current calculated in Question 3.5 compare to the current values recorded in Table 14.2?

3.7 Using the peak of the voltage waveform recorded in Graph 14.1 for the load resistor, calculate the theoretical load voltage (average or D.C. voltage).

3.8 How does the theoretical voltage calculated in Question 3.7 compare to the measured load voltage recorded in Table 14.3 for all three switches closed?

3.9 Using the RMS value you measured for V_{L1-LN} from Table 14.2, calculate the theoretical peak voltage for V_{L1-N}.

3.10 Based on your answer to Question 3.9, what is the smallest possible value you should specify for the V_{RRM} of the rectifier diodes in Figure 14.7?

3.11 Using a factor of safety of 2 and your answers to Questions 3.5 and 3.10, select a diode to be installed in the circuit shown in Figure 14.7. Specify the manufacturer's part number.

PART 4: Troubleshooting

4.1 Assume that you are troubleshooting a circuit like that shown in Figure 14.7. When displaying the load resistor waveform on your portable, hand-held oscilloscope, you get a waveform like that shown in Figure 14.8. What is the most likely problem? Explain how you arrived at your answer.

Figure 14.8: Load waveform for Troubleshooting Problem 4.1.

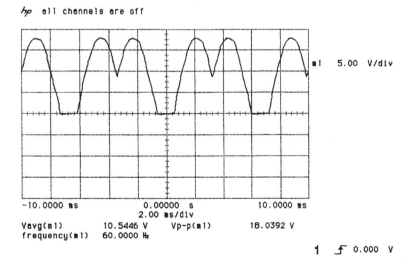

4.2 If one of the diodes in Figure 14.7 were to short because its V_{RRM} were exceeded, what would be the result or symptom of such a failure? Explain your thought processes (chain of logic) in arriving at your answer.

PART 5: Practice Problems

5.1 A circuit like that shown in Figure 14.7 has a line-to-neutral voltage of 270 volts RMS and a load resistor having a resistance of 50 Ohms. Answer the following questions. Show units where required.

a) What is the peak value of the line-to-neutral voltage?

b) Assuming a forward voltage drop, V_F, of 1.5 volts across each diode, what is the peak voltage across the load resistor?

c) Based on your answer to part b), what is the D.C. voltage across the load resistor?

d) Based on your answer to part c), what is the average current through the load resistor?

e) How much power does the load resistor dissipate?

f) Based on your answer to part d) what is the average current through each diode?

PART 6: Circuit Design

6.1 Design a three-phase, half-wave rectifier circuit like that shown in Figure 14.7 that will meet the following design requirements: a) provide an output voltage of 100 volts D.C. to the armature of a D.C. motor having a nominal resistance of 4 Ohms; b) the supply voltage is a 270 volts RMS line-to-neutral three-phase sinusoidal source; c) it will be necessary to step the source voltage down using a three-phase transformer connected wye-wye as shown in Figure 14.3; d) provide for a factor of safety of 3 for the rectifier diodes; e) the motor rotates at 1750 RPM and draws 2 amperes of current when operating at rated load conditions. Specify the transformer turns ratio and the part number for the rectifier diodes that you select. Write down any assumptions that you make. Show all calculations in the space provided below. Draw a schematic of your final design. Label all components with proper identification. Attach your schematic to your lab report. As a guide to help you complete your design, refer to Formulas 14.1 through 14.6.

Calculations:

PART 7: Three-Phase, Half-Wave Rectifier Formulas

$V_{RMS} = (0.707) \cdot (V_P)$ **14.1**

$V_{L-L} = \sqrt{3}(V_{L-N})$ **14.2**

$V_{DC} = (0.827) \cdot (V_P)$ **14.3**

$I_{DC(RESISTOR)} = V_{DC(RESISTOR)}/R$ **14.4**

$I_{AVERAGE(DIODE)} = I_{DC(RESISTOR)}/3$ **14.5**

$V_{PRV} = (1.5) \cdot (V_{PEAK(LINE-TO-NEUTRAL)})$ **14.6**

Three-Phase, Full-Wave Rectification: 15

INTRODUCTION:
Experiment 14 introduced you to the rectification of three-phase A.C. power supplies by examining the principle of operation of the three-phase, half-wave rectifier. This experiment will take the subject of three-phase rectification to its next natural step by examining the three-phase, full-wave rectifier.

SAFETY NOTE:
Before starting this experiment, have your instructor review the proper way to use an oscilloscope to display three-phase voltages. You will also be operating your circuit at high voltage, so take extra precaution to prevent shorts. Make sure that there are no exposed leads or connectors. Also, the power resistor that is to be used as a load will become very hot. Do not touch it after energizing your circuit. Before circuit disassembly, allow this resistor to cool down after turning off the power for the last time. As always, wear safety glasses, and remove jewelry from your fingers and wrists. Also have your instructor tell you what the maximum allowable power, current, and voltage ratings for your rectifier diodes and power resistor are so that you will not exceed any of these during the experiment.

OBJECT:
Upon successful completion of this experiment and all reading assignments, the student should be able to:
- construct an operational three-phase, full-wave rectifier circuit
- use an oscilloscope to display the voltage across the load resistor in a three-phase, full-wave rectifier circuit
- given a full-wave, three-phase rectifier circuit and given the line-to-line voltage, calculate the peak load voltage, average (D.C.) load voltage, peak load current, average load current, peak diode current, and average diode current

MATERIALS:
- 6 - rectifier diodes, 1N5407, or SK9009, or the equivalent
- 1 - 1500 Ohms, 25 watt power resistor
- 4 - switches, SPST, rated for 208 VAC or higher
- miscellaneous lead wires and connectors

EQUIPMENT:
- 1 - three-phase, variable A.C. power supply capable of 0 to 120 volts RMS single-phase and 0 to 208 volts line-to-line voltage with circuit breaker
- 1 - handheld multimeter
- 1 - oscilloscope
- 2 - oscilloscope probes
- 1 - isolation transformer

INSTRUCTOR'S NOTE: If your lab is not equipped with an adjustable three-phase power supply, you may still perform this lab if you have access to a fixed three-phase 120/208 A.C. supply with appropriate fuse and/or circuit breaker protection. You will <u>not</u> be able to use the 1500 Ohms, 25 watt power resistor. Because of the higher output voltage, and in turn higher power dissipation requirements, you should instead use a 2500 Ohms power resistor with a wattage rating of 50 watts. SK9009 rectifier diodes or diodes with an equivalent or higher rating should be used. Also provide your students with all of the necessary safety precautions and safety training required for working with three-phase voltages of this magnitude.

PART 1: The Three-Phase, Full-Wave Rectifier
Background:
Before discussing the three-phase, full-wave rectifier, let's review some three-phase circuit fundamentals. Recall that a three-phase A.C. power supply can be described as three single-phase A.C. sources with each sinusoidal voltage waveform separated from the other two by a phase shift of 120° as shown in Figure 15.1.

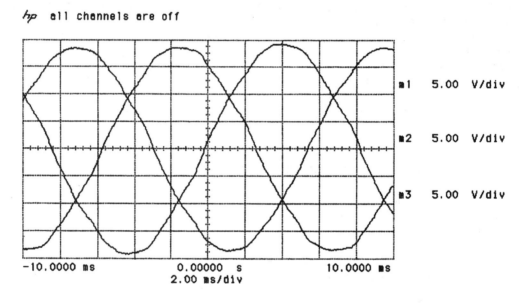

Figure 15.1: Sinusoidal voltage waveforms produced by a three-phase alternator displaced by 120°.

Remember also that the mathematical relationship between the peak voltage, V_P, of a single-phase sinusoidal A.C. waveform and its RMS value, V_{RMS} (V_{AC}), is as follows:

$$V_{RMS} = (0.707) \cdot (V_P) \qquad \textbf{15.1}$$

This formula also applies to three-phase, line-to-neutral and three-phase, line-to-line voltages. Figure 15.2 shows the relationship between the three line-to-neutral voltages, V_{L1-N}, V_{L2-N}, V_{L3-N}, and the three line-to-line voltages V_{L1-L2}, V_{L2-L3}, V_{L1-L3}. For example, if the peak of the line-to-neutral voltage of a particular three-phase A.C. supply is 392 volts, the RMS value would be

$$V_{RMS} = (0.707) \cdot (392)$$

$$= 277 \text{ volts RMS}$$

If the RMS value of the line-to-line voltage of a particular three-phase supply is 480 volts RMS, then the peak of this waveform would be

$$V_P = (V_{RMS})/(0.707)$$

$$= 480/(0.707)$$

$$= 679 \text{ volts peak}$$

The formula relating the line-to-neutral voltage to the line-to-line voltage of a sinusoidal three-phase A.C. supply is as follows:

$$V_{L-L} = \sqrt{3}(V_{L-N}) \qquad \textbf{15.2}$$

For example, if the RMS value of the line-to-neutral voltage of a particular A.C. supply is 277 volts RMS, then the RMS value of the line-to-line voltage is

$$V_{L-L} = \sqrt{3}(277 \text{ volts RMS})$$

$$= 480 \text{ volts RMS}$$

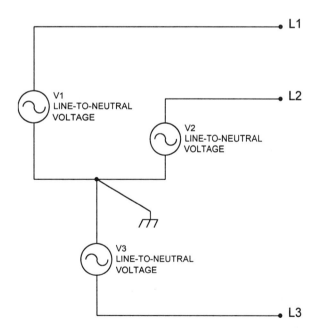

Figure 15.2: Simplified representation of a wye-connected, three-phase power supply.

Figure 15.2 is an example of a wye-connected power supply. Each of the sinusoidal supplies shares a common point referred to as the neutral. Figure 15.3 shows an example of a three-phase power supply that is delta-connected. Notice that there is no neutral connection in this circuit. Also note that each of the sinusoidal sources in Figure 15.3 represents the line-to-line voltage, whereas in Figure 15.2, each of the sinusoidal sources represents the line-to-neutral voltages.

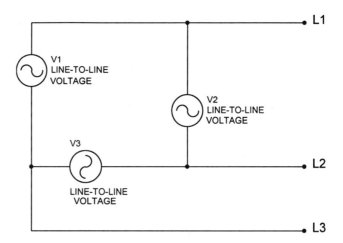

Figure 15.3: Simplified representation of a delta-connected, three-phase power supply.

Now let's go about explaining how a three-phase, full-wave rectifier works. Let's assume that the three-phase A.C. source shown in Figure 15.2 is connected to the circuit shown in Figure 15.4. Figure 15.4 is a schematic diagram of a three-phase transformer that is being used to step-down the three-phase supply voltage. A three-phase transformer works on the same principle as a single-phase transformer, but there are three sets of windings instead of one. Each set of windings steps down the voltage applied to the primary by a factor equal to the turns ratio for that particular set of windings. The primary and secondary of a three-phase transformer may be either wye- or delta-connected. In this example the primary is wye-connected and the secondary is delta-connected. The abbreviations H2 and H1 designate the primary connections, and X2 and X1 designate the secondary connections.

Each line coming out of the secondary of the transformer is connected to the anode of one rectifier diode and the cathode of another rectifier diode. In this circuit only two of the six diodes will be conducting at a time. This is similar to the operation of the single-phase, full-wave bridge rectifier in which only two out of the four diodes conduct at a time. Of diodes D4, D5, and D6, the one that conducts is the one that has the most positive instantaneous voltage applied to it. Of diodes D1, D2, and D3, the one that conducts is the one that has the most negative instantaneous voltage applied to it. The other four diodes are reverse-biased, acting as opens. Refer to Figure 15.5 for an example.

In Figure 15.5, line 3 has the most positive voltage and is forward biasing diode D6 while at the same instant in time line 2 is the most negative, causing diode D2 to be forward-biased. The four remaining diodes have been replaced by opens. The path of conventional current then is from line 3 through diode D6, into the positive terminal of the power resistor, out the negative terminal of the power resistor, through diode D2 and returning on line 2.

Notice an important difference between our three-phase, full-wave rectifier and the three-phase, half-wave rectifier in Experiment 14. Here the line-to-line voltage is applied across the resistor and the two forward-biased diodes that are in series with it. In the three-phase, half-wave rectifier circuit, the line-to-neutral voltage was applied across the load resistor and one of the forward-biased diode in series with it. Much like the single-phase, full-wave rectifier, our three-phase, full-wave rectifier effectively flips or reroutes each of the negative pulses shown in Figure 15.1 so that they each produce positive pulses across the load resistor. The resulting voltage waveform across the load for the circuit shown in Figure 15.4 is shown in Figure 15.6. Note how smooth this waveform is compared to the output of a three-phase, half-wave rectifier. Note also how much more area under the curve there is now, especially since the waveform does not drop to zero at any point along the x-axis.

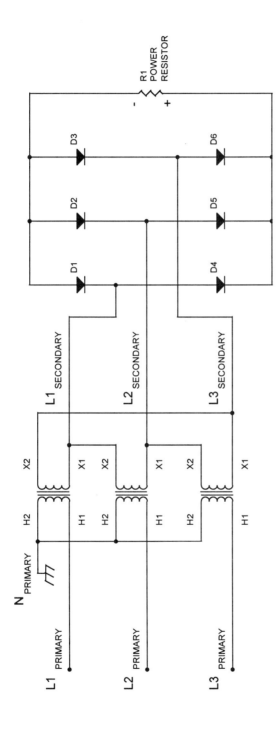

Figure 15.4: Schematic diagram of a wye-delta connected three-phase, step-down transformer. The three-phase secondary is connected to six diodes and a load resistor resulting in a three-phase, full-wave rectified power supply.

As with the three-phase, half-wave rectifier, the diodes in the three-phase, full-wave rectifier conduct for only one-third of the full cycle. However, there is an even higher peak reverse voltage across each of the diodes now. In this configuration, the peak of the voltage waveform across the load is equal to the peak of the applied line-to-line voltage minus the sum of the forward-bias voltage drops across each of the conducting diodes. Because we will be using rather high-power diodes for circuits such as this, the forward-bias drop will likely be more than 0.7 volt and may be as high as 2.0 volts or more due to the heavy doping of silicon-based power diodes and the high currents that will flow through each diode. However, this voltage drop is still relatively small compared to the peak of the supply voltage, which typically will be greater than 100 volts, usually far greater.

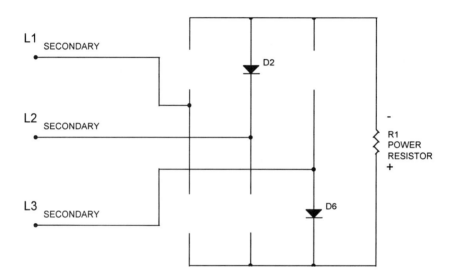

Figure 15.5: In the three-phase, full-wave rectifier, only two out of six diodes conduct at a time while the remaining four diodes act as opens.

The average value of the output voltage across the load resistor may be found by integrating the area under the voltage curve. If we assume that each of the peaks is sinusoidal with each pulse having a width of 60° ($\pi/3$), the area under the first pulse could be found by integrating from 60° to 120°. The interval from 60° to 120° represents the section of each three-phase waveform that is contributing to the positive voltage across the resistor. This area can then be divided by the length of that curve segment, ($\pi/3$). The result is the following formula for the D.C. or average value of the voltage waveform across the resistor:

$$V_{DC} = (0.955) \cdot (V_P) \qquad \textbf{15.3}$$

In this formula, V_P is the peak of the voltage waveform across the resistor. The average current then that flows through the resistor can be found using Ohm's Law:

$$I_{DC(RESISTOR)} = V_{DC(RESISTOR)}/R \qquad \textbf{15.4}$$

I know that all of you have a firm grasp of Ohm's Law by now so, it was not my intent to insult anyone's intelligence by including it amongst our list of formulas. Rather, I wanted to include it as an appropriate reminder of how I went about determining the current rating for each of the diodes. Because each of the diodes conducts for one-third of the time, the average current through each diode in Figure 15.4 can be found as follows:

$$I_{AVERAGE(DIODE)} = I_{DC(RESISTOR)}/3 \qquad \textbf{15.5}$$

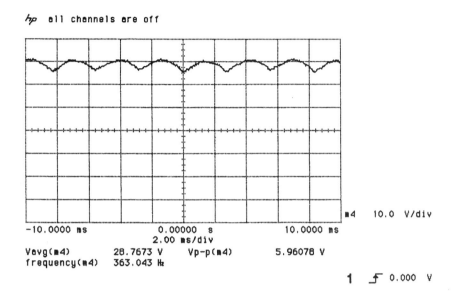

Figure 15.6: Three-phase, full-wave rectified load waveform for the circuit shown in Figure 15.4.

Another important parameter to specify when installing diodes in rectifier circuits is the peak repetitive voltage, V_{RRM}. If you study the three-phase waveform in Figure 15.1 closely, you will notice that when the first phase is at its most positive point, the other two waveforms are negative. At this point, diode D4 will be forward-biased and diode D2 will be starting to shut off while diode D3 will be coming on. Diodes D5 and D6 will be reverse-biased by the positive potential of the load resistor. At the same time, the anodes of each of these two diodes will be exposed to the negative voltages present on lines 2 and 3 respectively. The same positive voltage that forward biases diode D4 reverse biases diode D1. The differential voltage from the peak of the positive waveform to where the other two negative waveforms cross represents the magnitude of the reverse-bias voltage applied to diodes D5 and D6 in Figure 15.4. The positive peak of the first voltage waveform forward biases D4 causing a voltage drop across the load resistor. The magnitude of the voltage across the resistor is equal to the magnitude of the first sinusoidal waveform less the forward-bias voltage drop across D4. You must realize that the very same voltage waveform that appears across the load resistor causes diodes D5 and D6 to be reverse biased. At the very same instant in time, the remaining two negative sinusoidal waveforms (from lines L2 and L3) are also reverse biasing diodes D5 and D6. The total reverse bias voltage then is the sum of the absolute value of these two instantaneous voltages. This is similar to what happens in a voltage doubler circuit. This differential voltage is approximately equal to the square root of three times the peak of the line-to-line voltage. In other words the maximum peak reverse voltage, V_{PRV}, for any one of the diodes is

$$V_{PRV} = (\sqrt{3}) \cdot (V_{PEAK(LINE-TO-LINE)}) \qquad \textbf{15.6}$$

As always, choose an appropriate factor of safety when selecting rectifier diodes or any other electronic component for that matter. Never use a factor of safety of 1. At least double the expected load voltage, current, or power when determining electronic device specifications. Always consult with a professional engineer before making a final selection. Circuit failures associated with marginally designed components can be potentially catastrophic!

Procedure:

1.1 Obtain all materials and equipment required to complete this experiment.

1.2 Use your handheld multimeter to measure the resistance of the power resistor, and record your measurement in Table 15.1. If the resistor is out of tolerance, notify your instructor immediately and obtain a replacement.

Table 15.1: Nominal and measured load resistor values.

R1, NOMINAL RESISTANCE	R1, MEASURED RESISTANCE

1.3 With your power supply turned off and adjusted to produce zero volts when turned on, connect the circuit shown in Figure 15.7. Have your instructor check your work.

1.4 If your are using an analog meter in this experiment, be on the alert for excessive upscale deflection or pegging of the meter movement. Select an appropriate full-scale value to prevent this from happening. If you have any doubt about what range to select, ask your instructor for assistance.

1.5 Turn on the power supply. Slowly increase the supply voltage until M1, your handheld A.C. voltmeter, indicates 80 volts RMS. M1 indicates the voltage from line 1 to line 2, V_{L1-L2}. Close all four switches. Measure and record all of the voltages specified in Table 15.2.

Table 15.2: Measured A.C. voltages for the circuit shown in Figure 15.7.

V_{L1-L2}, VOLTS RMS	V_{L2-L3}, VOLTS RMS	V_{L1-L3}, VOLTS RMS

1.6 Using your handheld multimeter as a D.C. voltmeter, measure the voltage across the load resistor. Be sure to observe proper polarity when making your measurement, especially if using an analog voltmeter. Record your measurement in Table 15.3.

1.7 Use your oscilloscope to display the source voltage waveform from line 1 to line 2, V_{L1-L2}. Draw the waveform in the space provided on Graph 15.1.

1.8 Now use your oscilloscope to display the voltage waveform across the load resistor, V_{AB}. Draw the resulting waveform in the space provided on Graph 15.2.

1.9 Open switch S1. Using your handheld multimeter as a D.C. voltmeter, measure the voltage across the load resistor. Record your measurement in Table 15.3.

1.10 Use your oscilloscope to display the voltage waveform across the load resistor. Draw the resulting waveform in the space provided on Graph 15.3.

1.11 Open switch S2. Switch S1 should still be open. Using your handheld multimeter as a D.C. voltmeter, measure the voltage across the load resistor. Record your measurement in Table 15.3.

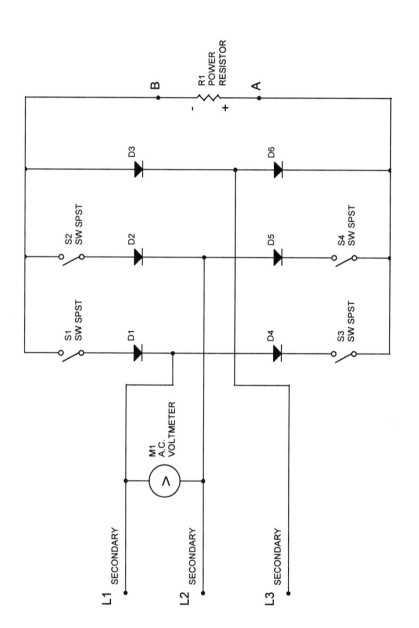

Figure 15.7: Three-phase, full-wave rectifier circuit.

1.12 Use your oscilloscope to display the voltage waveform across the load resistor. Draw the resulting waveform in the space provided on Graph 15.4.
1.13 Open switch S3. Switches S1 and S2 should still be open. Using your handheld multimeter as a D.C. voltmeter, measure the voltage across the load resistor. Record your measurement in Table 15.3.
1.14 Use your oscilloscope to display the voltage waveform across the load resistor. Draw the resulting waveform in the space provided on Graph 15.5.
1.15 Open switch S4. Switches S1, S2, and S3 should still be open. Using your handheld multimeter as a D.C. voltmeter, measure the voltage across the load resistor. Record your measurement in Table 15.3.
1.16 Use your oscilloscope to display the voltage waveform across the load resistor. Draw the resulting waveform in the space provided on Graph 15.6.
1.17 Reduce the power supply voltage to 0 volts. Turn off the power supply.

Table 15.3: Output voltage for different switch positions for the circuit shown in Figure 15.7.

SWITCH SETTINGS	OUTPUT VOLTAGE, D.C. VOLTS
ALL SWITCHES CLOSED	
S1 OPEN; S2, S3, AND S4 CLOSED	
S1 AND S2 OPEN; S3 AND S4 CLOSED	
S1, S2, AND S3 OPEN; S4 CLOSED	
ALL SWITCHES OPEN	

Graph 15.1: Source voltage waveform, V_{L1-L2}, for the circuit shown in Figure 15.7.

V_{L1-L2}: _____ volts/div   _____ sec/div

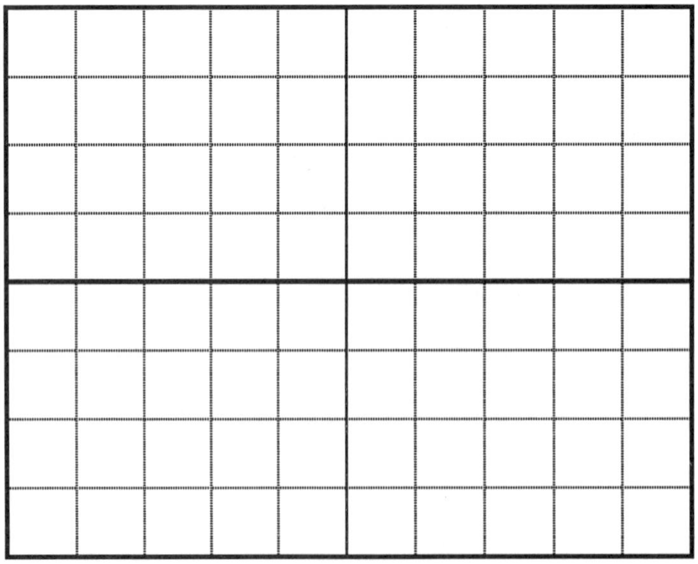

Graph 15.2: V_{R1} voltage waveform for the circuit shown in Figure 15.7.

V_{R1}: _____ volts/div   _____ sec/div

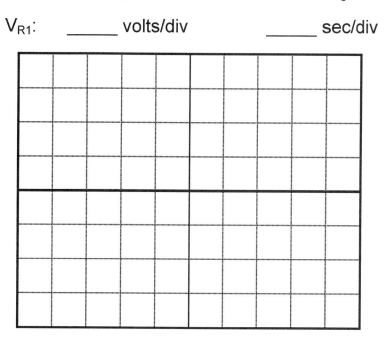

Graph 15.3: V_{R1} voltage waveform for the circuit shown in Figure 15.7 with switch S1 open.

V_{R1}: _____ volts/div   _____ sec/div

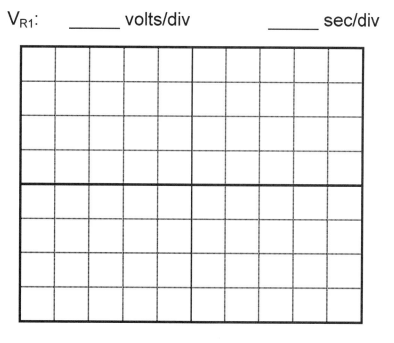

Graph 15.4: V_{R1} voltage waveform for the circuit shown in Figure 15.7 with switches S1 and S2 open.

V_{R1}: _____ volts/div   _____ sec/div

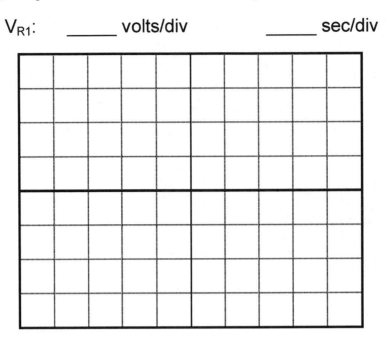

Graph 15.5: V_{R1} voltage waveform for the circuit shown in Figure 15.7 with switches S1, S2, and S3 open.

V_{R1}: _____ volts/div   _____ sec/div

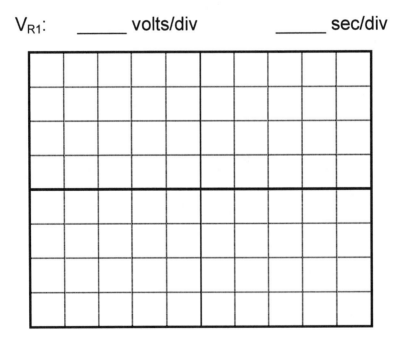

Graph 15.6: V_{R1} voltage waveform for the circuit shown in Figure 15.7 with all switches open.

V_{R1}: _____ volts/div   _____ sec/div

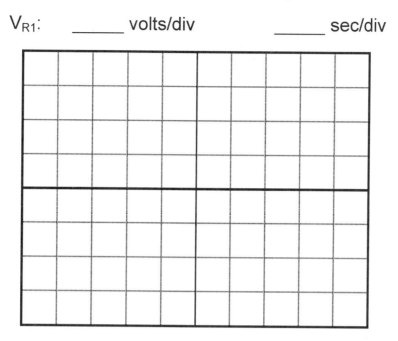

PART 2: Manufacturer's Ratings

2.1 Referring to the manufacturer's data manual or web site for your rectifier diodes, complete Table 15.4.

Table 15.4: Rectifier diode specifications.

Part Number	V_{RRM}, VOLTS	V_{RSM}, VOLTS	I_O, AMPS	V_F, VOLTS	I_R, uAmps

PART 3: Questions

3.1 Using the RMS value you measured for V_{L1-L2} from Table 15.2, calculate the theoretical line-to-neutral voltage.

3.2 Using the peak of the voltage waveform recorded in Graph 15.1, calculate the RMS value of the line-to-line voltage.

3.3 How does your answer to Question 3.2 compare to your measured line-to-line voltages recorded in Table 15.2? In other words, did your experimental results confirm the theoretical value predicted by Formula 15.1?

3.4 Referring to your experimental results in Table 15.3 and the measured resistance for the load resistor, calculate the current through R1 in the circuit shown in Figure 15.7 when all of the switches are closed.

3.5 Using the value for the load current calculated in Question 3.4, calculate the average value for the current through each diode in Figure 15.7.

3.6 Using the peak of the voltage waveform recorded in Graph 15.2 for the load resistor, calculate the theoretical load voltage.

3.7 How does the theoretical voltage calculated in Question 3.6 compare to the measured load voltage recorded in Table 15.3 for all four switches closed?

3.8 Using the RMS value you measured for V_{L1-L2} from Table 15.2, calculate the theoretical peak voltage for V_{L1-L2}.

3.9 Based on your answer to Question 3.8, what is the smallest possible value you should specify for the V_{RRM} of the rectifier diodes in Figure 15.7?

3.10 Using a factor of safety of 2 and your answers to Questions 3.5 and 3.9, select a diode to be installed in the circuit shown in Figure 15.7. Specify the manufacturer's part number.

PART 4: Troubleshooting

4.1 Assume that you work for a company that makes plastic products by injection molding. You are the lead maintenance technician. Your boss has just come to tell you that one of the injection molding units is not working properly because the plastic is not melting completely. This is indicative of the heater elements either not getting hot enough and/or not staying on long enough. You know that a circuit like that shown in Figure 15.7 is used to power the heating elements. First you check the signal from the microprocessor that controls the duty cycle of the D.C. power supply and find that it is working properly. All that is left to check now is the circuit shown in Figure 15.7. What do you think is wrong? Explain how you would go about confirming your suspicions.

4.2 If one of the diodes in Figure 15.7 were to short because its V_{RRM} was exceeded, what would be the result or symptom of such a failure? Explain your thought processes (chain of logic) in arriving at your answer.

PART 5: Practice Problems

5.1 A circuit like that shown in Figure 15.7 has a line-to-line voltage of 480 volts RMS and a load resistor having a resistance of 150 Ohms. Answer the following questions. Show units where required.

a) What is the peak value of the line-to-line voltage?

b) Assuming a forward voltage drop, V_F, of 2 volts across each diode, what is the peak voltage across the load resistor?

c) Based on your answer to part b), what is the D.C. voltage across the load resistor?

d) Based on your answer to part c), what is the average current through the load resistor?

e) How much power does the load resistor dissipate?

f) Based on your answer to part d), what is the average current through each diode?

PART 6: Circuit Design

6.1 Design a three-phase, full-wave rectifier circuit like that shown in Figure 15.7 that will meet the following design requirements: a) provide an output voltage of 150 volts D.C. to an electric powered heat-treating oven having a nominal resistance of 20 Ohms; b) the supply voltage is a 270 volts RMS line-to-neutral, three-phase sinusoidal source; c) it will be necessary to step the source voltage down using a three-phase transformer connected wye-delta as shown in Figure 15.4; and d) provide for a factor of safety of 3 for the rectifier diodes. Specify the transformer turns ratio and the part number for the rectifier diodes that you select. Write down any assumptions that you make. Show all calculations in the space provided below. Draw a schematic of your final design, and label all components with proper identification. Attach your schematic to your lab report. As a guide to help you complete your design, refer to Formulas 15.1 through 15.6.

Calculations:

PART 7: Three-Phase, Full-Wave Rectifier Formulas

$V_{RMS} = (0.707) \cdot (V_P)$ **15.1**

$V_{L\text{-}L} = \sqrt{3}(V_{L\text{-}N})$ **15.2**

$V_{DC} = (0.955) \cdot (V_P)$ **15.3**

$I_{DC(RESISTOR)} = V_{DC(RESISTOR)}/R$ **15.4**

$I_{AVERAGE(DIODE)} = I_{DC(RESISTOR)}/3$ **15.5**

$V_{PRV} = (\sqrt{3}) \cdot (V_{PEAK(LINE\text{-}TO\text{-}LINE)})$ **15.6**

Fundamental Op-Amp Circuits: 16

INTRODUCTION:

The operational amplifier (op-amp), or differential amplifier as it is sometimes called, has the flexibility to be used in many applications. The op-amp may be used as a buffer, summer, wave-shaper, comparator, inverting or non-inverting amplifier, current-to-voltage converter, voltage-to-current converter, and in many other applications. Op-amp technology has also been integrated into many other ICs, including instrumentation amplifiers, analog-to-digital (A/D) converters, and digital-to-analog (D/A) converters. Unlike the transistor, which has just one input (the base), the op-amp has two signal inputs. These two inputs are the non-inverting (+) and inverting (-) inputs. These are shown on the schematic symbol in Figure 16.1. The op-amp is referred to as a *differential amplifier* because it amplifies the difference between the voltage applied to the (+) and (-) inputs. The output of the op-amp may be either positive or negative. Therefore, most op-amps have connections for both a positive and negative power supply as shown in Figure 16.1. The connection of two D.C. power supplies of opposite polarity to the LM741 op-amp is shown in Figure 16.2. The op-amp is a device capable of very high voltage gains. The op-amp's gain, as with all amplifiers, is determined by its configuration. The op-amp may be connected in two basic configurations--open loop and closed loop. In the open-loop configuration, the op-amp has a very large gain, A_{VOL}. In the closed-loop configuration, all, or a portion of the op-amp's output, is connected back to one of the inputs. Connecting the output back to the (-) input is referred to as negative feedback. The resulting closed-loop gain, A_{CL}, is much less than A_{VOL}. The output may also be fed back to the (+) input--positive feedback. This particular configuration is referred to as a Schmitt trigger. In this experiment you will have the opportunity to construct op-amp circuits in each of the following configurations: 1) non-inverting amplifier, 2) inverting amplifier, 3) voltage-follower, 4) comparator.

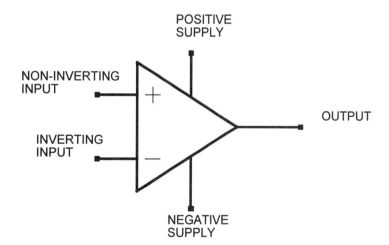

Figure 16.1: Op-amp symbol with pin identification.

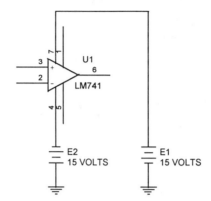

Figure 16.2: Op-amp power supply connections.

SAFETY NOTE:
While the op-amp is essentially a low-power device, it can still represent a safety hazard if not properly wired. It is literally possible to "blow the top off" of an op-amp. This is not recommended! This frequently happens when the power supply is improperly connected to the op-amp. Have your instructor review this important aspect of op-amp wiring before you proceed. As always, be sure to wear safety glasses while working in the lab, remove jewelry from fingers and wrists, and have your instructor check your work before you energize your circuit and each time you make changes thereafter.

OBJECT:
Upon successful completion of this experiment and all reading assignments, the student should be able to:
- construct an operational non-inverting amplifier circuit
- predict the output of a given non-inverting amplifier circuit
- construct an operational inverting amplifier circuit
- predict the output of a given inverting amplifier circuit
- construct an operational voltage-follower circuit
- predict the output of a given voltage-follower circuit
- construct an operational comparator circuit
- predict the output of a given comparator circuit

REFERENCES:
Chapter 8 of Maloney's Modern Industrial Electronics
www.national.com

MATERIALS:
- 1 - LM741 op-amp or its equivalent
- 1 - solderless breadboard
  - miscellaneous lead wires and connectors
- 1 - 10 Kohms, 1/4 watt resistor
- 1 - 47 Kohms, 1/4 watt resistor

EQUIPMENT:
- 1 - digital multimeter
- 1 - function generator
- 1 - frequency counter (optional)
- 1 - dual-outlet D.C. power supply
- 1 - dual-trace oscilloscope
- 2 - oscilloscope probes

PART 1: The Non-Inverting Op-Amp
Background:
A typical non-inverting op-amp configuration is shown in Figure 16.3. Note that part of the output is fed back to the (-) input. The signal is connected to the (+) input. The voltage gain for this configuration may be calculated as follows:

$$A_{CL} = (1 + R_F/R_1) \qquad \qquad 16.1$$

As you may have already studied, the general formula for the voltage gain, A_V, for any amplifier is

$$A_V = V_{OUT}/V_{IN} \qquad \qquad 16.2$$

In this case, V_{IN} is the 1 V_{P-P} sine wave signal, V_S. Combining Formulas 16.1 and 16.2 we have

$$V_{OUT} = V_S(1 + R_F/R_1) \qquad \qquad 16.3$$

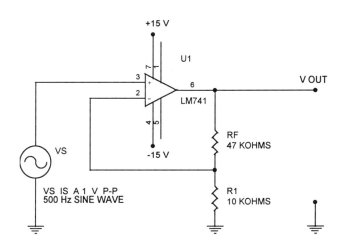

Figure 16.3: Op-amp configured as a non-inverting amplifier.

Procedure:
    1.1 Obtain all materials and equipment required to complete this experiment.
    1.2 Use your digital multimeter (DMM) to measure the resistance of each fixed resistor, and record in Table 16.1.

Table 16.1: Nominal and measured resistor values.

NOMINAL RESISTANCE	MEASURED RESISTANCE
47,000 OHMS	
10,000 OHMS	

1.3 Adjust your D.C. power supply so that each outlet produces 15 volts. Turn off your power supply.
1.4 Connect the D.C. power supply to the op-amp as shown in Figure 16.2. Have your instructor check your work. Verify the accuracy of your work by turning on your power supply and quickly and briefly touching the top of the op-amp to be sure that it is not getting hot. If the op-amp is getting hot, turn off the power supply immediately and re-check your work. If the op-amp is still cool, turn off the power supply and proceed to the next step.
1.5 Connect your function generator to Channel 1 of your oscilloscope. Adjust the function generator to provide a 1 volt peak-to-peak (V_{P-P}) sine wave with a frequency of 500 Hz. Once set, turn off the function generator.
1.6 Now connect the circuit shown in Figure 16.3. Note: You have already connected the +15 and -15 volt D.C. supplies. Use the same ground reference for your A.C. signal as was used for the D.C. supplies. Have your instructor check your work.
1.7 Turn on the D.C. power supply. Turn on the function generator. Use Channel 2 of the oscilloscope to display the output voltage waveform, V_{OUT}, in proper time phase with the signal, V_S. If you are using an oscilloscope with more than two channels, have your instructor indicate which two channels you should be using.
1.8 Draw both V_S and V_{OUT} in proper time phase with each other on Graph 16.1. Completely label the vertical and horizontal axes with voltage and time base values.

Graph 16.1: V_S and V_{OUT} for the non-inverting op-amp drawn in proper time phase.

V_S: _____ volts/div _____ sec/div
V_{OUT}: _____ volts/div _____ sec/div

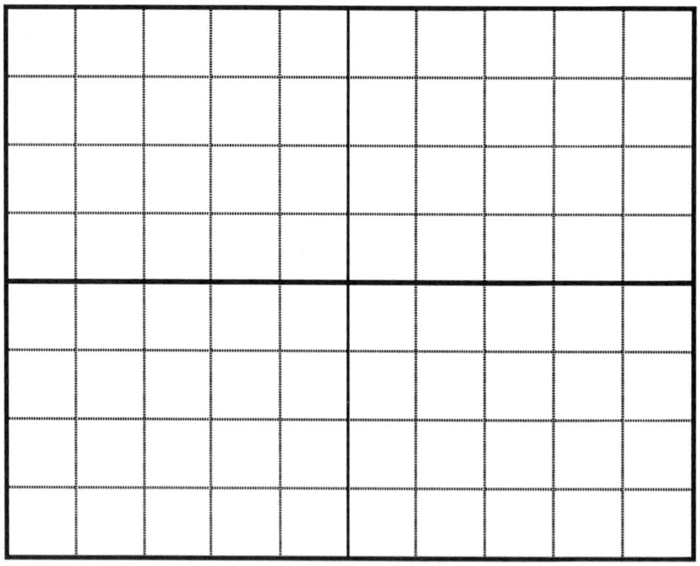

1.9 Use a voltmeter to measure the RMS value of V_{OUT} and V_S. Record in Table 16.2.
1.10 Turn off the function generator and your D.C. power supply.

Table 16.2: RMS values for V_{OUT} and V_S for the non-inverting op-amp.

V_{OUT}, VOLTS RMS	V_S, VOLTS RMS

PART 2: The Inverting Op-Amp

Background:

A typical inverting op-amp configuration is shown in Figure 16.4. Note that as with the non-inverting op-amp circuit, part of the output is fed back to the (-) input. However, now the source voltage is connected to the (-) input via R_1. The voltage gain for the inverting op-amp may be calculated as follows:

$$A_{CL} = -R_F/R_1 \qquad 16.4$$

Combining Formulas 16.2 and 16.4, we have the following relationship between V_{OUT} and V_S for the inverting op-amp:

$$V_{OUT} = V_S(-R_F/R_1) \qquad 16.5$$

The minus sign denotes the inversion. Some of you may have an invert function on your oscilloscope. When you select the invert function on your oscilloscope, you effectively multiply all points on the waveform by -1. Thus, all positive values become negative and all negative values become positive.

Figure 16.4: Op-amp configured as an inverting amplifier.

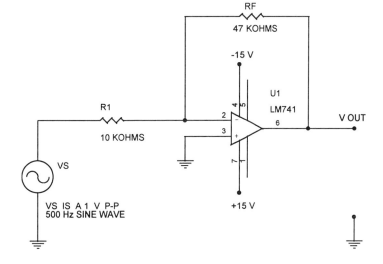

Procedure:

2.1 Modify your original circuit so that it appears as shown in Figure 16.4. The function generator and D.C. power supply should still have the same settings.

2.2 Have your instructor check your work. As before, verify the accuracy of your work by turning on your D.C. power supply and quickly and briefly touching the top of the op-amp to be sure that it is not getting hot. If the op-amp is getting hot, turn off the power supply immediately and re-check your work. If the op-amp is still cool, proceed to the next step.

2.3 Turn on the function generator.

2.4 Again display V_S on Channel 1 of your oscilloscope. Use Channel 2 of the oscilloscope to display the output voltage waveform, V_{OUT}, in proper time phase with the signal, V_S.

2.5 Draw both V_S and V_{OUT} in proper time phase with each other on Graph 16.2. Completely label the vertical and horizontal axes with voltage and time base values.

Graph 16.2: V_S and V_{OUT} for the inverting op-amp drawn in proper time phase.

V_S: _____ volts/div _____ sec/div
V_{OUT}: _____ volts/div _____ sec/div

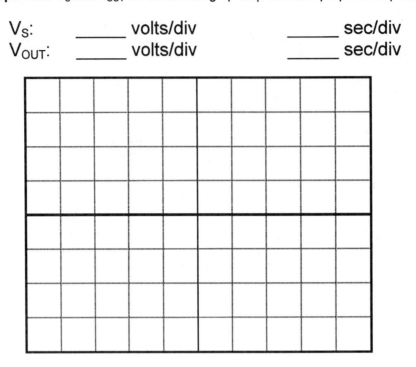

2.6 Use a voltmeter to measure the RMS value of V_{OUT} and V_S. Record in Table 16.3.
2.7 Turn off the function generator and your D.C. power supply.

Table 16.3: RMS values for V_{OUT} and V_S for the inverting op-amp.

V_{OUT}, VOLTS RMS	V_S, VOLTS RMS

PART 3: The Voltage-Follower
Background:
Referring to Figure 16.5, you will see an op-amp configured as a voltage follower. As you look at this circuit, think to yourself, "How is this circuit similar to and different from the non-inverting op-amp circuit shown in Figure 16.3?" Notice that the source signal, V_S, is connected to the non-inverting input for both of these circuits. Of course, the D.C. power supply connections are still the same. What's different? Well, in the voltage-follower circuit, all of the output voltage, V_{OUT}, is fed back to the inverting input, whereas in the non-inverting op-amp circuit, only a portion of the output is fed back to the inverting input via the voltage-divider network set up by R_F and R_1. As a result of feeding back 100% of the output to the inverting input, the op-amp's internal impedance, Z_{IN}, is increased significantly. As you may already know, Z_{IN} for any op-amp configuration is very high compared to a simple voltage-divider biased bipolar junction transistor circuit. Negative feedback, whether it be in an op-amp circuit or an emitter-follower or a source-follower circuit, increases the circuit's input impedance at the expense of reducing the circuit's voltage gain. However, the benefit of circuits such as the voltage-follower, emitter-follower, and source-

follower is that they are excellent *buffers*. Recall from your earlier studies that a buffer is used for *impedance matching*. The goal of impedance matching is to achieve maximum power transfer from a source to a load. A buffer is often installed between a small (or low-power) signal and a load requiring a current higher than the signal could otherwise provide without causing distortion. The voltage gain for the voltage-follower is approximately 1, or unity. In other words,

$$A_V = 1 \qquad \qquad 16.6$$

Combining Formulas 16.2 and 16.6, we have the following relationship between V_{OUT} and V_S for the voltage follower:

$$V_{OUT} = V_S \qquad \qquad 16.7$$

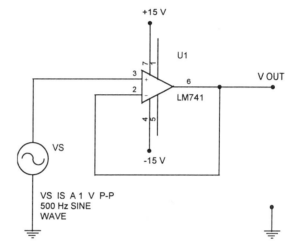

Figure 16.5: Op-amp configured as a voltage-follower.

Procedure:

3.1 Modify your original circuit so that it appears as shown in Figure 16.5. The function generator and D.C. power supply should still have the same settings.

3.2 Have your instructor check your work. As before, verify the accuracy of your work by turning on your D.C. power supply and quickly and briefly touching the top of the op-amp to be sure that it is not getting hot. If the op-amp is getting hot, turn off the power supply immediately and re-check your work. If the op-amp is still cool, proceed to the next step.

3.3 Turn on the function generator.

3.4 Again display V_S on Channel 1 of your oscilloscope. Use Channel 2 of the oscilloscope to display the output voltage waveform, V_{OUT}, in proper time phase with the signal, V_S. You might want to consider using different volts/div settings for each channel of the oscilloscope so that you may see the phase relationship between the two voltage waveforms.

3.5 Draw both V_S and V_{OUT} in proper time phase with each other on Graph 16.3. Completely label the vertical and horizontal axes with voltage and time base values.

Graph 16.3: V_S and V_{OUT} for the voltage-follower drawn in proper time phase.

V_S: _____ volts/div   _____ sec/div
V_{OUT}: _____ volts/div   _____ sec/div

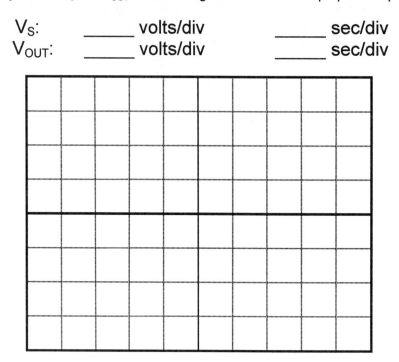

3.6 Use a voltmeter to measure the RMS value of V_{OUT} and V_S. Record in Table 16.4.
3.7 Turn off the function generator and your D.C. power supply.

Table 16.4: RMS values for V_{OUT} and V_S for the voltage-follower.

V_{OUT}, VOLTS RMS	V_S, VOLTS RMS

PART 4: The Op-Amp Comparator
Background:
Referring to Figure 16.6, you will see an op-amp configured as a comparator. As discussed in the introduction, this circuit will have a very large voltage gain because there is no negative feedback. In this configuration, the op-amp will amplify the voltage difference between the non-inverting and inverting inputs. In other words,

$$V_{OUT} = A_{VOL}(V_{(+)} - V_{(-)}) \qquad 16.8$$

where $V_{(+)}$ is the voltage applied to the non-inverting input and $V_{(-)}$ is the voltage applied to the inverting input. Because A_{VOL} is very large, even a small differential input voltage ($V_{(+)} - V_{(-)}$) will cause the output to go into *saturation*. In other words, the output will be *clipped* or limited to the maximum output voltage possible, which in this case would be approximately +15 or -15 volts. Due to internal voltage drops, the output may not quite swing to the full supply voltage.

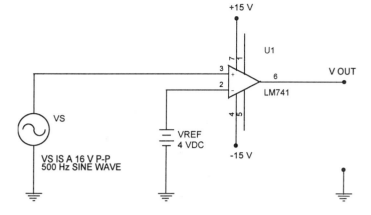

Figure 16.6: Op-amp configured as a comparator.

Procedure:

4.1 Modify your original circuit so that it appears as shown in Figure 16.6. If you do not have another D.C. source to provide the +4 volts to the inverting input, you may use your +15 volt supply and the voltage divider circuit shown in Figure 16.7 to obtain the +4 volt *reference* voltage. Note: V_S is now a 16 V_{P-P} sine wave with a frequency of 500 Hz.

4.2 Have your instructor check your work. As before, verify the accuracy of your work by turning on your D.C. power supply and quickly and briefly touching the top of the op-amp to be sure that it is not getting hot. If the op-amp is getting hot, turn off the power supply immediately and re-check your work. If the op-amp is still cool, proceed to the next step.

4.3 Turn on the function generator.

4.4 Again display V_S on Channel 1 of your oscilloscope. Use Channel 2 of the oscilloscope to display the output voltage waveform, V_{OUT}, in proper time phase with the signal, V_S.

4.5 Draw both V_S and V_{OUT} in proper time phase with each other on Graph 16.4. Completely label the vertical and horizontal axes with voltage and time base values.

4.6 Turn off the function generator and your D.C. power supply.

Figure 16.7: Op-amp comparator circuit.

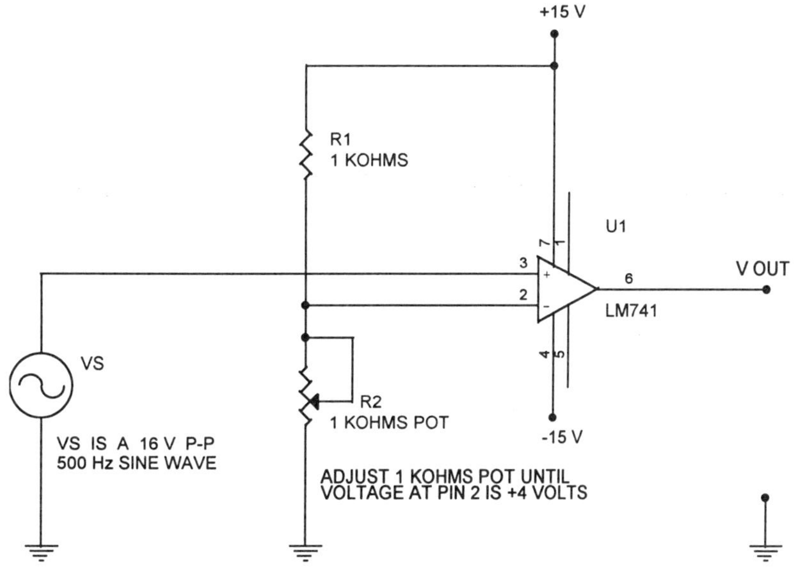

Graph 16.4: V_S and V_{OUT} for the op-amp comparator circuit drawn in proper time phase.

V_S: _____ volts/div _____ sec/div
V_{OUT}: _____ volts/div _____ sec/div

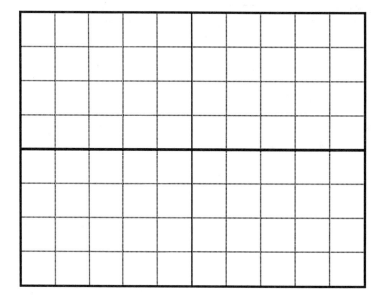

PART 5: Calculations

5.1 In the space provided below, calculate the *theoretical* voltage gain for the non-inverting and inverting op-amp circuits. In your calculations use the measured resistance values recorded in Table 16.1. Record your calculated results in Table 16.5. Also record the theoretical voltage gain for the voltage follower.

$A_{CL} = (1 + R_F/R_1) =$ **16.1**

$A_{CL} = -R_F/R_1 =$ **16.4**

5.2 In the space provided below, calculate the *actual* voltage gain for the non-inverting, inverting, and voltage-follower op-amp circuits. Use the following formulas and your measured voltage values from Tables 16.2, 16.3, and 16.4. Record your calculated results in Table 16.5.

$$A_{V(\text{NON-INVERTING})} = V_{OUT}/V_{IN} \qquad \textbf{16.2}$$

$$A_{V(\text{INVERTING})} = V_{OUT}/V_{IN} \qquad \textbf{16.2}$$

$$A_{V(\text{VOLTAGE-FOLLOWER})} = V_{OUT}/V_{IN} \qquad \textbf{16.2}$$

Table 16.5: Theoretical and actual voltage gains for the non-inverting, inverting, and voltage-follower op-amp circuits.

CIRCUIT	THEORETICAL VOLTAGE GAIN	ACTUAL VOLTAGE GAIN
NON-INVERTING		
INVERTING		
VOLTAGE-FOLLOWER		

PART 6: Questions

6.1 What changes could you make to the non-inverting op-amp circuit of Figure 16.3 to increase the voltage gain? Record your response in the space provided below.

6.2 What is the phase relationship between the source voltage and the output voltage in a non-inverting op-amp circuit?

6.3 What is the phase relationship between the source voltage and the output voltage in an inverting op-amp circuit?

6.4 What is the phase relationship between the source voltage and the output voltage in a voltage-follower op-amp circuit?

6.5 Referring to Graph 16.4 and your results for the op-amp comparator, at what point (voltage level) on the sine wave signal did the output voltage waveform switch from a negative value to a positive value?

PART 7: Practice Problems

7.1 What would be the value of the largest resistor that you could use for R_F in the inverting op-amp circuit and not produce clipping of the output voltage waveform? Assume that V_S is a 1 V_{P-P} sine wave and that R_1 is 10 Kohms. Show your calculations, and circle your final answer in the space provided below.

7.2 Referring to the circuit shown in Figure 16.8, what is the value for V_OUT?

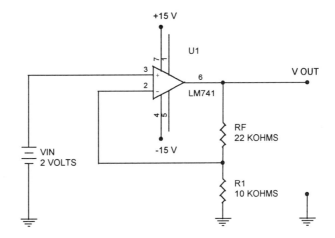

Figure 16.8: Op-amp circuit for Practice Problem 7.2.

7.3 Referring to the circuit shown in Figure 16.9, what is the value for V_OUT?

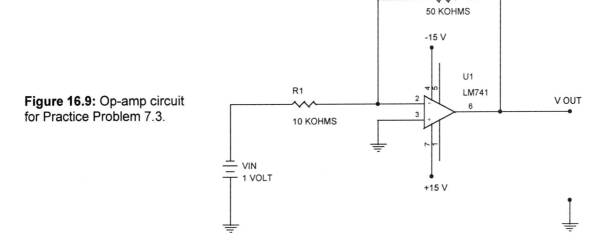

Figure 16.9: Op-amp circuit for Practice Problem 7.3.

7.4 Referring to the circuit shown in Figure 16.10, draw the input voltage waveform, V$_S$, and the output voltage waveform, V$_{OUT}$, in the space provided on Graph 16.5. Label the vertical and horizontal axes with voltage and time base information, respectively.

Calculations:

Figure 16.10: Op-amp circuit for Practice Problem 7.4.

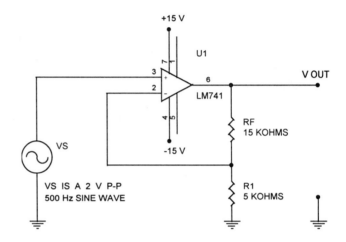

Graph 16.5: V$_S$ and V$_{OUT}$ for the op-amp circuit shown in Figure 16.10.

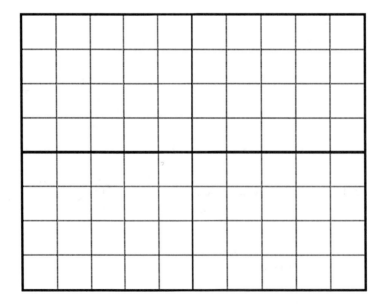

7.5 Referring to the circuit shown in Figure 16.11, draw the input voltage waveform, V_S, and the output voltage waveform, V_{OUT}, in the space provided on Graph 16.6. Label the vertical and horizontal axes with voltage and time base information, respectively.

Calculations:

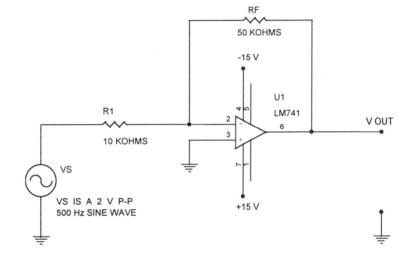

Figure 16.11: Op-amp circuit for Practice Problem 7.5.

Graph 16.6: V_S and V_{OUT} for the op-amp circuit shown in Figure 16.11.

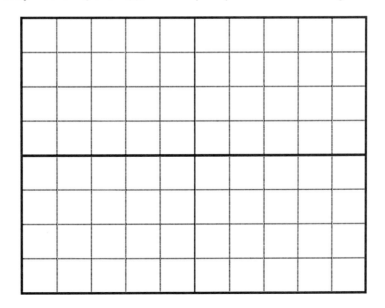

PART 8: Circuit Design

8.1 Modify the circuit shown in Figure 16.6 so that the output waveform will be inverted relative to the input voltage waveform. Sketch and label your new circuit in the space provided below. If a computer-aided drafting software package is available at your institution, use it to produce your drawing. Print a copy and attach to your report.

PART 9: Op-Amp Formulas

Non-inverting Op-Amp:

$A_{CL} = (1 + R_F/R_1)$ **16.1**

$V_{OUT} = V_S(1 + R_F/R_1)$ **16.3**

Inverting Op-Amp:

$A_{CL} = -R_F/R_1$ **16.4**

$V_{OUT} = V_S(-R_F/R_1)$ **16.5**

Voltage-Follower:

$A_V = 1$ **16.6**

$V_{OUT} = V_S$ **16.7**

Comparator:

$V_{OUT} = A_{VOL}(V_{(+)} - V_{(-)})$ **16.8**

General Amplifier Formula:

$A_V = V_{OUT}/V_{IN}$ **16.2**

Instrumentation Amplifier: 17

INTRODUCTION:
As you discovered in Experiment 16, the op-amp is a versatile integrated circuit. In addition to the applications in that lab, the op-amp has been used in applications requiring the amplification of small differential voltages produced by industrial sensors and transducers. As integrated circuits have grown smaller and more specialized, a single I.C. has been developed that will perform this very function--which requires three op-amps and several resistors. This device is the instrumentation amplifier. The schematic symbol and connection diagram of a common instrumentation amplifier--the AD620--are shown in Figure 17.1. Notice the similarity between the AD620 and the LM741 op-amp. Both have (+) and (-) power supply inputs. As with the op-amp, this will allow the output of the AD620 to be either (+) or (-). Likewise, both chips have (+) and (-) signal inputs. Do you remember how we set the gain of an op-amp circuit? With external resistors. Similarly, we will connect a resistor between pins 1 and 8 of the AD620 to set its differential voltage gain.

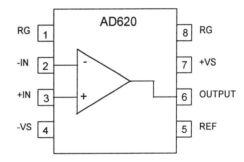

Figure 17.1: Top view and connection diagram of the AD620 instrumentation amplifier.

SAFETY NOTE:
Like the op-amp, the instrumentation amp is essentially a low-power device and can represent a safety hazard if not properly wired. It is literally possible to "blow the top off" of an instrumentation amplifier. This can happen when the power supply is improperly connected to the $+V_S$ and $-V_S$ pins. Have your instructor review this important aspect of amplifier wiring before you proceed. As always, be sure to wear safety glasses while working in the lab, remove jewelry from fingers and wrists, and have your instructor check your work before you energize your circuit and each time you make changes thereafter.

OBJECT:
Upon successful completion of this experiment and all reading assignments, the student should be able to:
- construct an operational instrumentation amplifier using the AD620
- predict the output of a given instrumentation amplifier circuit
- generate calibration data for a pressure transducer

REFERENCES:
www.analogdevices.com
www.omega.com

MATERIALS:
- 1 - instrumentation amplifier, AD620, or its equivalent
- 1 - solderless breadboard
- miscellaneous lead wires and connectors
- 5 - 10 Kohms, 1/4 watt resistors
- 1 - 5 Kohms, ten-turn potentiometer
- 1 - 470 Ohms, 1/4 watt resistor

EQUIPMENT:
- 1 - digital multimeter
- 1 - dual-outlet D.C. power supply
- 1 - pressure transducer, rated 0 – 500 psig, designed for hydraulic circuits
- 1 - hydraulic hand pump or source of adjustable hydraulic pressure with pressure gauge

PART 1: The Instrumentation Amplifier and the Wheatstone Bridge

Background:
The signal produced by a transducer is typically in the millivolt or milliamp range. This signal is commonly produced by a sensor installed as part of a wheatstone bridge. A simple wheatstone bridge circuit made from resistors is shown in Figure 17.2. The potentiometer represents the sensor or transducer whose electrical properties will change as a result of some physical stimulus. A thermistor is an example of a device whose resistance changes with temperature. A photoconductive cell changes resistance with change in light intensity. The change in the resistance of these devices will result in a change in the differential output voltage from the bridge (shown being measured by the voltmeter). However, this voltage change is so small that it must first be amplified before it can be displayed in a meaningful form. Additionally, small voltages are also susceptible to corruption by electromagnetic interference from nearby power lines or high-current carrying inductive devices.

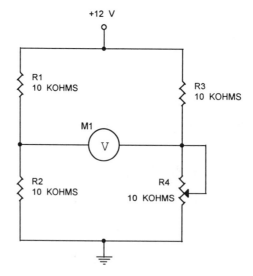

Figure 17.2: Wheatstone bridge circuit.

As noted in the introduction, the differential voltage gain of the AD620 is determined by the size of the resistor connected between pins 1 and 8 as shown in Figure 17.3. The AD620 will amplify the "difference" in the voltage applied to pins 3 and 2. If the voltage applied to the (+) input is 6.00 volts and the voltage applied to the (-) input is 5.98 volts, the amplifier will amplify the difference of +0.02 volt or 20 millivolts. Recall the general formula for the voltage gain, A_V, of an amplifier as introduced in Experiment 16. It is

$$A_V = V_{OUT}/V_{IN} \qquad \textbf{16.2}$$

Here, V_{IN} represents the differential input voltage. Assuming our amplifier gain is 100 and applying Formula 16.2, the output voltage would be found as follows

$$V_{OUT} = A_V(V_{IN})$$

$$V_{OUT} = 100(20 \text{ millivolts})$$

$$= 2000 \text{ millivolts}$$

The experimental voltage gain for this circuit may be calculated as follows:

$$\text{CIRCUIT VOLTAGE GAIN} = V_{OUTPUT}/(V_{+IN} - V_{-IN}) \qquad \textbf{17.1}$$

The theoretical voltage gain for an AD620 may be calculated using the following formula:

$$\text{AD620 GAIN} = 1 + (49.4 \text{ k}/R_G) \qquad \textbf{17.2}$$

Procedure:
1.1 Obtain all materials and equipment required to complete this experiment.
1.2 Use your digital multimeter (DMM) to measure the resistance of RG and record in Table 17.1.

Table 17.1: Nominal and measured resistor values for gain resistor, RG.

NOMINAL RESISTANCE	MEASURED RESISTANCE
470 OHMS	

1.3 Adjust your D.C. power supply so that each outlet produces 12 volts. Turn off your power supply.
1.4 Construct the wheatstone bridge shown in Figure 17.3. Do not connect it the AD620. Apply power to the bridge.
1.5 Measure the voltage V_{AB}. Adjust the potentiometer so that the voltage is as close to 0.0 volt as possible. Turn off the power supply.
1.6 Finish wiring the circuit shown in Figure 17.3. Have your instructor check your work. Verify the accuracy of your work by turning on your power supply and quickly and briefly touching the top of the amplifier to be sure that it does not get hot. If the amplifier is getting hot, turn off the power supply immediately and re-check your work. If the op-amp is still cool, proceed to the next step.
1.7 For each of the differential voltages shown in Table 17.2, measure the corresponding output voltage from pin 6 to ground. Record your results in Table 17.2.

Table 17.2: Experimental Data for Figure 17.3.

Differential Voltage	Output Voltage, volts	Voltage Gain, A_v
90 mV		
80 mV		
70 mV		
60 mV		
50 mV		
40 mV		
30 mV		
20 mV		
10 mV		
-10 mV		
-20 mV		
-30 mV		
-40 mV		
-50 mV		
-60 mV		
-70 mV		
-80 mV		
-90 mV		

1.8 For each set of data in Table 17.2, calculate and record the experimental voltage gain using Formula 17.1.

Calculations:

1.9 Use the measured value for R_G to calculate the theoretical voltage gain for the AD620. Record in the space provided below.

Calculations:

$A_v =$ _____

1.10 How does the theoretical value for the gain compare to the experimental voltage gains? If there is a significant difference, have your instructor check your work.

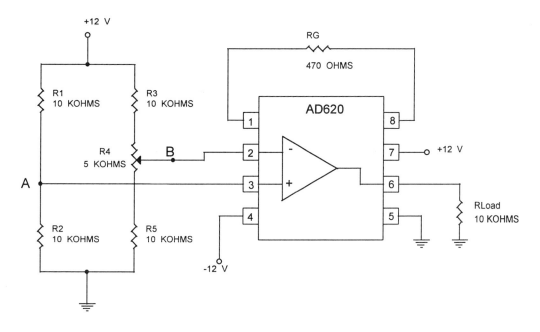

Figure 17.3: AD620 instrumentation amplifier being used to amplify the differential voltage from a wheatstone bridge.

PART 2: The Instrumentation Amplifier and the Pressure Transducer

Background:

An example of an industrial sensor that produces a millivolt signal from a wheatstone bridge circuit is the pressure transducer. A pressure transducer is used to convert hydraulic or pneumatic pressure into a corresponding millivolt (or milliamp) signal. This small signal will be amplified, then applied to a computer or some type of display for signal processing. The wheatstone bridge inside a pressure transducer is constructed from a strain gage or gages and precision resistors. The strain gage is a transducer that changes resistance when a tensile or compressive force is applied to it. In this part of the experiment you will obtain calibration data for a pressure transducer. Like the circuit shown in Figure 17.2 the pressure transducer must have an external voltage applied to power the internal wheatstone bridge. And like the circuit shown in Figure 17.3, a small differential output voltage (+ SIGNAL and − SIGNAL) will be amplified and connected to some type of numeric display. A pictorial of this is shown in Figure 17.4.

The process of transducer calibration involves applying a known physical input and measuring the corresponding electrical output. As a simple example, consider how you might calibrate an electronic bathroom scale. You would apply known, exact weights in incremental values (such as 50, 100, 150, 200 pounds, etc.) and check the digital readout for accuracy. If the digital reading did not match the applied weight, you would then adjust the electronic circuit that performed the conversion from weight to electrical signal.

To calibrate the pressure transducer, you must have a hydraulic hand pump or some other source of variable hydraulic pressure. Some mechanical labs may have a "dead-weight tester" used to calibrate pressure gauges. You will apply increasing levels of oil pressure to your transducer and measure the corresponding millivolt output signal. Once you reach a maximum pressure, you will then incrementally reduce the pressure, repeating your voltage measurements. This is done to check for any *non-linearity* or *hysteresis* in the response of the transducer.

INSTRUCTOR NOTE: You may choose to do the next activity as a classroom demonstration.

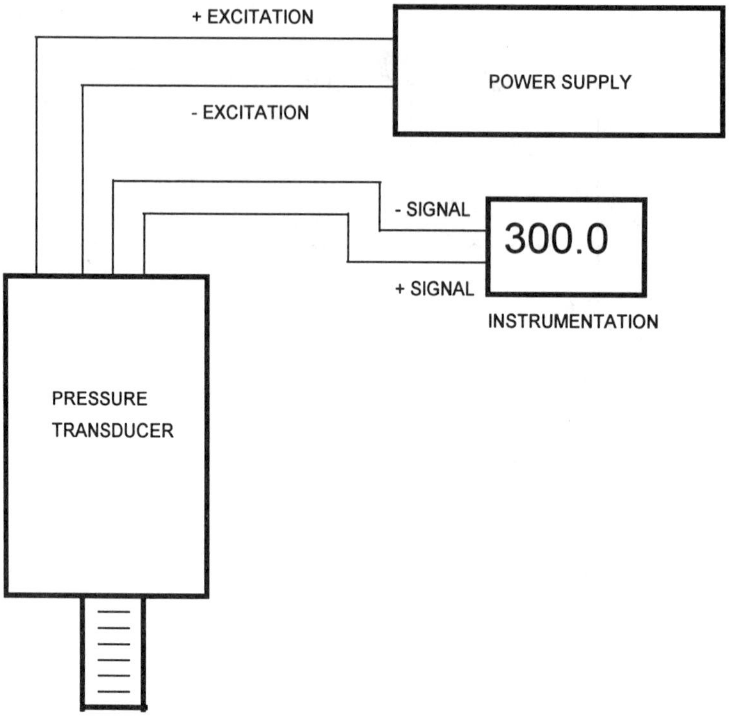

Figure 17.4: Pressure transducer with power supply and output signal.

Procedure:

    2.1 Attach your pressure transducer to a hydraulic hand pump. Do <u>not</u> use a motor driven pump. You will not be able to precisely control its output pressure. Use a precision hydraulic gauge to measure the pressure.

    2.2 Connect a power supply to the transducer to provide the desired bridge excitation as shown in Figure 17.4. Have your instructor help you identify the input pins.

    2.3 Connect a digital millivoltmeter to the output pins. Make sure that no hydraulic pressure is applied at this time.

    2.4 Apply power to the transducer's wheatstone bridge. Check for any *D.C. offset*. The voltmeter should read close to 0.0 volt with no pressure applied. If there is a D.C. offset, ask your instructor what to do about it. The D.C. offset can be removed by *bridge compensation*.

    2.5 Gradually increase the pressure to 50 psig. Record the corresponding voltmeter reading in Table 17.3. Increase the pressure in 50 psig increments. Record the corresponding voltages. Do not exceed the transducer's maximum pressure rating. Once you have reached 500 psig (or the maximum pressure, whichever is less), slowly reduce the pressure in 50 psig increments and record the corresponding voltages. Turn off the power supply.

    2.6 Use your data to plot a graph of output voltage as a function of applied pressure. Attach your graph to your report. Your graph should be nearly a straight line. Is it?

Table 17.3: Calibration Data for the Pressure Transducer.

Applied Pressure, psig	Ouput Voltage, mV
0	
50	
100	
150	
200	
250	
300	
350	
400	
450	
500	
450	
400	
350	
300	
250	
200	
150	
100	
50	
0	

PART 3: Practice Problems

3.1 If your replaced R_G in Figure 17.3 with a 1 Kohms resistor, what would be the theoretical voltage gain of the AD620?

3.2 What value for R_G would you have to select to produce a voltage gain of 200 from the AD620?

3.3 If the voltage applied to the (+) input of an AD620 is 5.00 volts and the voltage applied to the (-) input is 5.06 and the gain is 50, what is the output voltage relative to ground?

3.4 If the voltage applied to the (+) input of an AD620 is 4.03 volts and the voltage applied to the (-) input is 4.00 and the output voltage relative to ground is +6.00 volts, what is the circuit voltage gain?

PART 4: Circuit Design

4.1 Connect the (+) and (-) outputs of your pressure transducer to the corresponding (+) and (-) inputs of the instrumentation amplifier as shown in Figure 17.5. Select a value for R_G that will produce an output voltage from the AD620 in the range of +4 to +6 volts at a pressure of 300 psig. The output of the AD620 is to be connected to the op-amp comparator circuit shown in Figure 17.6. Select values for R1, R2, and R_B that will cause the solenoid of a hydraulic valve to turn on and off at a pressure specified by your instructor. Show all of your calculations in the space provided below.

INSTRUCTOR NOTE: If your school has hydraulic trainers, you may want to have the students control a hydraulic cylinder with the solenoid-operated control valve. Have the students connect the pressure transducer in an appropriate location in the hydraulic circuit.

Calculations:

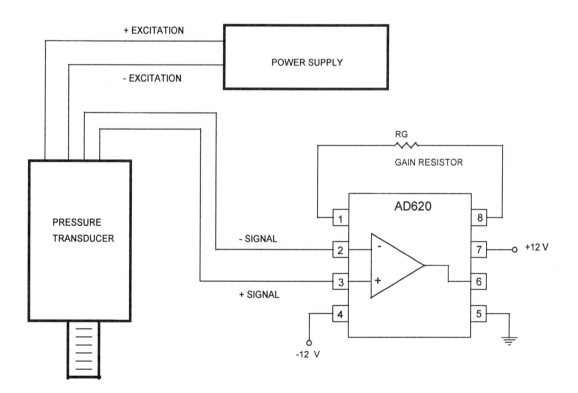

Figure 17.5: Pressure transducer with instrumentation amplifier.

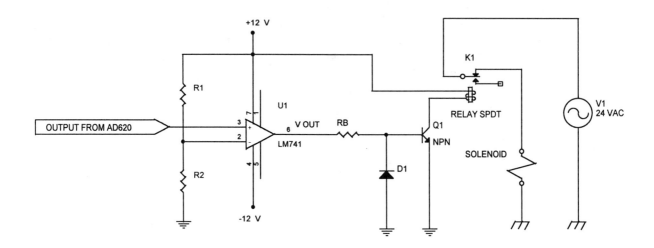

Figure 17.6: Op-amp comparator circuit. The output of the op-amp will switch from a negative output voltage to a positive output voltage when the pressure applied to a pressure transducer exceeds a certain set point.

PART 5: Instrumentation Amplifier Formulas

CIRCUIT VOLTAGE GAIN = $V_{OUTPUT}/(V_{+IN} - V_{-IN})$ **17.1**

AD620 GAIN = $1 + (49.4 \text{ k}/R_G)$ **17.2**

Op-Amp Summers: 18

INTRODUCTION:
In many instrumentation and control circuits, it is necessary to electronically monitor input signals from several transducers or other electronic devices with the sum of the inputs determining what action or result will take place--a relay being energized, a solenoid-operated valve being energized, the movement of the servo motor in a strip-chart recorder, etc. In addition to the many functions that it performed in Experiment 16, the operational amplifier (op-amp) is also an integral part of the electronic summing operation just described. The op-amp summer is essentially an electronic adder--summing the voltages applied to it and producing the sum or result at its output. A solid grasp of the operation of the op-amp summer is also fundamental to the understanding of the operation of the digital-to-analog converter (DAC). You will have the opportunity to study the DAC in detail in Experiment 30. In this experiment you will construct a basic op-amp summer circuit and a special version of the op-amp summer--the binary-weighted summer. The latter circuit can be used as a very basic DAC circuit.

SAFETY NOTE:
While the op-amp is essentially a low-power device, it can still represent a safety hazard if not properly wired. It is literally possible to "blow the top off" of an op-amp. This is not recommended! This frequently happens when the power supply is improperly connected to the op-amp. Have your instructor review this important aspect of op-amp wiring before you proceed. As always, be sure to wear safety glasses while working in the lab, remove jewelry from fingers and wrists, and have your instructor check your work before you energize your circuit and each time you make changes thereafter.

OBJECT:
Upon successful completion of this experiment and all reading assignments, the student should be able to:
- construct an operational op-amp summer circuit
- mathematically predict the output of a given op-amp summer circuit

MATERIALS:
- 4 - SPST switches or one dip switch pack
- 1 - LM741 op-amp or its equivalent
- miscellaneous resistors and/or potentiometers
- miscellaneous lead wires and connectors
- 5 - 1 Kohms, 1/4 watt resistors
- 1 - 10 Kohms, ten-turn trim pot
- 1 - solderless breadboard

EQUIPMENT:
- 1 - digital multimeter
- 1 - dual-outlet D.C. power supply

PART 1: The Op-Amp Summer
Background:
Before constructing your first op-amp summer circuit, let's review the operation of the inverting op-amp. Figure 18.1 shows the basic configuration of the inverting op-amp that you constructed in Experiment 16.

Recall that the relationship between the output voltage, V_{OUT}, and the input signal, V_S, for this circuit can be expressed as follows:

$$V_{OUT} = V_S(-R_F/R_1) \qquad 18.1$$

Figure 18.1: Op-amp configured as an inverting amplifier.

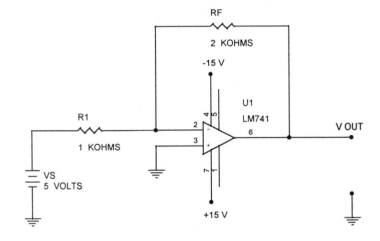

Now that you have some experience with op-amp circuits, in particular with the inverting op-amp, we can proceed to examine in detail how this mathematical relationship was developed. As discussed in Experiment 16, the input impedance, Z_{IN}, between the inverting and non-inverting terminals of the op-amp is very large. With the addition of negative feedback in the inverting configuration, Z_{IN} becomes even larger—on the order of several megohms. As a result, the current flow through the op-amp from the inverting input to the non-inverting input is zero. Therefore, the differential voltage (voltage drop) from the inverting input relative to the non-inverting input is also zero. Because the non-inverting input is connected to ground as shown in Figure 18.1, the net result is that the potential at the inverting input of the op-amp is 0.0 volt, or what is referred to as *virtual ground*. The resulting or equivalent circuit just described is shown in Figure 18.2. Just in case you are scratching your head at this point wondering just what in the world virtual ground is, let's review my earlier comments. I pointed out that there is no current flow from the inverting input to the non-inverting input. Now you might be thinking that this is equivalent to an open. Well, that is just not the case. In the circuit shown in Figure 18.1, there is a physical connection from the inverting input through the op-amp to the non-inverting input and in turn to ground. It just so happens that this connection has a very high resistance (impedance), but it is not an open! If it were truly an open, there would be a large voltage drop from the inverting input to the non-inverting input. In this case, however, we have a very large resistance through which no current can flow. According to Ohm's Law, zero amperes through a resistor of any size results in a voltage drop of zero volts. Because there is no voltage drop from the inverting input to the non-inverting input and since the non-inverting input is at ground potential, then the inverting input is in effect, or virtually, at ground potential. Hence the term virtual ground. The inverting input is not at true ground because no current can flow from the inverting input to the non-inverting input.

Now let's consider the ramifications of what I just discussed. Referring again to the circuit shown in Figure 18.2, we see that the node that was connected to the inverting input of the op-amp in Figure 18.1 has been replaced by a connection to ground representing virtual ground. Because this point is at virtual ground, the voltage drop from this point to true ground is 0.0 volt as shown in Figure 18.2. This results in a differential voltage across resistor R1 of 5.0 volts. As shown in Figure 18.3, this causes a current of 5 mA to flow through R1. Applying Kirchhoff's Current Law for the node established at the inverting input of the op-amp and keeping in mind that no current can flow through the op-amp, we have the resulting current distribution shown in Figure 18.4. Because the input current of 5 mA that flows through R1 must also flow through R_F, a voltage drop across R_F results with the polarity as shown in Figure 18.5. Next, applying Kirchhoff's Voltage Law (KVL) for a closed loop as shown in Figure 18.5, we see that the output voltage, V_{OUT}, relative to ground must be -10 volts. Take a moment to mentally review KVL, and convince yourself that the relationship of the voltages shown in Figure 18.5 is indeed correct. If you apply Formula 18.1 to the circuit shown in Figure 18.1, you will discover that indeed the output voltage is -10 volts.

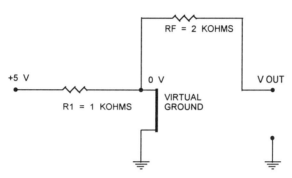

Figure 18.2: Equivalent circuit for the inverting op-amp circuit shown in Figure 18.1. The op-amp has been replaced by the virtual ground connection from the inverting to the non-inverting inputs of the op-amp.

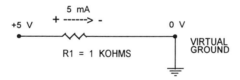

Figure 18.3: Equivalent circuit for the input stage of the circuit shown in Figure 18.1. The potential difference of +5 volts causes 5 mA of current (conventional) to flow from the +5 volts supply toward virtual ground.

All right, by now you might be saying, "What does all this have to do with an op-amp summer?" Well, a typical op-amp summer circuit is shown in Figure 18.6. Do you note any similarities between it and the circuit shown in Figure 18.1? Instead of just one input resistor, R1, there are now four input resistors, R1, R2, R3, and R4. Switches have been added so we can selectively control which of the inputs will "see" a potential of 1 volt. In real-world applications, each input will often see a different potential applied to it depending upon what type of process is being monitored and its current operating state. Let's now consider what will happen when just one of the switches is closed. Figure 18.7 shows the equivalent circuit that would result if switch S1 in Figure 18.6 was closed. Note that our analysis of the summer circuit can proceed just like that for the inverting amplifier. The 1.0 volt drop across R1 will cause 1 mA of current to flow through R_F. In turn, this causes a 1.0 volt drop across R_F. V_{OUT} then becomes -1.0 volt. Now let's consider what would happen if more than one switch is closed. Figure 18.8 shows the equivalent circuit that would result if switches S1 and S2 in Figure 18.6 were closed. As in Figure 18.7, 1 mA of current will flow through each resistor due to the 1.0 volt potential across each. Applying Kirchhoff's Current Law at the node joining R1, R2, and R_F, the resulting current through R_F must be the sum of the currents through R1 and R2 or 2 mA. This current causes 2.0 volts to be dropped across R_F. The resulting output voltage is then -2.0 volts. What do you think the output voltage would be if three or four switches were closed?

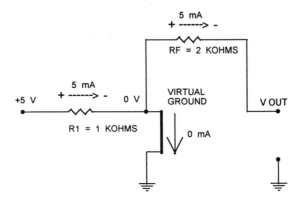

Figure 18.4: Equivalent circuit for Figure 18.1 showing current flow through R1 and R_F. Because of the very large input impedance of the op-amp, we can assume the current flow from the inverting input of the op-amp to the non-inverting input of the op-amp to be 0 mA.

Figure 18.5: Equivalent circuit of the output stage of the circuit shown in Figure 18.1.

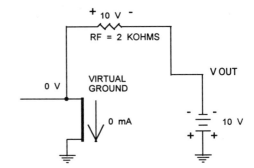

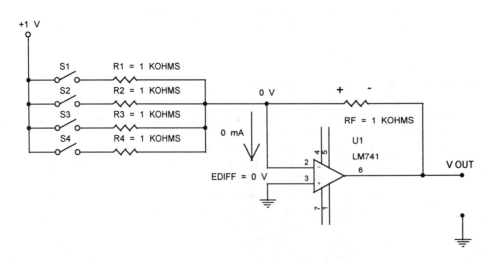

Figure 18.6: Op-amp summer circuit.

246

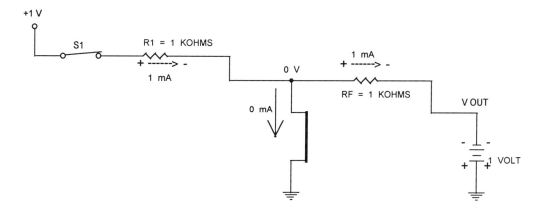

Figure 18.7: Equivalent circuit for Figure 18.6 with one switch closed.

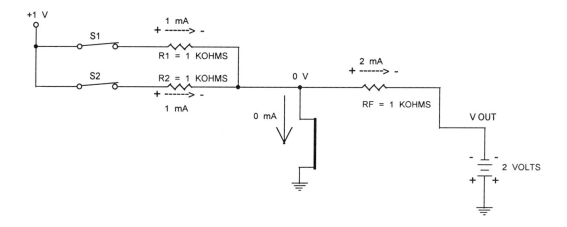

Figure 18.8: Equivalent circuit for Figure 18.6 with two switches closed.

Procedure:

1.1 Obtain the parts necessary to construct the circuit shown in Figure 18.9. In order to obtain accurate results, try to obtain resistors that are as close to 1 Kohms as possible. Or, obtain resistors that are as nearly equal to each other as possible. A final option would be to substitute 2 Kohms trim-pots adjusted to 1 Kohms.

1.2 Once you have obtained all of the necessary parts, you may construct the circuit shown in Figure 18.9.

1.3 For each of the switch settings shown in Table 18.1, measure and record the corresponding output voltage, V_{OUT}.

1.4 Turn off the power supply.

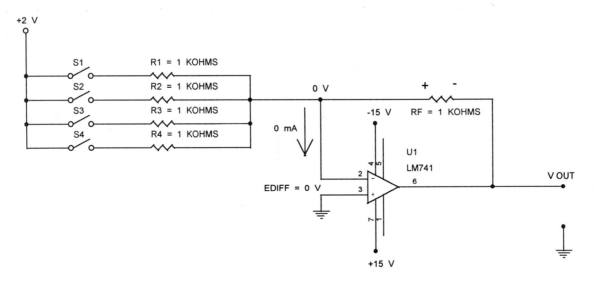

Figure 18.9: Op-amp summer circuit for Part 1 of the experiment.

1.5 Using the rated resistance values shown in Figure 18.9, calculate and record the theoretical output voltage for the op-amp summer circuit for each of the switch settings shown in Table 18.2. Show your work in the space provided below.

Calculations:

1.6 How do your measured values from Table 18.1 compare to the theoretical values shown in Table 18.2? If there are significant differences, check your work.

Table 18.1: Measured output voltages for different input combinations for the circuit shown in Figure 18.9.

SWITCH SETTINGS	OUTPUT VOLTAGE, VOUT, VOLTS
S1 CLOSED, ALL OTHERS OPEN	
S2 CLOSED, ALL OTHERS OPEN	
S3 CLOSED, ALL OTHERS OPEN	
S4 CLOSED, ALL OTHERS OPEN	
S1 AND S2 CLOSED, S3 AND S4 OPEN	
S1, S2, AND S3 CLOSED, S4 OPEN	
ALL SWITCHES CLOSED	

Table 18.2: Theoretical output voltages for different input combinations for the circuit shown in Figure 18.9.

SWITCH SETTINGS	OUTPUT VOLTAGE, VOUT, VOLTS
S1 CLOSED, ALL OTHERS OPEN	
S2 CLOSED, ALL OTHERS OPEN	
S3 CLOSED, ALL OTHERS OPEN	
S4 CLOSED, ALL OTHERS OPEN	
S1 AND S2 CLOSED, S3 AND S4 OPEN	
S1, S2, AND S3 CLOSED, S4 OPEN	
ALL SWITCHES CLOSED	

PART 2: The Binary-Weighted Summer
Background:
As was mentioned in the introduction to this experiment, a special version of the op-amp summer is the binary-weighted summer. The binary-weighted summer circuit that you will construct is shown in Figure 18.10. Rather than using resistors of equal value, as was the case for the circuit shown in Figure 18.9, five different resistor values are used. Note in particular the relationship of the values of the input resistors. Each succeeding resistor is twice the value of the preceding one. As a result, the current that will flow through R2 when S2 is closed will be twice the value of the current that will flow through R1 when S1 is closed. In turn, the voltage drop across R_F caused by closing S2 will be twice the voltage drop across R_F caused by closing S1. The "weight" or contribution of each input is then twice the preceding one. Because each input is related to the others by a factor of two, this configuration has been referred to as the binary-weighted summer. If we equate a switch closure to a binary 1 and an open switch to a binary 0, we have created a simple four-bit digital-to-analog converter. The input at switch S1 represents

the least significant bit (LSB) and the input at switch S4 represents the most significant bit (MSB). All switches open corresponds to a binary 0000, and all switches closed represents a binary 1111. Recall that $0000_2 = 0_{10}$ and that $1111_2 = 15_{10}$.

Procedure:

2.1 Obtain the parts necessary to construct the circuit shown in Figure 18.10. As in Part 1 of this experiment, it will be necessary to obtain resistance values as close as possible to those shown on the schematic diagram. Also, since 5 Kohms, 20 Kohms, and 40 Kohms are not standard resistance values, you may wish to use ten-turn trim pots to obtain these values.

2.2 Once you have obtained all of the necessary parts, you may construct the circuit shown in Figure 18.10. Note that the input voltage is now -5 volts and that the op-amp supply voltages have been changed! Adjust the ten-turn trim pot so that the value of R_F is 8 Kohms.

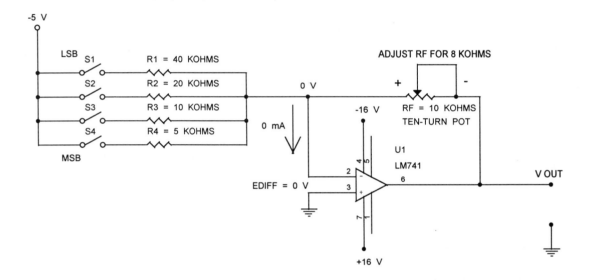

Figure 18.10: Binary-weighted summer circuit for Part 2 of the experiment.

2.3 For each of the switch settings shown in Table 18.3, measure and record the corresponding output voltage, V_{OUT}.

2.4 Turn off the power supply.

Table 18.3: Measured output voltages of different input combinations for the circuit shown in Figure 18.10.

SWITCH SETTINGS	OUTPUT VOLTAGE, VOUT, VOLTS
ALL SWITCHES OPEN	
S1 CLOSED, ALL OTHERS OPEN	
S2 CLOSED, ALL OTHERS OPEN	
S1 AND S2 CLOSED, S3 AND S4 OPEN	
S3 CLOSED, ALL OTHERS OPEN	
S1 AND S3 CLOSED, S2 AND S4 OPEN	
S2 AND S3 CLOSED, S1 AND S4 OPEN	
S1, S2, AND S3 CLOSED, S4 OPEN	
S4 CLOSED, ALL OTHERS OPEN	
S4 AND S1 CLOSED, S2 AND S3 OPEN	
S2 AN S4 CLOSED, S1 AND S3 OPEN	
S1, S2, AND S4 CLOSED, S3 OPEN	
S3 AND S4 CLOSED, S1 AND S2 OPEN	
S1, S3, AND S4 CLOSED, S2 OPEN	
S2, S3, AND S4 CLOSED, S1 OPEN	
ALL SWITCHES CLOSED	

2.5 Using the <u>rated</u> resistance values shown in Figure 18.10, calculate and record the theoretical output voltage for the binary-weighted summer for each of the switch settings shown in Table 18.4. Show your work in the space provided below.

Calculations:

Table 18.4: Theoretical output voltages of different input combinations for the circuit shown in Figure 18.10.

SWITCH SETTINGS	OUTPUT VOLTAGE, VOUT, VOLTS
ALL SWITCHES OPEN	
S1 CLOSED, ALL OTHERS OPEN	
S2 CLOSED, ALL OTHERS OPEN	
S1 AND S2 CLOSED, S3 AND S4 OPEN	
S3 CLOSED, ALL OTHERS OPEN	
S1 AND S3 CLOSED, S2 AND S4 OPEN	
S2 AND S3 CLOSED, S1 AND S4 OPEN	
S1, S2, AND S3 CLOSED, S4 OPEN	
S4 CLOSED, ALL OTHERS OPEN	
S4 AND S1 CLOSED, S2 AND S3 OPEN	
S2 AN S4 CLOSED, S1 AND S3 OPEN	
S1, S2, AND S4 CLOSED, S3 OPEN	
S3 AND S4 CLOSED, S1 AND S2 OPEN	
S1, S3, AND S4 CLOSED, S2 OPEN	
S2, S3, AND S4 CLOSED, S1 OPEN	
ALL SWITCHES CLOSED	

2.6 How do your measured values from Table 18.3 compare to the theoretical values shown in Table 18.4? If there are significant differences, check your work.

PART 3: Questions

3.1 In Figure 18.9, the theoretical value of the output voltage that would result if only one switch were closed is -2.0 volts. Describe three different ways that the circuit shown in Figure 18.9 could be changed to increase the value of V_{OUT} corresponding to a single switch closure. Record your responses in the space provided below.

3.2 Why were the values of the op-amp's (+) and (-) supply voltages increased to +16 and -16 volts, respectively, as shown in Figure 18.10?

3.3 How would you modify the circuit shown in Figure 18.10 so that the output voltage would be +1.5 volts with all four switches closed, i.e., binary 1111?

PART 4: Op-Amp Summer Formulas

$V_{OUT} = V_S(-R_F/R_1)$ **18.1**

SCRs: D.C. Characteristics: 19

INTRODUCTION:

Experiment 1 introduced you to types of mechanical switches. In Experiment 2, you worked with transistors configured to operate as switches. In Experiment 3, you investigated the operation of an electromagnetic switch--the control relay. Each of the types of switches studied in the first three experiments offers the designer of control circuitry its own unique set of advantages as well as limitations. Relays, for example, are subject to contact pitting due to the arcing that occurs when a relay turns off an inductive load. Transistors are limited to low- or medium-power applications. In recent years, the SCR has gained wide popularity in high-power control applications. The SCR is a member of the thyristor family. It is essentially a four-layer, three-terminal device as represented by the simplified model shown in Figure 19.1 a). The three terminals are identified as the anode, cathode, and gate. The schematic symbol for the SCR and its three terminals are shown in Figure 19.1 b). While the SCR is a three-terminal device like the transistor, the SCR does not operate in an active region like the transistor. Rather, the SCR acts like a switch, either on (closed) or off (open). While the control relay requires only a small current applied to the coil to close its contacts, which in turn control a much larger current, an SCR can likewise control a large amount of current flow from anode to cathode by the application of a relatively small current to its gate. In this experiment, you will construct an SCR circuit that will take a low-power (low-current) input and control a higher-power load.

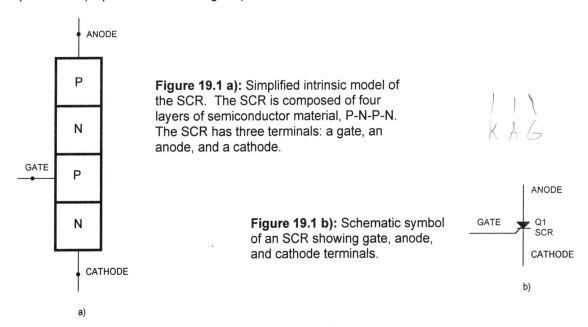

Figure 19.1 a): Simplified intrinsic model of the SCR. The SCR is composed of four layers of semiconductor material, P-N-P-N. The SCR has three terminals: a gate, an anode, and a cathode.

Figure 19.1 b): Schematic symbol of an SCR showing gate, anode, and cathode terminals.

As noted, an SCR is designed to act as a switch--either on or off. Figure 19.2 a) shows an SCR circuit with an open gate. If the gate of an SCR is open or reverse-biased, the SCR will be off, acting as an open from anode to cathode. As a result, the load (lamp in this case) will be off. This is shown in Figure 19.2 b). If the gate of an SCR is forward-biased, as shown in Figure 19.2 c), then the SCR will be on. Under these conditions, the SCR acts like a closed switch as shown in Figure 19.2 d). As a result, the load will be energized. Not only must the gate of the SCR be forward-biased to turn it on, there also must be a minimum amount of gate current for the given load voltage, V_{AA}. This gate current that is required to turn on an SCR is referred to as I_{GT}.

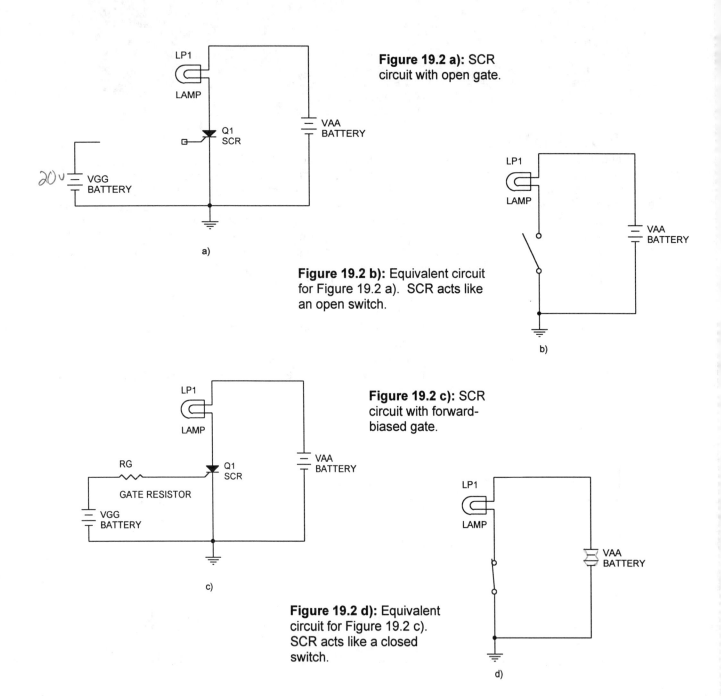

Figure 19.2 a): SCR circuit with open gate.

Figure 19.2 b): Equivalent circuit for Figure 19.2 a). SCR acts like an open switch.

Figure 19.2 c): SCR circuit with forward-biased gate.

Figure 19.2 d): Equivalent circuit for Figure 19.2 c). SCR acts like a closed switch.

SAFETY NOTE:
Read and familiarize yourself with the manufacturer's ratings for the SCR before wiring it in your circuit. As always, wear safety glasses, and remove all jewelry from your hands and fingers before the start of the experiment.

OBJECT:
Upon successful completion of this experiment and all reading assignments, the student should be able to:
- identify the anode, cathode, and gate of an SCR
- construct an operational SCR circuit that will control a D.C. powered load

REFERENCES:
Chapter 4 of Maloney's Modern Industrial Electronics
www.motorola.com

MATERIALS:
- 1 - SCR, MCR218-6FP or the equivalent
- 2 - normally-open push button switches
- 1 - 560 Ohms resistor, 1 watt
- 1 - solderless breadboard
- 1 - incandescent lamp, 24 to 28 volts rating
- 3 - 1.2 Kohms resistors, 1/2 watt
- 1 - 5 Kohms, ten-turn potentiometer
- miscellaneous lead wires and connectors

EQUIPMENT:
- 1 - adjustable D.C. power supply, 0 – 24 VDC (or 0 – 20 VDC if 24 VDC is not available)
- 2 - multimeters
- 1 - 24 VDC muffin-style fan

PART 1: The SCR D.C. Switch

Background:
While the SCR is similar to the control relay in that it is used as a switch, there is one very significant difference. Once the SCR is turned on, it stays on even after the removal of the gate current. Thus, the SCR is said to act like a latch. This is similar to a functional characteristic of flip-flops. Once set to a certain state a flip-flop will stay, or latch, in that state (output is high or low) until forced to change to the opposite state. Well then, you might ask, "How do you turn the SCR off?" There are two basic ways. One is to reverse bias the anode-cathode junction; this will be demonstrated in Experiment 20. The other way is to reduce the anode-cathode current, I_A, to a value below what is referred to as the *holding current*, I_H. This can be accomplished several different ways. A large switch could be placed in series with the SCR. Opening it would reduce I_A to zero. Another way would be to turn off the power supply. While either method is certainly possible, neither is practical from a controls standpoint. We would like to be able to electronically turn off the SCR. In Figure 19.3, momentarily pressing the normally-open push button switch S1 will divert the supply current from the SCR, causing it to turn off. It is not practical to turn off a high-power SCR this way; however, consider the circuit shown in Figure 19.4. In this circuit, the push button switch has been replaced by a transistor switch. If a momentary, active-high pulse from a digital control circuit is applied to the base of the transistor, it will momentarily go in to saturation, diverting the current away from the SCR just long enough to turn it off. While this will require a rather large transistor if turning off a high-power SCR, the transistor can have a lower current and power rating than the SCR because it will be on only momentarily, whereas the SCR must be sized to operate continuously under high-power conditions.

INSTRUCTOR'S NOTE: If the D.C. power supply in your lab cannot supply 24 volts D.C. as shown in Figure 19.3, the circuit will still work if you use a power supply rated for 20 volts. If using a 24 volt supply, warn the students not to touch the 1.2 Kohms resistors shown in Figure 19.5 since they will be operating near their maximum wattage rating when the SCR is fully turned on.

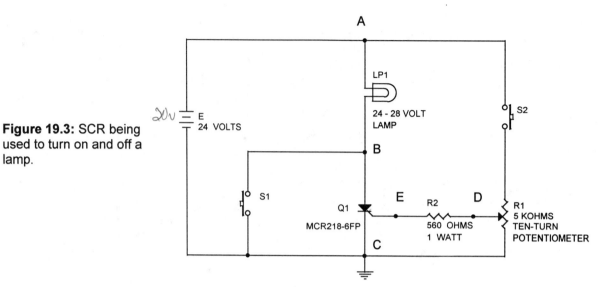

Figure 19.3: SCR being used to turn on and off a lamp.

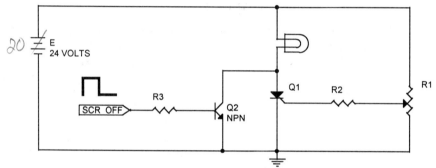

Figure 19.4: SCR circuit using a transistor to turn off the SCR.

Procedure:

1.1 Obtain all materials and equipment required to complete this experiment.
1.2 Adjust the output of your power supply for 24 volts. Turn off the power supply.
1.3 Make sure that you have correctly identified the terminals of the SCR. Once you have done so, you may construct the circuit shown in Figure 19.3.
1.4 Turn on the power supply.
1.5 While pressing and holding switch S2, measure the voltage from point D to ground, V_{DC}. Adjust the 5 Kohms pot until V_{DC} is 20 volts. Release switch S2.
1.6 Turn off the power supply.
1.7 Turn the power supply back on. Is the lamp on? It should not be. If it is on, immediately turn off the power supply and check your work.
1.8 Press and hold switch S2. Describe what happens in the space provided below.

The light turns on

1.9 Release switch S2. Describe what happens in the space provided below.

The light stays on

1.10 Press and hold switch S1. Describe what happens in the space provided below.

The light stays on

1.11 Release switch S1. Describe what happens.

Light turns off

1.12 Repeat steps 1.8 through 1.11 a number of times to see if you get consistent results.
1.13 With the light turned on, measure the voltages V_{AB} and V_{BC}. Record in Table 19.1.
1.14 Press S1 to turn the light off. With the light turned off, measure the voltages V_{AB} and V_{BC}. Record in Table 19.1.
1.15 Turn off the power supply.

Table 19.1: Lamp and SCR voltages for Figure 19.3.

CIRCUIT OPERATIONAL STATUS	LAMP VOLTAGE, V_{AB}, VOLTS	SCR VOLTAGE, V_{BC}, VOLTS
LAMP ON	19.24 v	.77 v
LAMP OFF	0 v	20.1 v

PART 2: D.C. Electrical Characteristics of the SCR
Background:
As pointed out in the introduction of this experiment, the SCR operates much like a switch, either on or off. If the gate of the SCR is at ground potential or the gate-cathode P-N junction is reverse-biased, the SCR will act as an open from anode to cathode. If the gate-cathode P-N junction is sufficiently forward-biased, the SCR will turn on and act like a closed switch with a small forward voltage drop from anode to cathode (since there is no such thing as a perfect switch). Well, just what does "sufficiently forward-biased" mean anyway? As you will see in Experiment 20, when controlling an A.C. source with an SCR, a certain combination of both gate current and anode-to-cathode voltage in combination are required to turn on an SCR. The larger the voltage applied to the anode, the smaller the gate current required to turn on an SCR. The smaller the anode voltage, the larger the gate current required to turn on an SCR. The exact amount of anode voltage and gate current required to turn on an SCR depends on the power rating of the SCR. Just as different transistors have different I-V curves that describe their operation, so do SCRs. As you might suspect, higher-power SCRs require larger anode voltages and gate currents to turn

them on, whereas lower-power SCRs can be turned on with smaller anode voltages and smaller gate currents. Manufacturers usually specify a minimum value for the gate current that will turn the SCR on at a very low anode voltage. This parameter is abbreviated as I_{GT}. Sometimes manufacturers specify the voltage drop from gate to cathode that will ensure SCR turn on. This parameter is abbreviated as V_{GT}. As you discovered in Part 1 of this experiment, once the SCR turns on, it latches on and stays on even after the removal of the gate current. To turn off the SCR, you must turn off the D.C. supply, reverse bias the anode-cathode junction (as will be seen in Experiment 20), or reduce the current through the SCR below the holding current, I_H. This can be accomplished by either diverting it away from the SCR or reducing the D.C. supply to near 0 volts. In this part of the experiment, you will have the opportunity to experimentally determine some of the SCR's electrical parameters.

Procedure:

2.1 Measure and record the resistance of each of the resistors shown in Table 19.2.

Table 19.2: Nominal and measured resistance values.

NOMINAL RESISTANCE	MEASURED RESISTANCE
R2 = 560 OHMS	560 Ω
R3 = 1.2 KOHMS	1.197 KΩ
R4 = 1.2 KOHMS	1.189 KΩ
R5 = 1.2 KOHMS	1.187 KΩ

2.2 Construct the circuit shown in Figure 19.5.

2.3 With the power supply set for 24 volts, adjust the potentiometer until V_{DC} (the voltage from point D to point C) is approximately 0.0 volt. This is necessary to ensure that the SCR will initially be in the off state. Turn off the power supply.

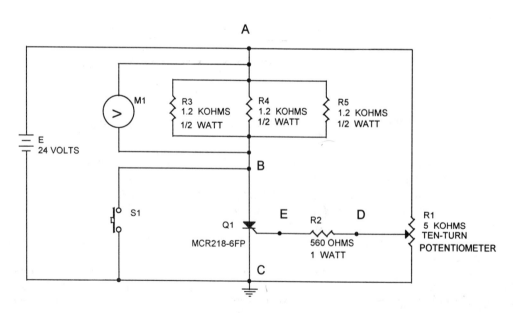

Figure 19.5: Circuit used to experimentally determine the D.C. electrical characteristics of an SCR.

2.4 Turn the power supply back on.
2.5 For each of the values for V_{DC} shown in Table 19.3, measure and record V_{AB}, V_{BC}, V_{EC}, and V_{DE}. Leave one of your voltmeters connected across the load as you adjust the potentiometer so that you can tell when the SCR fires--turns on. Slowly increase the voltage V_{DC} so that you can determine when the SCR just turns on. In the last row in Table 19.3 record the value for V_{DC} that caused the SCR to just turn on. Record the corresponding circuit voltages in Table 19.3. Once you complete Table 19.3, you may wish to repeat your measurements to ensure the accuracy of your results.
2.6 Use the measured values for R3, R4, R5, and the values for V_{AB} to calculate the corresponding SCR current, I_A, for each of the entries in Table 19.3. Record your calculated results in Table 19.3.

Calculations:

$$I = E/R$$

$$I = V_{AB} / \left[\left(R3^{-1} + R4^{-1} + R5^{-1} \right)^{-1} \right]$$

2.7 Use the measured values for R2 and V_{DE} to calculate the corresponding gate current, I_G, for each of the entries in Table 19.3. Record your calculated results in Table 19.3.

Calculations:

$$I = E/R$$

$$I = V_{DE} / 560$$

Table 19.3: SCR voltage and current data for Figure 19.5.

VDC, VOLTS	VAB, VOLTS	VBC, VOLTS	VEC, VOLTS	VDE, VOLTS	IA, mA	IG, mA
0	0 v	20.1 v	0 v	0 v	0 mA	0 mA
1	0 v	20.1 v	.3 v	.7 v	0 mA	1.25 mA
3	19.31 v	.75 v	.73 v	2.26 v	48.64 mA	4.04 mA
5	19.25 v	.75 v	.75 v	4.22 v	48.49 mA	7.54 mA
7	19.29 v	.75 v	.76 v	6.22 v	48.59 mA	11.11 mA
10	19.29 v	.75 v	.78 v	9.31 v	48.59 mA	16.63 mA
15	19.28 v	.75 v	.80 v	14.14 v	48.56 mA	25.25 mA
20	19.26 v	.75 v	.82 v	19.13 v	48.51 mA	34.16 mA
2.12 v*	19.07 v	.75 v	.72 v	1.41 v	48.04 mA	2.52 mA

* **VOLTAGE V_{DC} THAT CAUSES THE SCR TO JUST TURN ON.**

2.8 With the SCR turned on, open the gate-control circuit as shown in Figure 19.6. Do **not** turn off the power supply. The SCR must still be on to complete the next phase of the experiment.

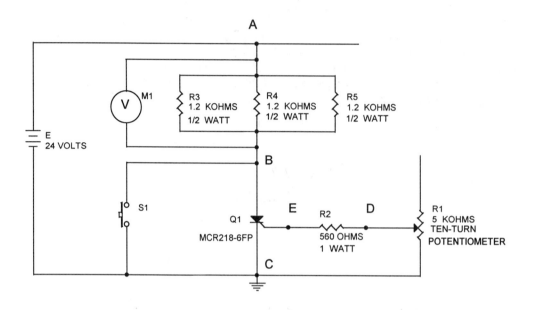

Figure 19.6: Circuit used to experimentally determine the D.C. electrical characteristics of an SCR with an open gate circuit.

2.9 With the SCR on, measure the voltage across the load, V_{AB}. Slowly reduce the supply voltage E until the SCR turns off (voltage across the load is 0.0 volt). Record in Table 19.4 the voltage V_{AB} that you observed just before the SCR shut off.

2.10 Use the measured values for R3, R4, R5, and the value for V_{AB} in Table 19.4 to calculate the current through the SCR just before shut off. This represents the holding current I_H. Record this value in Table 19.4.

Calculations:

$I = E/R$

$I = 2.1v / 560\Omega$

2.11 Turn off the power supply.

Table 19.4: Holding current data for the circuit shown in Figure 19.5.

CIRCUIT STATUS	V_{AB}, VOLTS	I_H, mA
SCR JUST ABOUT TO TURN OFF	2.1v	5.29 mA

PART 3: Application

Background:
While the next activity is a low-power circuit, it is intended to give you an idea of how an SCR might be used in a higher-power industrial control application.

Procedure:
3.1 Replace the resistors in the circuit shown in Figure 19.5 with a 24 VDC muffin fan or equivalent size 24 VDC motor.

3.2 Adjust potentiometer R1 for maximum resistance--SCR off. Turn on the power supply.

3.3 Slowly decrease, then increase the resistance of R1 over its entire range. Note the response (speed) of the fan as the resistance of the pot is decreased then increased.

PART 4: Manufacturer's Ratings
4.1 Referring to the web site www.motorola.com , or an electronic data base, or a set of data manuals, complete Table 19.5 for the MCR218-6FP SCR.

Table 19.5: Manufacturer's specifications for the MCR218-6FP SCR.

Part Number	V_{DRM}, VOLTS	$I_{T(RMS)}$, Amps	I_{GT}, mA	V_{GT}, VOLTS	I_H, mA
MCR218-6FP	400	10 A	15 mA	1.5 v	20 mA

PART 5: Questions
5.1 Based on your results to Part 1 (particularly steps 1.8 through 1.11) of this experiment, compare the operation of your SCR to a control relay. Discuss both similarities and differences.

5.2 Based on your results to Part 1 and Part 2 of this experiment, compare the operation of your SCR to a transistor. Discuss both similarities and differences.

5.3 Based on the results shown in Table 19.1, what general statement can you make about the voltage <u>across an SCR</u> when it is turned on?

There is very little voltage across the SCR

5.4 Based on the results shown in Table 19.1, what general statement can you make about the voltage <u>across the load</u> in an SCR-controlled circuit when the SCR is turned on?

Most of the voltage drops across the load.

5.5 Based on the results shown in Table 19.1, what general statement can you make about the voltage <u>across an SCR</u> when it is turned off?

All the voltage drops across the SCR

5.6 Based on the results shown in Table 19.1, what general statement can you make about the voltage <u>across the load</u> in an SCR-controlled circuit when the SCR is turned off?

There is no voltage across the load.

5.7 Referring to the results in Table 19.3, what effect does increasing gate current have on the operation of the SCR once it is turned on?

It has no effect. Voltage across the SCR stays the same. A very low voltage.

5.8 How does the experimental data in Table 19.3 and 19.4 compare to the manufacturer's data for V_{GT}, I_{GT}, and I_H?

They were not very close in values but can be reasonable.

5.9 What happened to the speed of the fan in Part 3 of this experiment as the resistance of the potentiometer was decreased? Were you able to vary the speed of the fan?

Didn't do

5.10 In Part 3 of this experiment, once the fan turned on, what did you have to do to turn it off?

Didn't do

SCRs: A.C. Characteristics: 20

INTRODUCTION:
Experiment 19 provided you with an opportunity to observe the operation of an SCR when powered by a D.C. source and to determine some of its electrical characteristics. Now you will have the opportunity to observe the operational characteristics of an SCR when powered by an A.C. supply. SCRs are widely used in single-phase and three-phase circuits to control (vary) the magnitude of the voltage applied to a load. For example, in applications requiring the speed control of large D.C. motors, steel processing plants used (some still do) very high horsepower, constant-speed synchronous motors to drive large D.C. generators that would in turn serve as the D.C. supply to variable-speed D.C. motors. The speed of the D.C. motors would be varied by changing the applied voltage from the D.C. generator. The output of the D.C. generator would be varied by varying its field excitation. Doesn't this seem rather inefficient to have three large pieces of rotating machinery to perform one task? Well, in some respects it is. However, this arrangement (referred to as a motor-generator set) was very reliable, being capable of operating under extremely heavy loads for hundreds or even thousands of hours without a major failure. In recent years, the motor-generator set has been replaced by banks of computer-controlled SCRs that are powered by the same three-phase source that used to energize the synchronous motor. By controlling the timing of the firing of the SCRs, the magnitude of the D.C. voltage applied to the D.C. drive motor could be varied, in turn varying the motor's speed. This is just one example of the application of SCRs in an industrial setting. By no means, however, are SCRs limited to industrial control circuits. SCRs have also been integrated into medical instrumentation, laboratory test equipment, and numerous applications requiring an adjustable D.C. source or over-voltage protection.

SAFETY NOTE:
Before starting this experiment, have your instructor review with you the proper use of the oscilloscope to display the secondary voltages of a center-tapped transformer. It is important that you know how to use your oscilloscope to display the true voltage waveforms that appear across the secondary terminals of a center-tapped transformer. Incorrect use of the oscilloscope will result in either an incorrect waveform such as displaying a floating voltage or worse, causing a short circuit to ground through your oscilloscope. Read and familiarize yourself with the manufacturer's ratings for the SCR before wiring it in your circuit. As always, wear safety glasses, and remove all jewelry from your hands and fingers before the start of the experiment.

OBJECT:
Upon successful completion of this experiment and all reading assignments, the student should be able to:
- construct an A.C. powered SCR circuit
- use an oscilloscope to display the load and SCR voltage waveforms in an A.C. powered SCR circuit
- determine the delay angle and conduction angle for an A.C. powered SCR circuit

REFERENCES:
Chapter 4 of Maloney's Modern Industrial Electronics
www.motorola.com

MATERIALS:
- 1 - 120:12.6 V$_{RMS}$ center-tapped transformer
- 2 - 1N4003 diodes or the equivalent
- 1 - 1 µF capacitor
- 1 - 10 µF capacitor
- 1 - 1 Kohms resistor, 1/2 watt
- miscellaneous lead wires and connectors
- 1 - 1/8 amp slow-blow fuse
- 1 - in-line fuse holder
- 1 - solderless breadboard
- 1 - SCR, MCR218-6FP or the equivalent
- 1 - 5 Kohms, ten-turn potentiometer
- 1 - 220 Ohms resistor, 1 watt

EQUIPMENT:
- 1 - handheld multimeter
- 1 - dual-trace oscilloscope
- 2 - oscilloscope probes
- 1 - 24 VDC muffin-style fan

PART 1: The SCR in an A.C. Powered Circuit
Background:
When operated in an A.C. circuit, the SCR will act much like a diode in that it will conduct (be turned on) when forward-biased (if sufficient gate current is applied) and act like an open (be turned off) when reverse-biased. Because of the SCR's ability to rectify an A.C. waveform, it can be used to power loads requiring a D.C. voltage. However, it is much more flexible than a simple rectifier diode. As was pointed out in Experiment 19, an SCR will fire, turn on, when a certain combination of anode voltage and gate current are applied. While it is possible to turn an SCR on without the application of gate current, this is not practical because the voltage required to turn on an SCR under these conditions is very large. Recall from Experiment 19 that the greater the gate current, the smaller the anode voltage required to turn on an SCR. The advantage of this property of the SCR is the circuit designer's ability to vary the point along the A.C. supply voltage at which the SCR will turn on. Figure 20.1 shows the voltage waveform across an SCR when being powered by a single-phase A.C. source. This waveform shows that the SCR is fully turned on. Note that this waveform looks very much like a half-wave rectified waveform. The span of time that this waveform is flat--SCR conducting--is from approximately 0° to 180°. The span of time that the SCR is on is referred to as the *conduction angle*. By lowering the gate current to our SCR, we can delay the firing--turn on--of the SCR. For example, in Figure 20.2 the firing of the SCR has been delayed until the A.C. supply voltage reaches its peak. As a result, less voltage is available to the load. In effect, by varying the gate current to an A.C. powered SCR, we can vary the D.C. voltage applied to a load. In Figure 20.2, the SCR does not fire until approximately 90° after the sinusoidal supply voltage crosses zero and goes positive. The span of time that the SCR is off during the positive alternation of the supply voltage is referred to as the *delay angle*. In this example, the conduction angle is approximately 90°.

By now you might be thinking, "Gee, I can do the same thing with a potentiometer without a lot of complex control circuitry. Why should I bother with an SCR?" Well, one reason is efficiency. Consider the circuit shown in Figure 20.3. A potentiometer will dissipate a lot of power to control the voltage across the series load resistor, whereas an SCR is either fully on or fully off regardless of the point along the A.C. supply voltage at which it is fired or triggered into conduction. When fully on, the voltage drop across the SCR is very small, and hence the power dissipation (power loss) is very small. When the SCR is off, it acts like an open. Of course, no current flows through an open. No current means no power dissipation. So, the SCR is much more efficient. Also, we can design a digital circuit that can automatically control the firing of the SCR. A potentiometer must be manually adjusted. Figure 20.4 shows the circuit you will construct. The transformer is used to step the 120 volts RMS single-phase source down to 12.6 volts RMS. The diodes, potentiometer, fixed resistor, and capacitor make up the gate control circuitry. Varying the resistance of the potentiometer will change the conduction angle of the SCR.

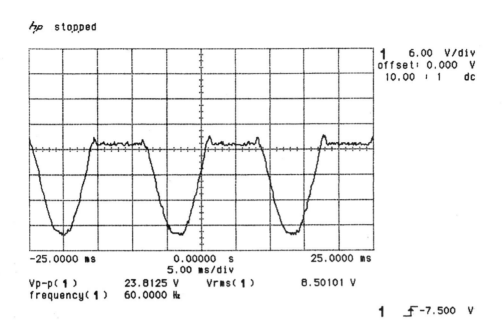

Figure 20.1: Voltage waveform across an SCR (from anode to cathode) when powered by a single-phase sinusoidal source. The SCR is fully turned on.

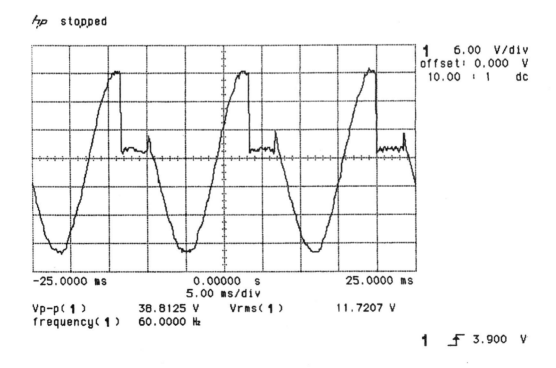

Figure 20.2: SCR voltage waveform with a 90° delay angle.

Figure 20.3: Potentiometer used as a lamp dimmer.

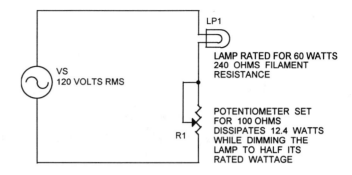

Procedure:

1.1 Obtain all materials and equipment required to complete this experiment.
1.2 Make sure that you have correctly identified the terminals of the SCR. Once you have done so, you may construct the circuit shown in Figure 20.4.

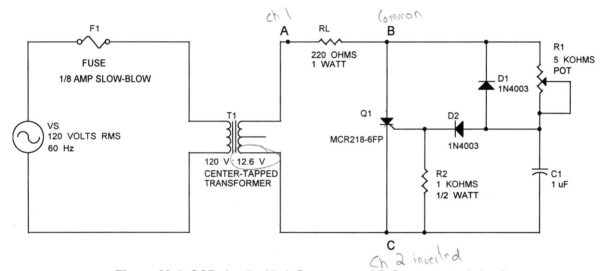

Figure 20.4: SCR circuit with A.C. source and R-C gate control circuit.

1.3 Adjust the potentiometer for maximum resistance.
1.4 Plug in the transformer.
1.5 Use the oscilloscope to display the transformer's secondary voltage, V_{AC}. Record this waveform in the space provided on Graph 20.1. Label both the vertical and horizontal axes with voltage and time base information.
1.6 With the potentiometer adjusted for maximum resistance, use the oscilloscope to display the voltage across the SCR, V_{BC}. On Graph 20.1, record this waveform in proper time phase with the voltage waveform V_{AC}. Label both the vertical and horizontal axes with voltage and time base information. You may wish to select a different volts/div setting for this waveform to distinguish it from V_{AC}.

Graph 20.1: V_{AC} and V_{BC} voltage waveforms for the circuit shown in Figure 20.4 with the potentiometer adjusted for maximum resistance.

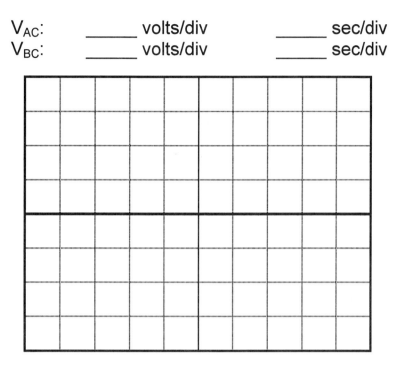

V_{AC}: _____ volts/div _____ sec/div
V_{BC}: _____ volts/div _____ sec/div

1.7 Based on the voltage waveforms drawn on Graph 20.1, what can you conclude about the present status (operating condition) of the SCR? Explain how you arrived at your answer.

With pot at max resistance it doesn't allow some of the voltage across the load.

1.8 Decrease the resistance of the potentiometer until the SCR just fires. Do this slowly!
1.9 Use the oscilloscope to display the voltage across the SCR, V_{BC}. Record this waveform on Graph 20.2. Label both the vertical and horizontal axes with voltage and time base information.
1.10 With your instructor's assistance, display the voltage waveform across the load resistor, V_{AB}. Record this waveform on Graph 20.3. Label both the vertical and horizontal axes with voltage and time base information.
1.11 Decrease the resistance of the potentiometer until the SCR is fully on.
1.12 Use the oscilloscope to display the voltage across the SCR, V_{BC}. Record this waveform on Graph 20.4. Label both the vertical and horizontal axes with voltage and time base information.

1.13 Display the voltage waveform across the load resistor, V_{AB}. Record this waveform on Graph 20.5. Label both the vertical and horizontal axes with voltage and time base information.
1.14 For each of the SCR operating conditions described in Table 20.1, use your handheld voltmeter to measure the D.C. voltages specified. Make sure that your voltmeter is set to measure D.C. voltage, not A.C. voltage.
1.15 Unplug the transformer.
1.16 Replace the 1 µF capacitor with a 10 µF capacitor.
1.17 Plug in the transformer.
1.18 Adjust the resistance of the potentiometer until the SCR just turns on.
1.19 Use the oscilloscope to display the voltage across the SCR, V_{BC}. Record this waveform on Graph 20.6. Label both the vertical and horizontal axes with voltage and time base information.
1.20 Unplug the transformer.

Graph 20.2: V_{BC} voltage waveform for the circuit shown in Figure 20.4 with the SCR just turned on.

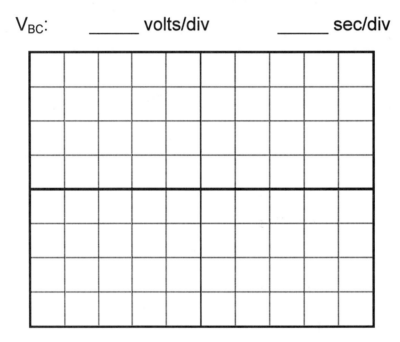

V_{BC}: _____ volts/div   _____ sec/div

Table 20.1: D.C. voltages for three different SCR operating conditions for the circuit shown in Figure 20.4.

SCR OPERATIONAL STATUS	LOAD VOLTAGE, VAB, VOLTS D.C.	SCR VOLTAGE, VBC, VOLTS D.C.
SCR FULLY OFF	2.63v	2.63v
SCR JUST TURNED ON	5.65v	5.70v
SCR FULLY ON	5.68v	5.74v

Graph 20.3: V_{AB} voltage waveform for the circuit shown in Figure 20.4 with the SCR just turned on.

V_{AB}: _____ volts/div _____ sec/div

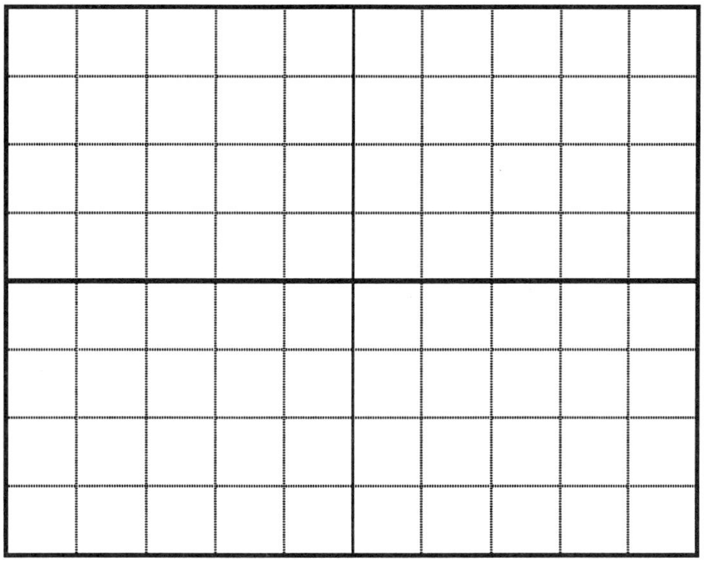

Graph 20.4: V_{BC} voltage waveform for the circuit shown in Figure 20.4 with the SCR fully turned on.

V_{BC}: _____ volts/div _____ sec/div

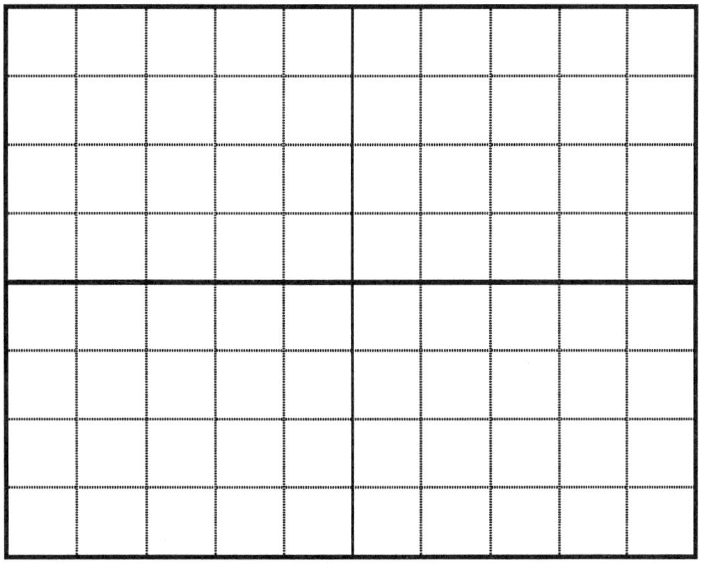

Graph 20.5: V_{AB} voltage waveform for the circuit shown in Figure 20.4 with the SCR fully turned on.

V_{AB}: _____ volts/div   _____ sec/div

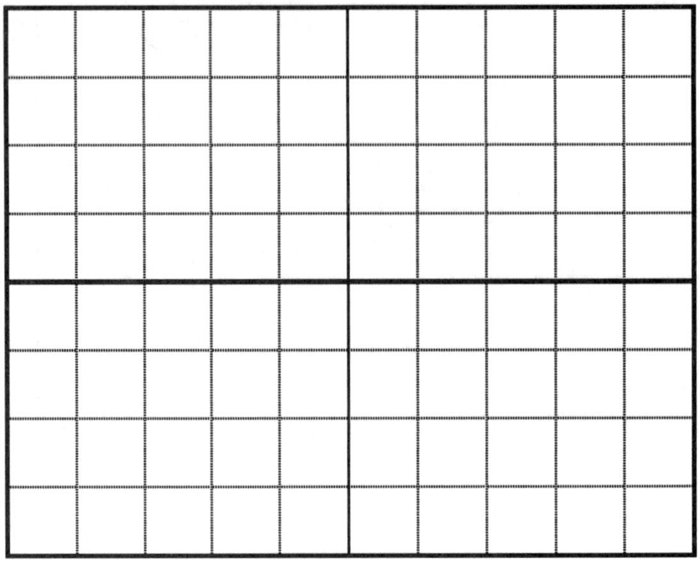

Graph 20.6: V_{BC} voltage waveform for the circuit shown in Figure 20.4 with the SCR just turned on with a 10 µF capacitor installed in the gate circuit.

V_{BC}: _____ volts/div   _____ sec/div

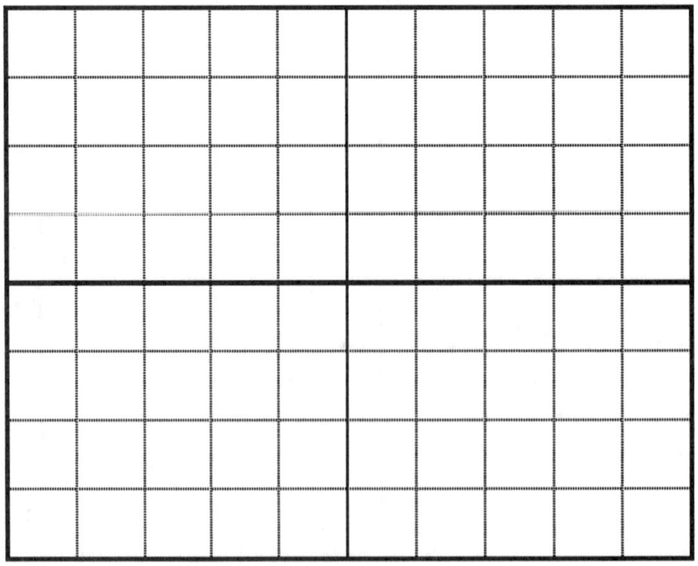

PART 2: Application
Background:
This activity will allow you to compare the response of a load in an A.C. powered circuit to its response in a D.C. powered circuit. In Experiment 19 you observed the operation of a small fan in a D.C. powered circuit. Now you will have the opportunity to see how the speed of a D.C. motor can be varied using an SCR in an A.C. powered circuit by installing the same fan in your SCR circuit constructed in this lab.

Procedure:
2.1 Replace R_L, the 220 Ohms load resistor, in the circuit shown in Figure 20.4 with a 24 VDC muffin fan or equivalent size 24 VDC motor.
2.2 Adjust potentiometer R1 for maximum resistance--SCR off. Turn on the power supply.
2.3 Use a D.C. voltmeter to measure V_{AK} (V_{BC}).
2.4 <u>Slowly</u> decrease, then increase the resistance of R1 over its entire range. Note the response (speed) of the fan as the resistance of the pot is decreased then increased. Also note the corresponding change in V_{AK}.

PART 3: Questions
3.1 Referring to the web site www.motorola.com, an electronic data base, or a set of data manuals, what is V_{RRM} for the MCR218-6FP SCR?

3.2 Based on your answer to Question 2.1, what is the largest A.C. voltage that can be applied to the MCR218-6FP SCR? Express your answer in <u>RMS</u> units.

3.3 Based on your experimental results from Part 1 of this experiment, compare the <u>operation</u> of your SCR to a rectifier diode. Discuss both similarities and differences.

3.4 Based on your experimental results, what effect did decreasing the resistance of the potentiometer in the gate control circuit have on the D.C. voltage across the load?

It allowed more voltage through the load.

3.5 Referring to your data in Table 20.1, explain how you can have a negative value for V_{BC}?

Reverse the leads of the meter.

3.6 In Graph 20.2, what is the conduction angle?

90° – 180°

3.7 In Graph 20.4, what is the conduction angle?

0° – 180°

3.8 In Graph 20.6, what is the conduction angle?

110° – 125°

3.9 In Graph 20.4 what is the delay angle?

180° – 0°

3.10 In Graph 20.6, what is the delay angle?

125° – 110°

3.11 Changing capacitor C1 in the circuit shown in Figure 20.4 changed the circuits *range of control*. Range of control is defined as the range of adjustment over which the area under the SCR's voltage curve can be adjusted (changed) by the operator. The operator is not able to control the area under the curve from the point where the SCR is fully off to the point where the SCR suddenly turns on (snaps on). What affect did the size of the capacitor have on the range of control of the SCR's conduction angle?

3.12 Figure 20.5 shows experimental results from an SCR circuit like that shown in Figure 20.4. Which component's voltage is being displayed? What conclusions could an electronics technician draw about the operation of this circuit based upon the <u>shape</u> of the waveform?

Figure 20.5: Experimental SCR data.

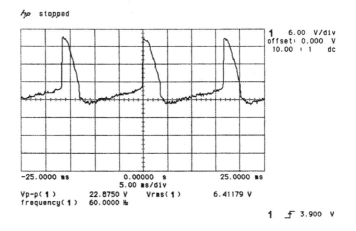

3.13 Figure 20.6 shows the voltage waveform <u>across the SCR</u> from a circuit like that shown in Figure 20.4. What conclusions could an electronics technician draw about the operation of this circuit based upon the <u>shape</u> of this waveform?

Figure 20.6: Experimental SCR data.

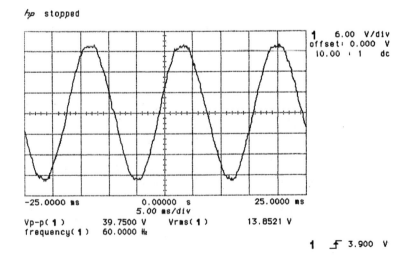

3.14 What happened to the speed of the fan in Part 2 of this experiment as the resistance of the potentiometer was decreased? Explain why this happened by referring to the resulting change in I_G and V_{AK}.

3.15 Compare the operation of the fan in this experiment's SCR circuit to that in Experiment 19.

PART 4: Electronics Workbench®

4.1 Use Electronics Workbench® to create the circuit shown in Figure 20.4. You do not have to include the fuse. Select a transformer turns ratio of 120:12.6. Select an SCR that has electrical properties as close as possible to the one used in your lab experiment. Use the oscilloscope function to display the voltage waveforms at points A and B with R1 set for maximum resistance, minimum resistance, and at a value that will cause the SCR to just turn on. Print a copy of your schematic diagram and oscilloscope waveforms. How do these waveforms compare to the graphs of your experimental data? List possible causes for any differences.

Triacs and Diacs: 21

INTRODUCTION:

In Experiment 20, you learned that the D.C. voltage to a load in an SCR-controlled circuit can be adjusted by varying the current to the gate of the SCR. For those applications requiring an adjustable A.C. voltage, another member of the thyristor family can be selected. That device is the triac. The SCR is a unidirectional device like a diode in that it allows current (conventional) to flow from the anode through the SCR to the cathode but acts like an open when reverse-biased. The triac is an example of a bi-directional device in that it will allow current to flow through it in either direction. Like the SCR, the triac acts like a switch and is either on or off. Also like the SCR, the triac has a gate terminal. Figure 21.1 shows the schematic symbol for the triac and identifies the gate terminal and the other two leads of the triac--MT2 and MT1.

Figure 21.1: Schematic symbol for a triac showing the gate, MT2, and MT1 terminals.

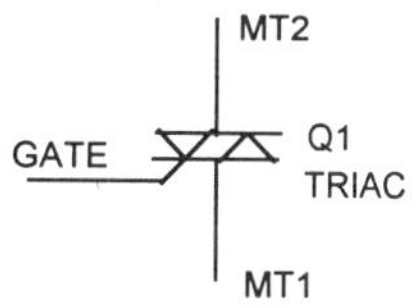

By varying the current to the gate of the triac, the voltage at which the triac breaks over (turns on) can be varied. Experiment 20 provided you with an opportunity to observe the operation of an SCR when powered by an A.C. source. In this experiment, you will have the opportunity to observe the operational characteristics of a triac when powered by an A.C. supply. Triacs are used in devices as common as the light dimmer found in most homes today and in applications such as variable-speed, electronically controlled drills. Triacs are also used in business office equipment such as photocopiers that require a heat source to *fuse* or melt print powder to the surface of the page that you are now reading. The triac is used in this equipment to turn on and off a high-wattage, glass-enclosed heating coil. The triac is also used in photocopiers to vary the intensity of the high-intensity lamp that *scans* the original document that is being copied. Triacs are primarily used in low- to medium-power applications. In applications requiring the control of high-voltage A.C. supplies such as electric heat treating ovens and other high- voltage A.C. loads, SCRs will be used instead. I know, you're thinking, "Didn't he say earlier that SCRs are used in circuits to control or provide varying D.C. voltages?" That's right, I did. But SCRs can be installed in what is known as the inverse-parallel configuration to control or adjust the magnitude of the voltage to an A.C. powered load. The inverse-parallel configuration is shown in Figure 21.2. SCR1 will allow current to flow during the positive alternation of the A.C. sinusoidal supply, and SCR2 will allow current to flow in the opposite direction through the load when the supply voltage becomes negative. Based on your work in Experiment 20, recall how the waveform across an SCR appears when it has a 90° delay angle. Such a waveform is shown in Figure 21.3. To understand the operation of SCR2, imagine how SCR1's waveform would look if it were inverted. Applying the principle of superposition and putting these two waveforms together, you have the waveform shown in Figure 21.4. This is exactly how a triac operates, just like two SCRs wired in inverse-parallel.

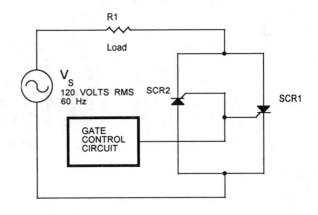

Figure 21.2: A.C. power control circuit using two SCRs connected in inverse-parallel. This configuration operates like a triac.

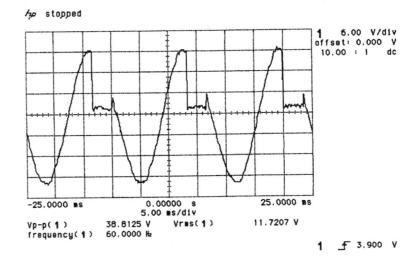

Figure 21.3: SCR voltage waveform with a 90° delay angle.

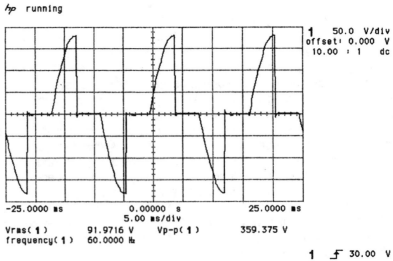

Figure 21.4: Triac voltage waveform with a 90° delay angle.

Another member of the thyristor family will be used in your triac circuit. That device is the *diac*. The diac is a low-power device (so be careful when installing it and making subsequent measurements) that will be used in the triac's gate control circuitry. You will have the opportunity to experimentally determine how the diac improves the performance of the triac circuit. Like the triac, the diac is an A.C. device. However, the diac is a two-terminal device like a diode, whereas the triac is a three-terminal device. The schematic symbols for the diac are shown in Figure 21.5. Diacs come in two basic configurations--symmetrical and nonsymmetrical. The diac will conduct or breakdown in both directions when powered by an A.C. source. The breakdown voltage of symmetrical diacs is the same in both the forward and reverse directions. Nonsymmetrical diacs have different forward and reverse breakdown voltages. The diac improves the firing of the triac as it will suddenly "snap on," allowing a sudden "rush" of electrons to flow into the gate of the triac. This results in a much more consistent--repeatable--point at which the triac turns on.

Figure 21.5: Schematic symbols for a diac.

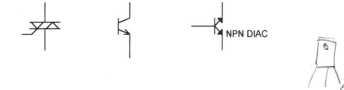

SAFETY NOTE:

In this experiment, you will be working with a high-voltage source, so be extremely cautious as you make measurements. Be careful not to touch any exposed circuit leads. Before starting this experiment, have your instructor review with you the proper use of the oscilloscope to display the A.C. line, load, and triac voltage waveforms. You must electrically isolate your circuit from your oscilloscope and any other grounded measuring instruments. Incorrect use of the oscilloscope will result in either an incorrect waveform, such as displaying a floating voltage, or worse, causing a short circuit to ground through your oscilloscope. Read and familiarize yourself with the manufacturer's rating for the triac before wiring it in your circuit. As always, wear safety glasses, and remove all jewelry from your hands and fingers before the start of the experiment. Once power is applied, do not touch the triac, it will be hot!

OBJECT:

Upon successful completion of this experiment and all reading assignments, the student should be able to:
- construct an operational triac-controlled circuit
- use an oscilloscope to display the load and triac voltage waveforms in an A.C. circuit
- determine the delay angle and conduction angle for a triac-controlled circuit
- experimentally determine if a diac is symmetrical or nonsymmetrical
- determine the range of control of the conduction angle of a triac circuit
- identify the gate, MT2, and MT1 leads of a triac
- observe and describe the phenomenon referred to as hysteresis
- observe the phase shifting of the voltage waveforms in an R-C gate-controlled triac circuit
- experimentally determine the effect that each component has on the operation of a triac circuit
- observe the phenomenon of radio-frequency interference (RFI) induced by the firing of a triac

REFERENCES:

Read Chapter 6 of Maloney's Modern Industrial Electronics

MATERIALS:

- 1 - triac, MAC210A8 or the equivalent
- 1 - 60 watt incandescent light bulb
- 1 - 1 Kohms resistor, 1/2 watt
- 1 - SPST switch, rated for 120 VAC or higher
- 2 - 0.1 µF capacitors, 400 WVDC
  (any capacitor in the range of 0.1 to 0.5 µF will work satisfactorily)
- 1 - diac, ST2, or SK3523, or the equivalent
- 1 - lamp socket
- 1 - 100 Kohms potentiometer, 2 watts
- 1 - solderless breadboard
- miscellaneous lead wires and connectors

EQUIPMENT:
- 1 - handheld multimeter
- 1 - dual-trace oscilloscope
- 2 - oscilloscope probes
- 1 - inexpensive A.M. radio
- 1 - hand-held electric drill rated for 120 VAC with universal motor

PART 1: The Triac Circuit
Background:
The circuit you are to construct is shown in Figure 21.7. The method used to control the triggering (or firing) of the triac in this circuit is an example of what is referred to as phase-shift control. Take a moment to mentally review the relationship between the voltage and current in a capacitive circuit. Remember ELI the ICE man? The current leads the voltage in a capacitive circuit. Notice the R-C network set up by the potentiometer and C1. By varying the resistance of the potentiometer, the voltage waveform at C1 (effectively the voltage applied to the gate of the triac) can be varied in phase relative to the voltage applied to the MT2 terminal of the triac. The resulting waveforms are shown in Figure 21.6. If the voltage waveform across C1 is 90° out of phase with the voltage waveform present across the triac (from point B to point C), the triac will be off--acting like an open. This will occur when the potentiometer is adjusted for maximum resistance. Reducing the resistance of the potentiometer will cause the voltage waveform that is applied to the triac's gate to start to come into phase with the voltage at MT2 of the triac. This will cause the triac to turn on. When the waveform applied to the gate is in phase with the waveform at MT2, the triac will be fully on. By varying the resistance of the potentiometer, we can vary the delay and conduction angles of the triac and load waveforms much like that in Experiment 20 with the A.C. powered SCR. By varying the delay angle, we can vary the area under the load waveform. This has the effect of varying the RMS value of the voltage across the load.

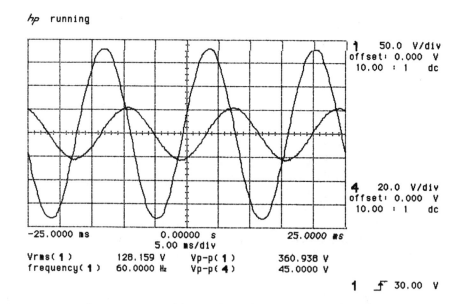

Figure 21.6: Voltage waveforms from the R-C network in the gate control portion of the triac circuit shown in Figure 21.7. The larger waveform represents the voltage across the open triac, and the smaller waveform is the voltage across C1. Note that the two waveforms are 90° out of phase.

Procedure:
1.1 Obtain all materials and equipment required to complete this experiment.
1.2 Make sure that you have correctly identified the terminals of the triac. You do not have to worry about identifying the terminals of the diac. You may install it in either direction as you

will be reversing its orientation later in the experiment.

1.3 The A.C. power supply should be off. Wire the circuit shown in Figure 21.7. Include a fast-blow fuse in your circuit between the power supply and lamp rated for 1 amp.
1.4 Switch S1 should be open at this time.
1.5 Adjust the potentiometer for maximum resistance.
1.6 Turn on the A.C. power supply.
1.7 Close switch S1. If you notice (smell) any components overheating, open S1 immediately.
1.8 Use the oscilloscope to display the voltage waveform V_{AC}. Record this waveform in the space provided on Graph 21.1. Label both the vertical and horizontal axes with voltage and time base information.
1.9 With the potentiometer adjusted for maximum resistance, use the oscilloscope to display the voltage across the triac, V_{BC}. On Graph 21.1 record this waveform in proper time phase with the voltage waveform V_{AC}. Label both the vertical and horizontal axes with voltage and time base information. You may wish to select a different volts/div setting for this waveform to distinguish it from V_{AC}.

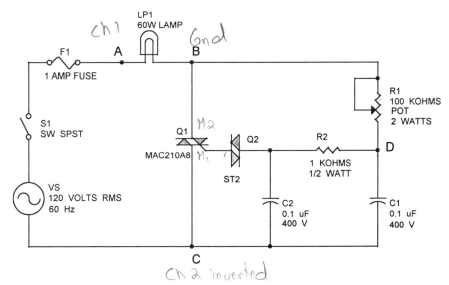

Figure 21.7: Triac circuit with single-phase A.C. source and R-C gate control circuit.

1.10 Based on the voltage waveforms drawn on Graph 21.1, what can you conclude about the present status (operating condition) of the triac? Explain how you arrived at your answer.

It's off, all voltage through the pot.

Graph 21.1: V_{AC} and V_{BC} voltage waveforms for the circuit shown in Figure 21.7 with the potentiometer adjusted for maximum resistance.

V_{AC}: _____ volts/div _____ sec/div
V_{BC}: _____ volts/div _____ sec/div

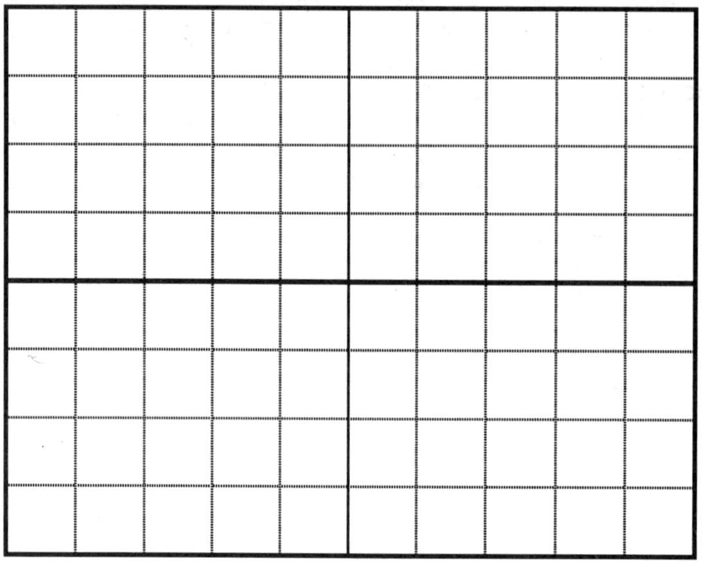

1.11 Decrease the resistance of the potentiometer until the triac just fires. Do this slowly! When the triac just turns on, the light bulb should be glowing dimly.

1.12 Use the oscilloscope to display the voltage across the triac, V_{BC}. Record this waveform on Graph 21.2. Label both the vertical and horizontal axes with voltage and time base information.

1.13 With your instructor's assistance, display the voltage waveform across the light bulb, V_{AB}. Record this waveform on Graph 21.3. Label both the vertical and horizontal axes with voltage and time base information.

1.14 Now with the triac just on, slowly increase the resistance of the potentiometer, noting what happens to the voltage waveform V_{BC} shown on the oscilloscope and what happens to the intensity of the lamp. Continue increasing the resistance of the potentiometer until the triac again turns off. Slowly decrease the resistance of the potentiometer once more until the triac just fires. Again slowly increase the resistance of the potentiometer, but do not increase the resistance enough to cause the triac to turn off. In the space provided below, describe the difference in operation of the triac with increasing versus decreasing potentiometer resistance. The phenomenon that you just observed is referred to as *hysteresis*.

When increasing the pot the light goes dimmer & triac is off, when decreasing the pot the light goes brighter and triac is fully on.

1.15 Decrease the resistance of the potentiometer until the triac is fully on.
1.16 Use the oscilloscope to display the voltage across the triac, V_{BC}. Record this waveform on Graph 21.4. Label both the vertical and horizontal axes with voltage and time base information.
1.17 Display the voltage waveform across the load resistor, V_{AB}. Record this waveform on Graph 21.5. Label both the vertical and horizontal axes with voltage and time base information.
1.18 For each of the triac operating conditions described in Table 21.1, use your handheld voltmeter to measure the A.C. voltages specified. Make sure that your voltmeter is set to measure A.C. voltage, not D.C. voltage.
1.19 With the potentiometer set to produce a delay angle of approximately 90°, bring an A.M. radio tuned to a local radio station within a few inches of the triac circuit. In the space below, describe what happened.

The circuit is putting out noise that interferes with the AM signal.

1.20 Open switch S1, and turn off the A.C. power supply.
1.21 Reverse the orientation of the diac in the circuit.
1.22 Turn on the A.C. power supply, and close S1.
1.23 Adjust the potentiometer from maximum resistance to minimum resistance. Mentally compare the operation of the circuit now to the operation in the preceding steps.
1.24 Open switch S1, and turn off the A.C. power supply.
1.25 Place a jumper wire in parallel with the diac so that it is shorted out.
1.26 Turn on the A.C. power supply, and close S1.
1.27 Adjust the potentiometer until the triac just fires.
1.28 Use the oscilloscope to display the voltage across the triac, V_{BC}. Record this waveform on Graph 21.6. Label both the vertical and horizontal axes with voltage and time base information.
1.29 Open switch S1. Remove the jumper wire that you placed in parallel with the diac.
1.30 Using a second scope probe, display the voltage waveform V_{DC} in proper time phase with the voltage waveform V_{BC}. Vary the pot from maximum resistance to minimum resistance. What is the phase angle between waveforms V_{BC} and V_{DC} when the pot is set for maximum resistance? Record your answer in Table 21.2. What is the phase angle between waveforms V_{BC} and V_{DC} when the pot is set for minimum resistance?
1.31 Open switch S1, and turn off the power supply.
1.32 Remove capacitor C1 from the circuit so it simulates an open circuit. The potentiometer should still be connected to resistor R2.
1.33 Turn on the A.C. power supply, and close S1.
1.34 Adjust the potentiometer from maximum resistance to minimum resistance. Mentally compare the operation of the circuit now to the operation in the preceding steps.

Graph 21.2: V_{BC} voltage waveform for the circuit shown in Figure 21.7 with the triac just turned on.

V_{BC}: _____ volts/div   _____ sec/div

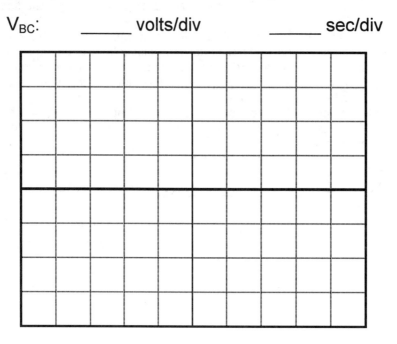

Graph 21.3: V_{AB} voltage waveform for the circuit shown in Figure 21.7 with the triac just turned on.

V_{AB}: _____ volts/div   _____ sec/div

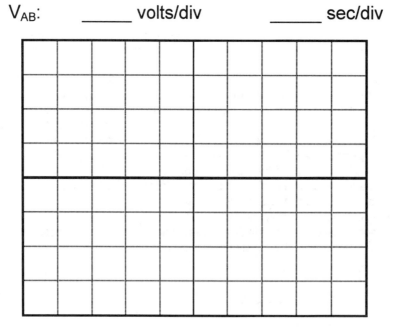

Graph 21.4: V_{BC} voltage waveform for the circuit shown in Figure 21.7 with the triac fully turned on.

V_{BC}: _____ volts/div _____ sec/div

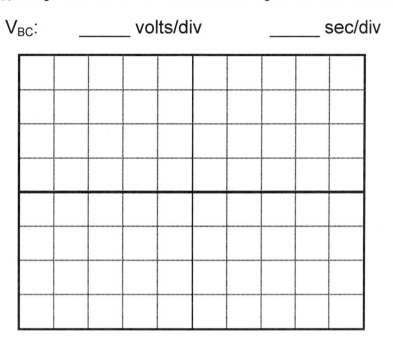

Graph 21.5: V_{AB} voltage waveform for the circuit shown in Figure 21.7 with the triac fully turned on.

V_{AB}: _____ volts/div _____ sec/div

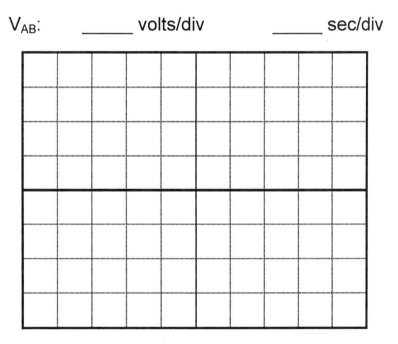

Table 21.1: A.C. voltages for three different triac operating conditions for the circuit shown in Figure 21.7.

TRIAC OPERATIONAL STATUS	LAMP VOLTAGE, VAB, VOLTS RMS	TRIAC VOLTAGE, VBC, VOLTS RMS
TRIAC FULLY OFF	0 v	120.7 v
TRIAC ON WITH A 90 DEGREE DELAY ANGLE	61.3 v	65.1 v
TRIAC FULLY ON	118.3 v	2.55 v

Graph 21.6: V_{BC} voltage waveform for the circuit shown in Figure 21.7 with the diac replaced by a short.

V_{BC}: _____ volts/div   _____ sec/div

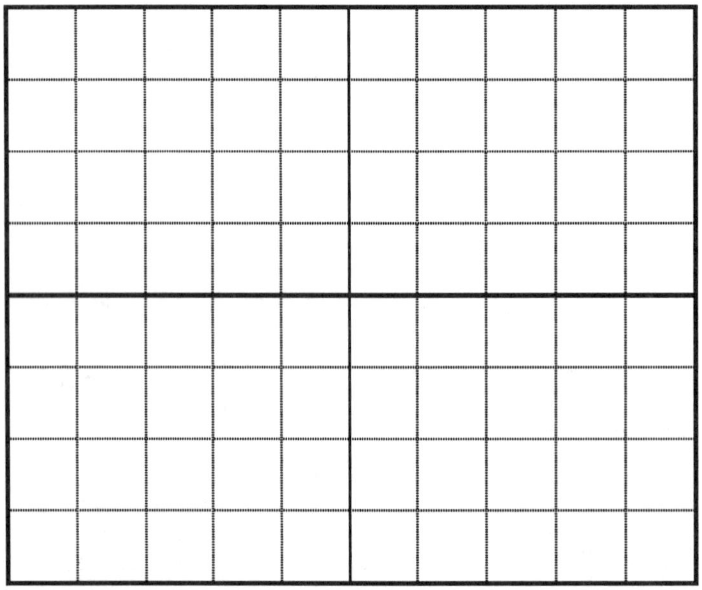

Table 21.2: Phase angle between voltage waveforms V_{BC} and V_{DC} with the potentiometer set for maximum and minimum resistance.

POTENTIOMETER SETTING	PHASE ANGLE BETWEEN VOLTAGE WAVEFORMS VBC AND VDC IN DEGREES
MINIMUM RESISTANCE	180°
MAXIMUM RESISTANCE	90°

PART 2: Application

Background:
While the next activity is a relatively low-power circuit, it is intended to give you an idea of how a triac might be used in a medium-power industrial control application.

Procedure:
2.1 Replace the light bulb in the circuit shown in Figure 21.7 with a hand-held electric drill rated for 120 VAC. Use a drill constructed with a universal motor (one that has brushes). You will need to replace the 1 amp fuse with one of a higher rating. Ask your instructor.

2.2 Adjust potentiometer R1 for maximum resistance--triac off.

2.3 Use your oscilloscope to monitor the source and triac voltage waveforms. Turn on the power supply.

2.4 Slowly decrease, then increase the resistance of R1 over its entire range. Note the response (speed) of the drill as the resistance of the pot is decreased then increased.

2.5 Turn off all power.

NOTE: The triac will dissipate a large amount of power. After applying power, do not touch the triac. After removing power, allow it to cool before touching it. Operate the drill only long enough to complete these steps; otherwise, you may overheat the triac.

PART 3: Manufacturer's Ratings
3.1 Referring to an electronic data base or a set of data manuals, complete Table 21.3 for the triac, MAC210A8.

Table 21.3: Manufacturer's ratings for the triac, MAC210A8.

V_{DRM} VOLTS	I_T AMPS RMS	V_{GT} VOLTS	$I_{GT(1)}$ mA	$I_{GT(2)}$ mA	$I_{GT(3)}$ mA	$I_{GT(4)}$ mA	V_{TM} VOLTS	I_H mA
400v	10 A	2.5v	50	50	50	—	—	50

PART 4: Questions
4.1 How did removing capacitor C1 change the operation of the triac circuit?

4.2 How did replacing the diac with a short affect the operation of the triac circuit?

4.3 What effect did reversing the orientation of the diac have on the operation of the triac circuit? Based on your answer to this question, is the diac symmetrical or nonsymmetrical?

4.4 Based on the manufacturer's ratings recorded in Table 21.3, what is the largest A.C. voltage that can be applied to the SK3658 triac? Express your answer in RMS units.

4.5 Based on your experimental results, what effect did decreasing the resistance of the potentiometer in the gate control circuit have on the A.C. voltage across the lamp?

4.6 In Graph 21.2, what is the conduction angle?

4.7 In Graph 21.4, what is the conduction angle?

4.8 In Graph 21.6, what is the conduction angle?

4.9 In Graph 21.4, what is the delay angle?

4.10 In Graph 21.6, what is the delay angle?

4.11 What caused the noise that you heard from your A.M. radio when you brought it in close proximity to the triac circuit? You must be more specific than just saying RFI. Exactly what caused the RFI?

4.12 Figure 21.8 shows the source and triac voltage waveforms for the electric drill control circuit. Based on the shape of the triac waveform, is the drill motor completely off, turning at a fast speed, or turning at a slow speed? Also, explain how you arrived at your answer. Make reference to the triac and drill voltages. If necessary, make a sketch of the drill's voltage waveform in the space below.

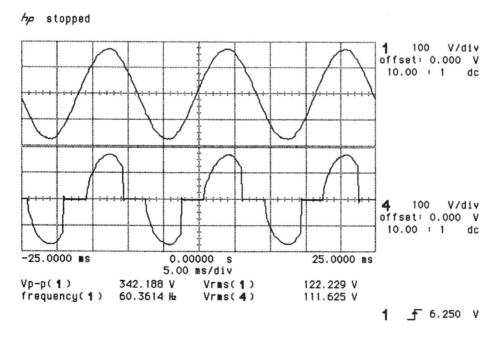

Figure 21.8: Source and triac voltage waveforms for the electric drill control circuit.

PART 5: Electronics Workbench®

5.1 Use Electronics Workbench® to create the circuit shown in Figure 21.7. You do not have to include the fuse. Select a triac that has electrical properties as close as possible to the one used in your lab experiment. Use the oscilloscope function to display the voltage waveforms at points A and B with R1 set for maximum resistance, minimum resistance, and at a value that will cause the triac to just turn on. Print a copy of your schematic diagram and oscilloscope waveforms. How do these waveforms compare to the graphs of your experimental data? List possible causes for any differences.

Unijunction Transistors: 22

INTRODUCTION:

The unijunction transistor (UJT) gets its name from the fact that it has one (uni) junction where P- and N-type semiconductor material have been brought together as a single unit. NPN and PNP transistors have two junctions or boundaries where the N and P layers of semiconductor material come together. Figure 22.1 a) shows the semiconductor model of the UJT. As you can see from the figure, a region of P-type semiconductor material is doped onto a base or substrate of N-type semiconductor material. The UJT does have three terminals, as does the bipolar junction transistor (BJT), but that's where the similarities end. The three terminals or connections are referred to as the emitter (E), base 1 (B1), and base 2 (B2). The terminals are labeled in Figures 22.1 a) and 22.1 b). The equivalent circuit of the UJT is shown in Figure 22.1 b). We will refer to this figure to explain the principle of operation of the UJT. The schematic symbols for the UJT are shown in Figure 22.1 c). Note that there are two symbols, one for the N-type UJT and one for the P-type UJT. Recall how field-effect transistors (FETs) are either N-channel or P-channel. Well, N-type UJTs--like the one shown in Figure 22.1 a)--are constructed from a small bar of N-type material to which is doped an even smaller layer of P-type material. P-type UJTs, as you probably have already guessed, are constructed from a small bar of P-type material to which is doped a layer of N-type material. In this experiment, you will determine some of the electrical characteristics of the UJT and construct a UJT-based relaxation oscillator.

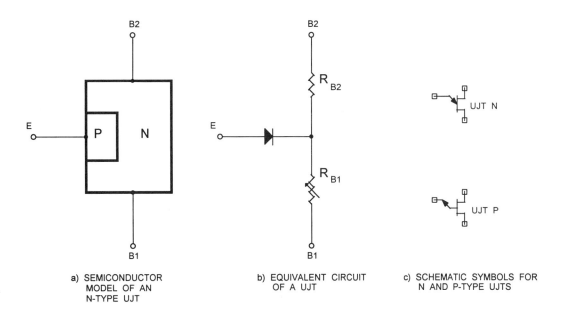

a) SEMICONDUCTOR MODEL OF AN N-TYPE UJT

b) EQUIVALENT CIRCUIT OF A UJT

c) SCHEMATIC SYMBOLS FOR N AND P-TYPE UJTS

Figure 22.1: Semiconductor model, equivalent circuit, and schematic symbols for the UJT.

SAFETY NOTE:
UJTs are low-power devices. Familiarize yourself with the manufacturer's ratings before designing or constructing a circuit that requires a UJT. As always, wear safety glasses, and remove all jewelry from your hands and fingers before the start of the experiment.

OBJECT:
Upon successful completion of this experiment and all reading assignments, the student should be able to:
- experimentally determine the peak voltage, V_P, required to trigger a UJT into the "on" state
- construct an operational relaxation oscillator circuit using a UJT

REFERENCES:
Read Chapter 5 of Maloney's <u>Modern Industrial Electronics</u>

MATERIALS:
- 1 - UJT, 2N2646, or SK9123, or the equivalent
- 1 - 470 Ohms resistor, 1/2 watt
- 1 - 100 Kohms ten-turn trim potentiometer
  - miscellaneous lead wires and connectors
- 1 - capacitor, 0.1 µF
- 1 - capacitor, 0.047 µF
- 1 - 100 Ohms resistor, 1/2 watt
- 1 - solderless breadboard

EQUIPMENT:
- 1 - D.C. power supply
- 2 - digital multimeters
- 1 - oscilloscope
- 2 - oscilloscope probes

PART 1: Electrical Characteristics of the UJT
Background:
The UJT is more like an SCR than a transistor. The UJT, like an SCR or triac, has an "on" state and an "off" state. However, these two states cannot be truly described as a short circuit for the on state nor as an open for the off state. Such a descriptive model is accurate for the SCR and triac. Rather, the UJT can be thought of as being in a high-resistance state from terminal B2 to B1 when it is in the off state and in a low-resistance state from terminal B2 to B1 when in the on state. This is why Figure 22.1 b) depicts the UJT as having an internal resistance, RB1, that varies much like an adjustable resistor. When the UJT is off, RB1 is in its high-resistance state. When the UJT is triggered into the on state, RB1 rapidly changes to the low-resistance state. Please keep in mind when referring to Figure 22.1 that RB2 and RB1 represent the UJT's <u>internal</u> resistance. These resistance values are not to be confused with the external resistors shown in Figures 22.2 and 22.3. In effect, the internal and external resistors form a voltage divider network for the output stage of the circuits shown in Figures 22.2 and 22.3. When the UJT is in the high-resistance state, there is a small voltage drop across R1. When the voltage applied to the emitter of the UJT becomes high enough, the UJT will switch into the low-resistance state, allowing more current to flow through it and R1, causing an increase in the voltage drop across R1. This characteristic is often taken advantage of by incorporating UJTs in the design of SCR gate-trigger circuitry. Just how high does the voltage applied to the emitter have to be to trigger the UJT into the on state? The formula for this trigger voltage, V_P, is as follows:

$$V_P = \eta \cdot V_{B2B1} + 0.6 \text{ volts} \qquad \frac{V_P - .6v}{V_{B2B1}} = \eta \qquad \textbf{22.1}$$

The 0.6 number represents the voltage required to overcome the potential barrier presented by the P-N junction. The Greek letter η is the symbol for the UJT's *intrinsic standoff ratio*. This unitless parameter represents the fraction or percent (this number is specified as a decimal) of the voltage from B2 to B1 that acts to oppose (as in counter-EMF) the turn on of the UJT. Now apply your knowledge of algebra to this formula. As the value of V_{B2B1} increases, the value of V_P increases linearly with it. As you will see later in this experiment, the value of V_{B2B1} is dependent on the value of the external supply voltage as well as the value of the internal resistances, RB2 and RB1.

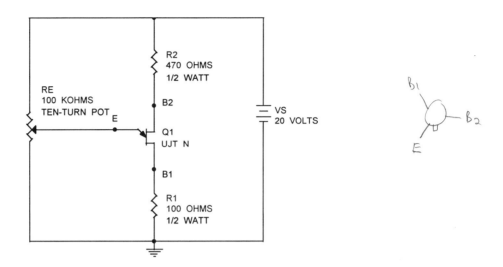

Figure 22.2: Test circuit to determine the voltage required at the emitter (E) of the UJT to trigger it into the "on" state.

Procedure:

1.1 Obtain all materials and equipment required to complete this experiment.
1.2 If your lab has one, use a capacitance tester to check the capacitance values of your capacitors. If either of them are out of tolerance, inform your instructor and obtain replacements. Record the measured capacitance values in Table 22.1.

Table 22.1: Nominal and measured capacitance values.

RATED CAPACITANCE	0.1 uF	0.047 uF
MEASURED CAPACITANCE	.101 µF	.048 µF

1.3 Adjust the output of your power supply to provide 20 volts. Turn off the power supply.
1.4 Construct the circuit shown in Figure 22.2. Adjust the potentiometer for minimum resistance.
1.5 Turn on the power supply.
1.6 Measure the voltage at point E to ground. It should be at or near 0.0 volt. The UJT should now be in the high resistance state.
1.7 While monitoring the voltage from point E to ground, use your other voltmeter to measure the voltage across R1.
1.8 While monitoring the two voltmeter readings, slowly increase the resistance of the potentiometer.
1.9 You will know when the UJT is triggered into the low-resistance on state by a small but sudden increase in the voltage across R1. The value of the voltage from point E to ground that causes this to happen is the V_P value for this circuit. Record the value for V_P in Table 22.2.
1.10 Reduce the resistance of the potentiometer until the UJT is again in the high-resistance off state.
1.11 Now use your second voltmeter to measure the voltage across the UJT, V_{B2B1}. Record this

value in Table 22.2.

1.12 Use the values recorded in Table 22.2 and Formula 22.1 to calculate the value for the UJT's intrinsic standoff ratio, η. Record your value for η in Table 22.2.

Calculations:

$$\eta = \frac{V_P - .6v}{V_{B2B1}} \qquad \frac{12.9v - .6v}{19.64v} = .63$$

1.13 Reduce the supply voltage to 10 volts.
1.14 Repeat steps 1.6 through 1.12 for this new supply voltage. Record your results in Table 22.3. Turn off the power supply.

Calculations:

$$\frac{6.5v - .6v}{9.15v} = .64$$

Table 22.2: V_P, V_{B2B1}, and η for the circuit shown in Figure 22.2 with a supply voltage of 20 volts.

V_P, VOLTS	V_{B2B1}, VOLTS	INTRINSIC STANDOFF RATIO
12.90 v	19.64 v	.63

Table 22.3: V_P, V_{B2B1}, and η for the circuit shown in Figure 22.2 with a supply voltage of 10 volts.

V_P, VOLTS	V_{B2B1}, VOLTS	INTRINSIC STANDOFF RATIO
6.5 v	9.15 v	.64

PART 2: The UJT-Based Relaxation Oscillator
Background:
The test circuit you just constructed can be converted to a relaxation oscillator by adding a capacitor in series with the potentiometer as shown in Figure 22.3. In this circuit, the R-C network produced by the potentiometer and C_E will continuously charge and then discharge through the E-B1 junction of the UJT. Assuming that the capacitor is initially discharged, it will charge up until it reaches the UJT turn-on voltage, V_P. At this point, the capacitor will discharge through the UJT, causing a momentary rush of current through R_E as the UJT switches into its low-resistance state. The voltage across the capacitor will suddenly fall, dropping below the UJT's trigger voltage. At this point, the UJT will switch back to its high-resistance state. The capacitor will start charging again, and the cycle will repeat itself until the power supply is turned off. The period, T, of the resulting voltage waveform across capacitor C_E can be approximated according to the following formula:

$$T = (R_E) \cdot (C_E) \qquad \qquad 22.2$$

It follows that the frequency of the waveform can be found using the formula:

$$f = 1/T = 1/[(R_E) \cdot (C_E)] \qquad 22.3$$

In Experiment 35, a UJT-based relaxation oscillator circuit like that in Figure 22.3 will be used as part of a solid-state timer circuit.

Procedure:

2.1 Construct the circuit shown in Figure 22.3.
2.2 Adjust the potentiometer for a resistance of 50 Kohms. Turn on the power supply. The supply voltage should be 20 volts.

Figure 22.3: UJT-based relaxation oscillator circuit.

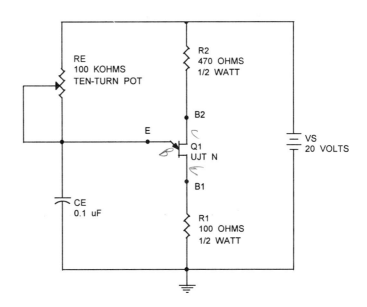

2.3 Use an oscilloscope to display the voltage waveform at the emitter of the UJT relative to ground. This represents the capacitor voltage. Draw the resulting waveform on Graph 22.1. Label both X and Y-axes.
2.4 Record the period of the capacitor's waveform in Table 22.4. Use this value to calculate the frequency of the capacitor's waveform. Record the frequency in Table 22.4.

Calculations: $(4 \text{ msec})^{-1} = 250 \text{ Hz}$

2.5 Turn off the power supply. Replace the 0.1 µF capacitor with a 0.047 µF capacitor.
2.6 Turn on the power supply, and display the capacitor's voltage waveform on the oscilloscope. Use the same volts/div and sec/div. Record the resulting waveform on Graph 22.2. Label both X and Y axes.

Graph 22.1: Capacitor voltage waveform for the relaxation oscillator circuit shown in Figure 22.3 with $C_E = 0.1$ µF, R_E set for 50 Kohms, and a supply voltage of 20 volts.

V_{CE}: _____ volts/div   _____ sec/div

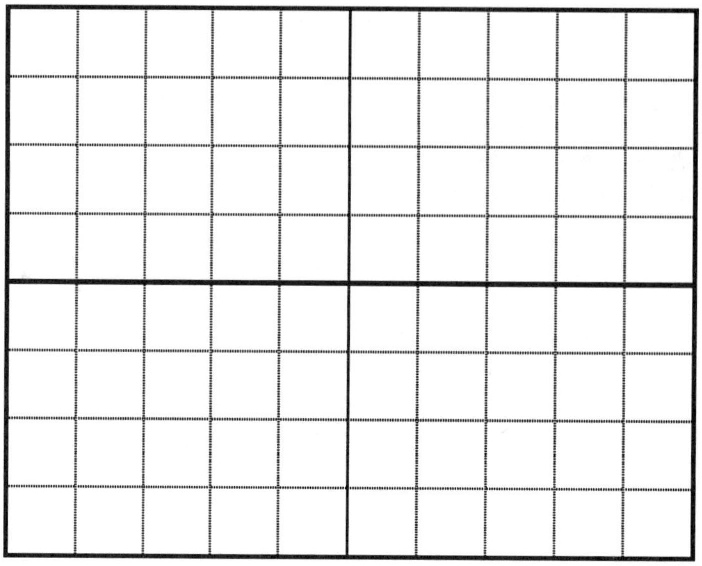

Table 22.4: Period and frequency of the capacitor voltage waveform shown on Graph 22.1 for the circuit shown in Figure 22.3 with $C_E = 0.1$ µF, R_E set for 50 Kohms, and a supply voltage of 20 volts.

PERIOD, T, IN MILLISECONDS	FREQUENCY, F, IN HERTZ
4 msec	250 Hz

2.7 Record the period of the capacitor's waveform in Table 22.5. Use this value to calculate the frequency of the capacitor's waveform. Record the frequency in Table 22.5.

Calculations: $(2.2 msec)^{-1} = 454.5 Hz$

2.8 Now place your second scope probe at point B1 on the circuit, and display the voltage waveform across R1. Look closely. There is a voltage spike in phase with the discharge of the capacitor. Draw this waveform in the space provided on Graph 22.3. Label both X and Y axes.

2.9 Decrease the value of the supply voltage to 10 volts. Draw the resulting capacitor voltage waveform on Graph 22.4. Label both X and Y axes.

2.10 Record the period of the capacitor's waveform in Table 22.6. Use this value to calculate the frequency of the capacitor's waveform. Record the frequency in Table 22.6.

Calculations: $(2 msec)^{-1} = 500 Hz$

2.11 Slowly increase the resistance of the potentiometer. Notice what happens to the period of the waveform. What effect does increasing the resistance of R_E have on the frequency of the capacitor waveform?

It decreases the frequency

2.12 Did the peak (highest point) of the capacitor voltage change any as a result of changing the value of R_E?

No

Graph 22.2: Capacitor voltage waveform for the relaxation oscillator circuit shown in Figure 22.3 with $C_E = 0.047\ \mu F$, R_E set for 50 Kohms, and a supply voltage of 20 volts.

V_{CE}: _____ volts/div   _____ sec/div

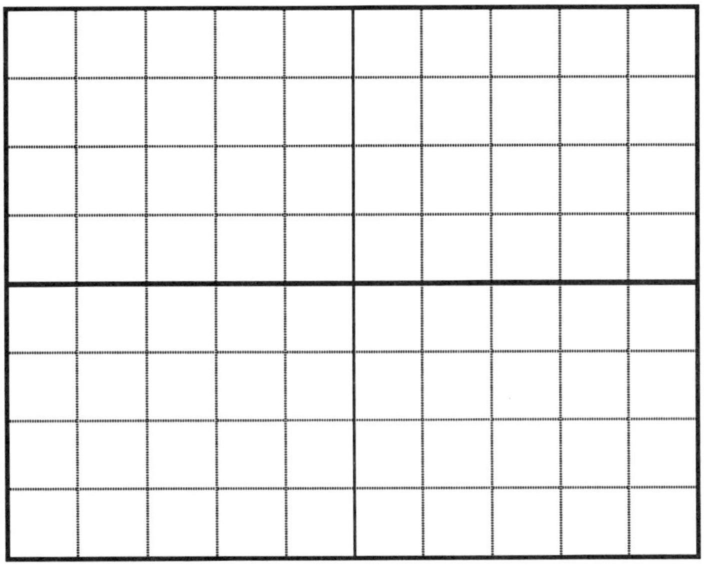

Table 22.5: Period and frequency of the capacitor voltage waveform shown on Graph 22.2 for the circuit shown in Figure 22.3 with $C_E = 0.047\ \mu F$, R_E set for 50 Kohms, and a supply voltage of 20 volts.

PERIOD, T, MILLISECONDS	FREQUENCY, F, HERTZ
2.2 msec	454.55 Hz

Graph 22.3: V_{R1} voltage waveform for the relaxation oscillator circuit shown in Figure 22.3 with $C_E = 0.047\ \mu F$, R_E set for 50 Kohms, and a supply voltage of 20 volts.

V_{R1}: _____ volts/div   _____ sec/div

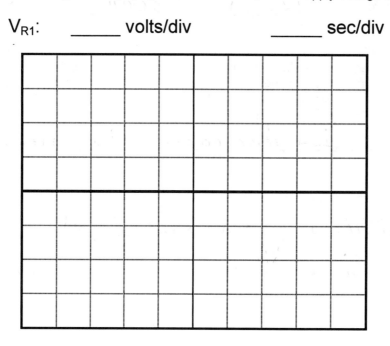

Graph 22.4: Capacitor voltage waveform for the relaxation oscillator circuit shown in Figure 22.3 with $C_E = 0.047\ \mu F$, R_E set for 50 Kohms, and a supply voltage of 10 volts.

V_{CE}: _____ volts/div   _____ sec/div

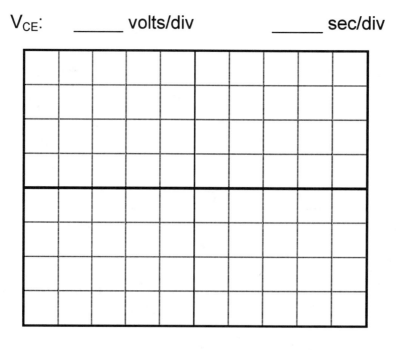

Table 22.6: Period and frequency of the capacitor voltage waveform shown on Graph 22.4 for the circuit shown in Figure 22.3 with C_E = 0.047 µF, R_E set for 50 Kohms, and a supply voltage of 10 volts.

PERIOD, T, MILLISECONDS	FREQUENCY, F, HERTZ
2 msec	500 Hz

PART 3: Manufacturer's Ratings

3.1 Referring to the manufacturer's data manual for your UJT, complete Table 22.7.

Table 22.7: Manufacturer's data for the UJT.

P_T, mW	V_{B2B1} MAXIMUM, V	INTRINSIC STANDOFF RATIO	R_{BB}, KOHMS	I_E MAXIMUM, mA
450 mW	35 V	.56 – .75	4.7 KΩ – 9.1 KΩ	50 mA

PART 4: Questions

4.1 Referring to your data in Tables 22.2 and 22.3, did the value of the supply voltage have a significant effect on the value of the intrinsic standoff ratio?

No

4.2 Did the value of the supply voltage in Figure 22.2 affect the value of the voltage V_{B2B1}?

Yes

4.3 Why did the frequency of the capacitor voltage waveform in the relaxation oscillator circuit decrease when the potentiometer's resistance was increased?

With an increase in resistance

4.4 Referring to the data in Tables 22.4 and 22.5, why did the period of the capacitor voltage waveform decrease when the 0.1 µF capacitor was replaced by the 0.047 µF capacitor?

4.5 Explain why changing the value of R_E in Figure 22.3 did not change the value of the peak of the capacitor voltage waveform, but changing the value of the supply voltage did.

4.6 How do your values for η recorded in Tables 22.2 and 22.3 compare to the range of values for η specified by the manufacturer?

My values were within the range specified by the manufacturer.

PART 5: Unijunction Transistor Formulas

$$V_P = \eta \cdot V_{B2B1} + 0.6 \text{ volts} \tag{22.1}$$

$$T = (R_E) \cdot (C_E) \tag{22.2}$$

$$f = 1/T = 1/[(R_E) \cdot (C_E)] \tag{22.3}$$

Photoelectric Devices: 23

INTRODUCTION:
In general, industrial control electronics circuits serve either one, or both, of two major functions. These two primary functions are to either turn on or off an energy source (typically electrical, though not always) or to control the magnitude of the energy source (increase or decrease it). Whether the primary energy source is to be turned on, off, increased, or decreased depends on the present condition of the process being controlled. To determine the status or current condition of any industrial process requires the use of one or more sensors or *transducers*. A transducer is any device that converts some type of mechanical or physical parameter or stimulus into a current or voltage whose magnitude is in proportion to the magnitude of the stimulus. For example, a pressure transducer converts pressure from a hydraulic or pneumatic device into a small voltage. A flowmeter is a device that converts fluid flow in gallons/minute or liters/hour into a high-frequency square wave. Transducers fall into one of two major categories--active or passive. Active transducers generate their own source of voltage (though it is usually small). Two examples are thermocouples and solar cells. One generates a small voltage when heated; the other generates a small voltage when exposed to light. Passive transducers require a separate voltage or current source to operate in a control circuit. Examples include strain gages, thermistors, and photoconductive cells. The type of sensor or transducer chosen depends on the type of process or product being monitored. In this experiment you will study the electrical characteristics of two photoelectric devices--the photoconductive cell and the solar cell.

SAFETY NOTE:
The solar cell is a very brittle device and is easily broken. It is extremely fragile, so handle with care! If it does break, use caution in picking up and disposing of the broken pieces as they can be very sharp. Most photoconductive cells, such as the one you will work with today, are low-power devices. Therefore, you should never apply a large voltage source, say, 120 volts RMS, across the terminals of this device. In this experiment, you will be working with a 28-volt incandescent lamp. Do not be deceived by this low voltage. This small lamp will become very hot. Do not touch it or any conductive surface that it comes in contact with. Also, do not expose it to any material that is flammable (such as paper) or to any material with a low melting point. As always, wear safety glasses, and remove all jewelry from your hands and fingers before the start of the experiment.

OBJECT:
Upon successful completion of this experiment and all reading assignments, the student should be able to:
- construct a circuit to generate calibration data for a photoconductive cell
- construct a circuit to generate calibration data for a solar cell

REFERENCES:
Chapter 10 of Maloney's Modern Industrial Electronics

MATERIALS:
- 1 - silicon solar cell 2 X 4 cm or larger, Radio Shack® Catalog No. 276-124A or the equivalent
- 1 - photoconductive cell, VACTEC® VT-204, VT935G, VT90D782G, or the equivalent
- 1 - solderless breadboard

1 - incandescent lamp, 28-volt rating
1 - lamp socket
1 - 10 Kohms resistor, 1/4 watt
1 - 50 Kohms, ten-turn trim potentiometer
  - miscellaneous lead wires and connectors
  - rigid, opaque, nonconductive tube with an inner diameter of approximately 0.5" and a length of at least 7" to 8"

INSTRUCTOR'S NOTE: It is not a requirement to use a 28-volt lamp to complete this experiment. Any conventional incandescent lamp with a voltage rating in the range of 14.4 volts to 28 volts will yield satisfactory results. If you do use a lamp having a rating of less than 28 volts, just have the students substitute this value whenever a reference is made to the 28-volt rating.

EQUIPMENT:
3 - digital multimeters
1 - adjustable D.C. power supply, 0 - 28 volts

PART 1: The Photoconductive Cell
Background:
The schematic symbols for the photoconductive cell and solar cell are shown in Figures 23.1 a) and 23.1 b), respectively. The photoconductive cell, also known as photoresistor, is a conductor whose resistance varies with the amount of light that is applied to the light-sensitive material used in its construction. For many years, the compound cadmium sulfide, CdS, was used in the construction of photoconductive cells. For this reason, you will often hear the term *CdS cell* in reference to this particular device. Photoconductive cells have been replaced in many industrial control applications by infrared emitters and detectors, which are not sensitive to ambient sunlight or other light sources within a factory. Photoconductive cells continue to be used in applications where it is necessary to sense the magnitude of white light or sunlight, such as in the control circuitry for home security lights.

Figure 23.1: Schematic symbols for the photoconductive cell and solar cell.

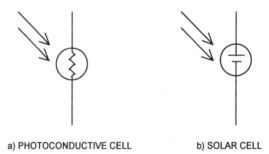

a) PHOTOCONDUCTIVE CELL   b) SOLAR CELL

Procedure:
1.1 Obtain all materials and equipment required to complete this experiment.
1.2 The electrical circuit that you will connect is shown in Figure 23.2. In this part of the experiment, you will determine the effect that distance between a light source and a photoconductive cell has on the response (change in resistance) of the cell.
1.3 One way to do this is to arrange the lamp and photoconductive cell as shown in Figure 23.3. In this figure, the lamp and photoconductive cell have been inserted in opposite ends of a piece of rigid, opaque tubing. Anything that will serve as a barrier to ambient light will be sufficient. However, do not use heat shrink tubing. Its melting point it too low! Your goal is to try to provide a light source of constant intensity while varying the distance between it and

the light-sensitive device, then record the change in resistance. A piece of tubing constructed from a high-melting point, nonflammable, rigid, opaque material at least 7" to 8" in length should be selected as a container for the lamp and the photoconductive cell so that the cell may be shielded from ambient light, or select a tube made from transparent material and wrap it with electrician's tape. The diameter of the tubing should be slightly larger than the diameter of the lamp. I suggest that you crimp the tubing around the end of the photoconductive cell so that no ambient light is able to enter that end, or consider using electrician's tape to cover that end. I also suggest that you keep the photoconductive cell stationary and move the lamp incrementally closer to it as shown in Figure 23.4.

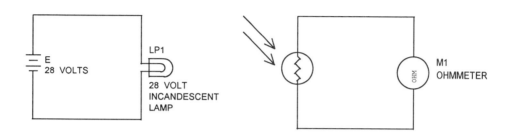

Figure 23.2: Test circuit to obtain calibration data for the photoconductive cell.

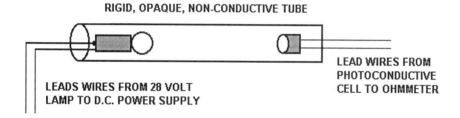

Figure 23.3: Experimental setup to determine a photoconductive cell's response to a light source of varying intensity.

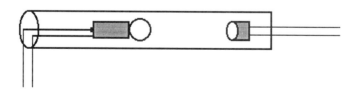

Figure 23.4: Light source moved closer to the photoconductive cell to determine the effect that distance between the light source and the photoconductive cell has on the response of the cell.

1.4 With the lamp operating at rated voltage (28 volts as shown on the schematic, though you may be using a lamp of lower voltage rating), vary the distance between the lamp and the photoconductive cell according to the distances shown in Table 23.1 and measure the corresponding cell resistance. Record your resistance values in Table 23.1.

1.5 Modify your original circuit to appear as shown in Figure 23.5.

1.6 Position your lamp within 1" of the photoconductive cell while still inside the tube.
1.7 Adjust the voltage source so that rated voltage is applied to the lamp (M2 indicates the lamp voltage). Record the corresponding lamp current and photoconductive cell resistance in Table 23.2. Repeat these measurements for each of the remaining lamp voltage values in Table 23.2. Note that the lamp voltages are expressed as a percent of the full-scale or rated lamp voltage. For example, if you are using a lamp rated for 28 volts, adjust the power supply for $(0.8) \cdot (28\ V) = 22.4$ volts when operating at 80% of rated voltage.
1.8 Turn off the power supply.

Table 23.1: Photoconductive cell resistance as a function of distance from a light source of a fixed intensity.

DISTANCE BETWEEN LAMP AND PHOTOCONDUCTIVE CELL IN INCHES	PHOTOCONDUCTIVE CELL RESISTANCE IN OHMS
5.0"	
4.5"	
4.0"	
3.5"	
3.0"	
2.5"	
2.0"	
1.5"	
1.0"	
0.5"	

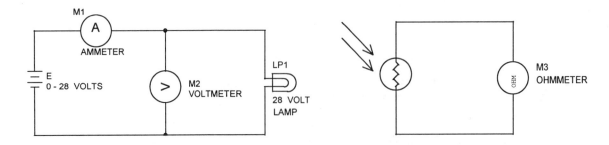

Figure 23.5: Test circuit to determine a photoconductive cell's resistance as a function of lamp voltage and lamp current.

Table 23.2: Photoconductive cell resistance as a function of lamp voltage and lamp current with a fixed distance between the lamp and photoconductive cell.

LAMP VOLTAGE AS A PERCENT OF RATED VOLTAGE	LAMP VOLTAGE, VOLTS	LAMP CURRENT, MILLIAMPS	PHOTOCONDUCTIVE CELL RESISTANCE, OHMS
100%			
90%			
80%			
70%			
60%			
50%			
40%			
30%			
20%			

PART 2: The Solar Cell

Background:

Solar cells--also known as *photovoltaic* cells--convert light energy into electrical energy. Like thermocouples, solar cells are composed of two dissimilar metals or semiconductor materials. When thermocouples are heated, they produce a small output voltage. In a similar fashion, solar cells produce a small voltage when exposed to light. I am sure that you have seen pictures of solar panel arrays extending like wings from the main body of an orbiting satellite. Or, closer to home (maybe even right beside you as you are reading this), you have used or seen solar-powered calculators. Other uses of photovoltaic cells include the sensor in camera light meters. Solar cells serve as the power source for emergency roadside telephones and as power sources for small, remote weather recording stations. And of course, scientists are trying to improve the efficiency of solar cells so that they may be used as a possible alternative energy source in the future. In the experimental procedure that follows, you will have the opportunity to determine the characteristics of the photovoltaic cell under varying load conditions. The physical outline of a photovoltaic cell similar to the one that you will be using is shown in Figure 23.7.

Procedure:

2.1 Take a small metric ruler and record the length and width of the active region of your solar cell. Record the cell's dimensions (centimeters) in Table 23.3. Use these dimensions to calculate the cross-sectional area of the solar cell. Record the area (cm^2) in Table 23.3.

Table 23.3: Photovoltaic cell dimensions and cross-sectional area.

LENGTH, CM.	WIDTH, CM.	CROSS-SECTIONAL AREA, SQ. CM.

2.2 Connect the leads of your solar cell to a digital voltmeter as shown in Figure 23.6. **Measure the output voltage when the solar cell is exposed to bright light.** If possible, bring the solar cell near a window so that you may expose it to sunlight. Record the voltage in Table 23.4.

2.3 Complete Table 23.4 by measuring the solar cell voltage when it has increasingly more of its cross-sectional area covered by an opaque material as shown in Figure 23.8.

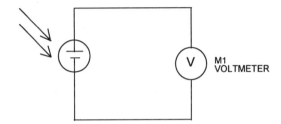

Figure 23.6: Test circuit to determine the no-load voltage characteristics of the solar cell as a function of the surface area of the cell that is exposed to light.

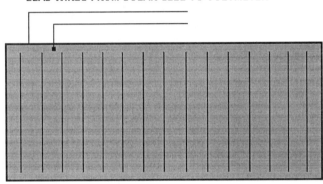

Figure 23.7: Physical outline of a typical experimenter's solar cell.

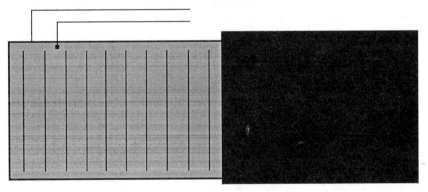

Figure 23.8: Solar cell covered with an opaque material to determine solar cell voltage as a function of solar cell area exposed to light.

Table 23.4: Photovoltaic cell voltage as a function of cross-sectional area exposed to light.

% OF SOLAR CELL AREA EXPOSED TO LIGHT	AREA OF SOLAR CELL EXPOSED TO LIGHT, SQ. CM.	SOLAR CELL VOLTAGE, MILLIVOLTS
100%		
90%		
80%		
70%		
60%		
50%		
40%		
30%		
20%		
10%		

2.4 Connect your solar cell as shown in Figure 23.9. For each of the load resistance settings shown in Table 23.5, measure and record the solar cell voltage. Use Ohm's Law to calculate the corresponding load current. Record the current values in Table 23.5.

2.5 Disconnect your circuit upon completion of your measurements.

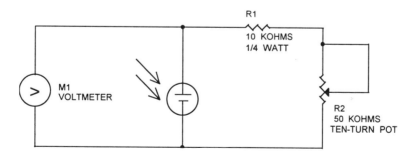

Figure 23.9: Test circuit to determine the voltage characteristics of the solar cell with a constant light source and a varying load.

Table 23.5: Photovoltaic cell voltage as a function of load resistance.

LOAD RESISTANCE	SOLAR CELL VOLTAGE, MILLIVOLTS	LOAD CURRENT, MICROAMPS
55 KOHMS		
50 KOHMS		
45 KOHMS		
40 KOHMS		
35 KOHMS		
30 KOHMS		
25 KOHMS		
20 KOHMS		
15 KOHMS		
10 KOHMS		

PART 3: Questions

3.1 Using the data shown in Table 23.1, construct a graph of photoconductive cell resistance as a function of distance from the lamp. Remember from algebra that Y = f(X). In this case, R = f(D), where R is the resistance of the photoconductive cell and D is the distance of the photoconductive cell from the lamp. Attach a copy of your graph to your report. Label both vertical and horizontal axes on your graph. Label your graph with the title: <u>Graph 23.1: Resistance as a function of distance from the lamp</u>.

3.2 Using the data shown in Table 23.2, construct a graph of resistance as a function of lamp current. Remember from algebra that Y = f(X). In this case, R = f(I), where R is the resistance of the photoconductive cell and I is the current through the lamp. Attach a copy of your graph to your report. Label both vertical and horizontal axes on your graph. Label your graph with the title: <u>Graph 23.2: Resistance as a function of lamp current</u>.

3.3 Using the data shown in Table 23.2, construct a graph of resistance as a function of lamp voltage. Remember from algebra that Y = f(X). In this case, R = f(V), where R is the resistance of the photoconductive cell and V is the voltage applied to the lamp. Attach a copy of your graph to your report. Label both vertical and horizontal axes on your graph. Label your graph with the title: <u>Graph 23.3: Resistance as a function of lamp voltage</u>.

3.4 Based on the graphs of your data, does the resistance of your photoconductive cell increase or decrease with increasing levels of light?

3.5 Which graph, Graph 23.2 or Graph 23.3, seemed to depict a more linear response? Speculate on why one graph was more linear than the other.

3.6 Using the data from Table 23.4, construct a graph of solar cell output voltage as a function of the cell area (cm^2) exposed to light. This assumes that the amount of light applied to each square centimeter of the photovoltaic cell was a constant. Remember from algebra that Y = f(X). In this case, E = f(A), where E is the voltage produced by the photovoltaic cell and A is the cross-sectional area of the photovoltaic cell exposed to light. Attach a copy of your graph to your report. Label both vertical and horizontal axes on your graph. Label your graph with the title: Graph 23.4: Voltage as a function of photovoltaic cell area.

3.7 Using the data from Table 23.5, construct a graph of solar cell output voltage as a function of load current. This assumes that the solar cell area exposed to light and the amount of light applied to each square centimeter of the photovoltaic cell was a constant. Remember from algebra that Y = f(X). In this case, E = f(I), where E is the voltage produced by the photovoltaic cell and I is the current through the load resistor. Attach a copy of your graph to your report. Label both vertical and horizontal axes on your graph. Label your graph with the title: Graph 23.5: Voltage as a function of load current.

3.8 Assume for the moment that you are designing a counter circuit that is to be used to count the number of filled soda pop bottles passing by an inspection point on a bottle filling line. What would be the disadvantages of using a white light source on one side of the bottling line and a photoconductive cell on the other to sense the presence of filled bottles passing by this sensor circuit. The filled bottles would block and unblock the path of light to the photoconductive cell as each one passed by. Where possible, try to incorporate your experimental results in your response.

3.9 Figure 23.10 shows a light-sensitive control circuit constructed from three series-connected photovoltaic cells like the one used in this experiment. From experience, you know that the maximum total series output voltage from the three cells is 1.2 volts (0.4 V + 0.4 V + 0.4 V). Determine the range of output voltage, V_{OUT}, as the voltage applied to the non-inverting input of the op-amp from the photovoltaic cells varies from a minimum of 0.0 volt in the dark to a maximum of 1.2 volts when exposed to bright light. In other words, calculate $V_{OUT(MINIMUM)}$ when the solar cells are in the dark and determine $V_{OUT(MAXIMUM)}$ when the solar cells are operating at their maximum potential.

Calculations:

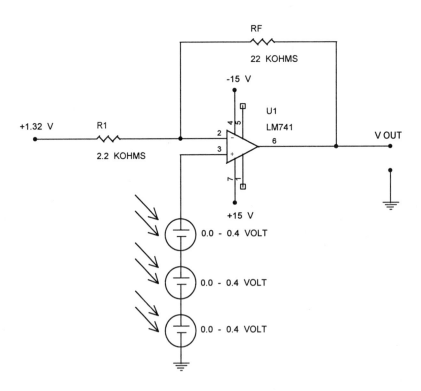

Figure 23.10: Circuit for Question 3.9.

PART 4: Circuit Design

4.1 Design an op-amp comparator circuit similar to the one shown in Figure 23.11 that will act as the control circuitry for an outdoor home security lighting system. The op-amp circuit will compare the input from the voltage divider (composed of the fixed resistor, R3, and photoconductive cell) to the reference or set point established by the other two resistors. The output of the op-amp should drive a current amplifier (remember, an op-amp is a low-current output device) that will power the coil of a control relay. The contacts of the relay are used to turn on and off the security light. Your design should provide a means to adjust the sensitivity of the circuit. In other words, provide some way to adjust the circuit to turn on or off at different ambient light levels. Draw a schematic diagram of your circuit, and label all components, including the power supply for the op-amp, control relay, and 120 volt RMS security light. Attach a copy of your schematic to your report. If your instructor wishes, construct the circuit and demonstrate it. Show all calculations in the space provided below.

Calculations:

4.2 What should be the physical location of the photoconductive cell relative to the security light?

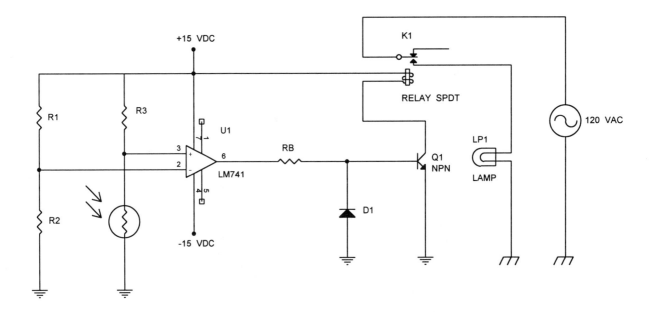

Figure 23.11: Op-amp comparator circuit connected to a Wheatstone bridge circuit. The output of the op-amp is used to turn on or off the transistor depending on the resistance of the photoconductive cell. The transistor switches the relay on and off. The relay should turn the lamp on just before sunset and turn it off just after sunrise.

Optoelectronic Devices: 24

INTRODUCTION:
In Experiment 23, you were introduced to two devices (the solar cell and the photoconductive cell) whose electrical characteristics depend on the amplitude and wavelength of light that they are exposed to. In this experiment, we will continue the study of light-sensitive and light-emitting devices by examining several optoelectronic devices. Electronic devices that fall into this category include light-emitting diodes (LEDs), infrared-emitting diodes (IREDs), photodiodes, phototransistors, light-activated SCRs (LASCRs), light-activated triacs, optoisolators, and photointerrupters. LASCRs and light-activated triacs will be covered in Experiment 25. In this experiment, you will have the opportunity to investigate the electrical characteristics of LEDs, seven-segment displays, phototransistors, and optoisolators (also referred to as optocouplers).

SAFETY NOTE:
All of the devices in this experiment have a low power dissipation rating; therefore, make sure that you look up the maximum current, voltage, and power rating for each of the electronic devices in this experiment before you proceed. Before constructing any circuit you should make sure that its design will not result in any device having its maximum ratings exceeded. In one part of this experiment, you will be working with a 28-volt incandescent lamp. Do not be deceived by this low voltage. This small lamp will become very hot. Do not touch it or any conductive surface that it comes in contact with. Also, do not expose it to any material that is flammable (such as paper) or to any material with a low melting point. As always, wear safety glasses, and remove all jewelry from your hands and fingers before the start of the experiment.

OBJECT:
Upon successful completion of this experiment and all reading assignments, the student should be able to:
- construct a circuit to generate calibration data for a light-emitting diode
- construct a circuit to identify each of the cathodes and the corresponding segments of a common anode, seven-segment display
- construct a circuit to generate calibration data for a phototransistor
- construct a circuit to generate calibration data for an optoisolator

REFERENCES:
Read Chapter 10 of Maloney's <u>Modern Industrial Electronics</u>

MATERIALS:
- miscellaneous red, green, and amber LEDs
- 1 - common anode, seven-segment display
- 1 - optoisolator, 4N26, or SK2040, or the equivalent
- 1 - photointerrupter, H21A2 or SK4930
- 1 - phototransistor, ECG3031 or SK2031
- 1 - incandescent lamp, 28 volt rating
- 1 - control relay, 12 VDC coil
- miscellaneous lead wires and connectors
- 1 - 1.5 Kohms, 1/2 watt resistor
- 1 - 330 Ohms, 1/2 watt resistor
- 1 - 1 Kohms, 1/2 watt resistor
- 1 - 100 Kohms, 1/4 watt resistor
- 1 - solderless breadboard
- 1 - lamp socket
- 1 - NPN transistor, 2N3905

- rigid, opaque, nonconductive tube with an inner diameter of approximately 0.5" and a length of at least 7" to 8"

INSTRUCTOR'S NOTE: It is not a requirement to use a 28 volt lamp to complete this experiment. Any conventional incandescent lamp with a voltage rating in the range of 14.4 volts to 28 volts will yield satisfactory results. If you do use a lamp having a rating of less than 28 volts, just have the students substitute this value whenever a reference is made to the 28 volt rating.

EQUIPMENT:
1 - dual-outlet, adjustable D.C. power supply
2 - digital multimeters

PART 1: Light-Emitting Diodes
Background:
If you have already worked with LEDs in other electronics classes, your instructor may have you skip over this section of the experiment. LEDs have been used for years in stand-alone applications as indicator lights in control panels, watches, calculators, and many types of consumer electronics. You are probably familiar with LEDs that emit a wavelength of light in the visible spectrum (such as red, green, amber, and even blue). However, in the past several years, the infrared-emitting diode (IRED) has been incorporated in many electronic devices designed for use in business, industry, and the home. If you have a remote control for your television or VCR, chances are that it has an IRED mounted inside it to act as the source of the electronic signal from the remote to the detection circuitry in your TV or VCR. In many public restrooms, manually operated faucets and hand-dryers have been replaced by units with IRED emitters and detectors that control the turn on and off of these devices. IREDs are used in opotoisolators and solid-state relays (SSRs). At this point, a word about safety is in order. When working with any light source--visible or invisible--never look directly into it. When working with any light source that may cause damage to the eye, the technician should wear eye protection meeting OSHA and/or ANSI standards for the particular activity being performed. The package outline and schematic symbol for an LED are shown in Figure 24.1.

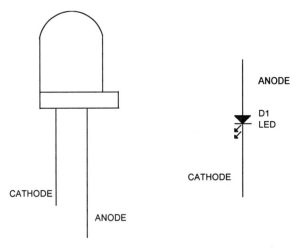

Figure 24.1: Package outline of a typical LED showing cathode and anode leads. The flat spot on the left of the plastic lens identifies the cathode, and the longer of the two leads identifies the anode. The schematic symbol for the LED is shown on the right.

As with rectifier diodes, LEDs have an anode and a cathode. Before installation, a new LED has two leads of different length. The longer lead is the anode. Even if the leads have been shortened for installation, you can still determine which lead is the anode and which is the cathode. Most LEDs have a flat spot on one side of the plastic lens. The lead wire next to the flat spot is the cathode. When an LED is forward-biased as shown in Figure 24.2, it will emit light. The brightness of the light will depend on the amount of current flowing through the LED. Manufacturer's specify the amount of forward-bias current required to produce rated brightness. This will vary from LED to LED. There are also some other points to keep in mind when working with LEDs. When forward-biased, LEDs will have a higher voltage drop from anode to cathode than will rectifier diodes. Typical voltage drops will be in the range of 1.6 volts to 2.3 volts. Also, care must be taken never to reverse bias LEDs. The reverse breakdown voltage of LEDs is very small compared to rectifier diodes. Figure 24.3 shows two possible techniques that designers will use to prevent LED failure (blowing their tops off) due to inadvertent reverse biasing. In the figure on the left, if the power supply polarity is accidentally reversed, the rectifier diode will be forward-biased, limiting the reverse-bias voltage across the LED to approximately 0.7 volt. In Figure 24.3, if the LED is accidentally installed backwards or exposed to an alternating voltage (one that goes negative), the rectifier diode will protect the LED from failure due to reverse breakdown. These are two simple examples. Other protective methods are also available.

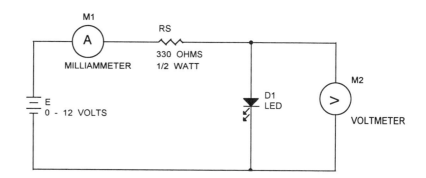

Figure 24.2: Test circuit to determine the I-V characteristics of an LED.

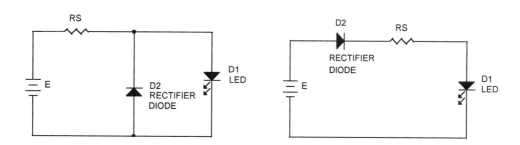

Figure 24.3: Two methods to protect an LED against accidental reverse-biasing or excessive reverse-biasing.

Procedure:

    1.1 Obtain all materials and equipment required to complete this experiment.
    1.2 The first circuit that you will connect is shown in Figure 24.2. Select a red LED for the first phase of the experiment.
    1.3 Vary the supply voltage according to the values specified in Table 24.1. Measure and record the corresponding LED voltage and current. Turn off the power supply.
    1.4 Replace the red LED with a green LED. Repeat step 1.3.
    1.5 Replace the green LED with an amber LED. Repeat step 1.3.
    1.6 If your lab has a curve tracer, use it to produce the I-V curve for each LED.

Table 24.1: LED voltage and current data obtained from the circuit shown in Figure 24.2.

SUPPLY VOLTAGE, VOLTS	RED LED CURRENT, mA	RED LED VOLTAGE, VOLTS	GREEN LED CURRENT, mA	GREEN LED VOLTAGE, VOLTS	AMBER LED CURRENT, mA	AMBER LED VOLTAGE, VOLTS
2						
3						
4						
5						
6						
7						
9						
12						

PART 2: The Seven-Segment Display

Background:

When it is necessary to have a numeric display, LEDs can be arranged in a group as shown in Figure 24.4, where each segment of the seven-segment display is an individual LED. The combination of LED segments lit will determine the number displayed. The letters A through G shown on the seven-segment display are the means used to identify each segment. Seven-segment displays come in one of two forms, common cathode or common anode. In order to minimize the number of external pins on each seven-segment display package and minimize the number of external connections, all of the cathodes in a common-cathode display are internally connected and brought out with just one common connection as shown in Figure 24.5. Typically, this connection is tied to ground as shown. Then each anode is connected to a voltage source through a current-limiting resistor as shown for the leftmost LED. Similarly, the anodes in a common-anode package are internally connected. The common external connection is then generally connected directly to a voltage source as shown in Figure 24.6. The LED segments are then lit by individually grounding each cathode through a current-limiting resistor as shown for the leftmost LED. This type of configuration is commonly used when an active-low display driver chip is used to control the seven-segment display.

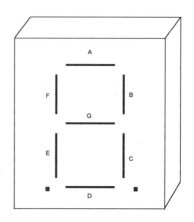

Figure 24.4: Package outline and segment identification for a seven-segment display showing left- and right-hand decimal points.

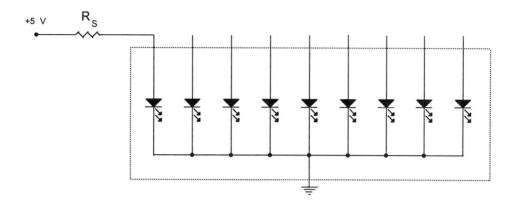

Figure 24.5: Internal connection of LEDs for a common cathode seven-segment display.

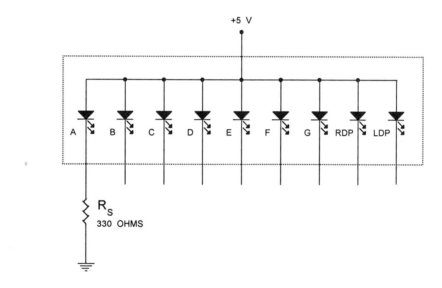

Figure 24.6: Internal connection of LEDs for a common anode seven-segment display.

Procedure:

2.1 Obtain a common-anode, seven-segment display. With your instructor's assistance, first identify the pin that is the common-anode connection. On many displays this is pin 3. However, never assume! Always ask someone who knows or consult the manufacturer's data sheet.

2.2 Connect a +5 volts supply to the common-anode connection as shown in Figure 24.6.

2.3 One at a time, ground each of the remaining pins through a current-limiting resistor to determine which pin number corresponds to the segment that lights (A,B,C,D,E,F,G, RDP, or LDP). Note RDP and LDP are abbreviations for right decimal point and left decimal point, respectively. Some seven-segment displays may not have a left decimal point.

2.4 In Table 24.2, record the pin number that corresponds to the segments shown.

Table 24.2: LED segments and corresponding pin numbers for the common anode, seven-segment display shown in Figure 24.6.

SEGMENT LIT	A	B	C	D	E	F	G	RDP	LDP
PIN NUMBER									

PART 3: The Phototransistor

Background:

Much like the photoconductive cell, the phototransistor can be used in applications where it is necessary to monitor light levels or in counting applications where an object passes between the phototransistor and a light source. Phototransistors can be doped to respond to white light or to wavelengths of light outside the visible spectrum. As pointed out in the background information for the LED, phototransistors can act as the detector in applications such as TV and VCR remote control circuits. Phototransistors are also used as the detector in safety light curtains that sense the presence of a person entering a hazardous area in a plant or manufacturing setting. In this application, a bank or array of phototransistors will shut off the machinery in a potentially hazardous work area to prevent injury to a human operator. When selecting LEDs as emitters and phototransistors as detectors, the designer must match the response of the phototransistor to the wavelength of light emitted by the source. In other words, a phototransistor designed to respond to white light will not respond to the output from an IRED. In this phase of the experiment, you will determine the response of a phototransistor to a white light source of varying intensity. This will be accomplished in two different ways. First, you will vary the distance of the light source from the phototransistor. This is being done to simulate a situation where you might have to design some type of emitter-detector circuit for use in a manufacturing setting where the emitter and detector are separated by a known distance and you must determine if the circuitry you have specified will operate properly at that distance. In the second test circuit, you will vary the light intensity to the phototransistor by varying the voltage applied to the light. The collector current through a conventional NPN or PNP transistor can be increased by increasing the base current. In a phototransistor, the collector current can be increased by increasing the amount of light energy applied to the light-sensitive surface of the phototransistor.

Procedure:

3.1 In this part of the experiment, you will determine the effect that distance between a light source and a phototransistor has on the response of the phototransistor by constructing the circuit shown in Figure 24.7.

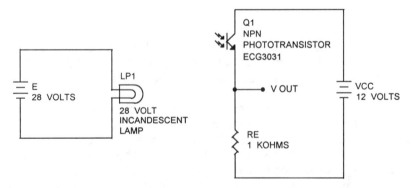

Figure 24.7: Test circuit to obtain calibration data for a phototransistor. The two circuits shown are electrically isolated. In other words, they do not share a common ground.

3.2 One way to do this is to arrange the lamp and phototransistor as shown in Figure 24.8. In this figure, the lamp and phototransistor have been inserted in opposite ends of a piece of rigid, opaque tubing. Your goal is to try to provide a light source of constant intensity while varying the distance between it and the phototransistor, then record the change in output voltage. A piece of tubing constructed from a high-melting point, nonflammable material at least 7" to 8" in length should be selected as a container for the lamp and the phototransistor so that the phototransistor may be shielded from ambient light. Do not use heat shrink tubing. Its melting point is too low. The diameter of the tubing should be slightly larger than the diameter of the lamp. I suggest that you crimp the tubing around the end of the phototransistor so that no ambient light is able to enter that end. I also suggest that you keep the phototransistor stationary and move the lamp incrementally closer to it as shown in Figure 24.9.

3.3 With the lamp operating at rated voltage (28 volts as shown on the schematic, though you may be using a lamp of lower voltage rating), vary the distance between the lamp and the phototransistor according to the distances shown in Table 24.3 and measure the corresponding output voltage, V_{OUT}. Record your data in Table 24.3.

3.4 Now repeat step 3.3 with a 100 Kohms resistor connected between the base and ground of the output stage as shown in Figure 24.10. Record your results in Table 24.3.

3.5 Modify your original circuit to appear as shown in Figure 24.11.

3.6 Position your lamp within 1" of the phototransistor while still inside the tube.

3.7 Adjust the voltage source so that rated voltage is applied to the lamp (M2 indicates the lamp voltage). In Table 24.4, record the corresponding lamp current and output voltage, V_{OUT}, from the phototransistor circuit. Repeat these measurements for each of the remaining lamp voltage values in Table 24.4. Note that the lamp voltages are expressed as a percent of the full-scale or rated lamp voltage. For example, if you are using a lamp rated for 28 volts, adjust the power supply for $(0.8) \cdot (28 \text{ V}) = 22.4$ volts when operating at 80% of rated voltage.

3.8 Turn off the power supply.

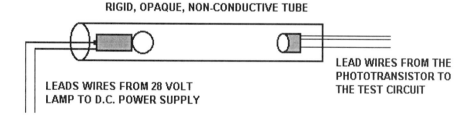

Figure 24.8: Experimental setup to determine a phototransistor's response to a light source of varying intensity.

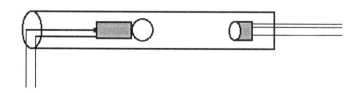

Figure 24.9: Light source moved closer to the phototransistor to determine the effect that distance between the light source and the phototransistor has on the response of the phototransistor.

Table 24.3: Phototransistor circuit output voltage as a function of distance from light source of a fixed intensity with and without a base resistor.

DISTANCE BETWEEN PHOTOTRANSISTOR AND LAMP IN INCHES	PHOTOTRANSISTOR OUTPUT VOLTAGE WITHOUT BASE RESISTOR, VOLTS	PHOTOTRANSISTOR OUTPUT VOLTAGE WITH BASE RESISTOR INSTALLED, VOLTS
5.0"		
4.5"		
4.0"		
3.5"		
3.0"		
2.5"		
2.0"		
1.5"		
1.0"		
0.5"		

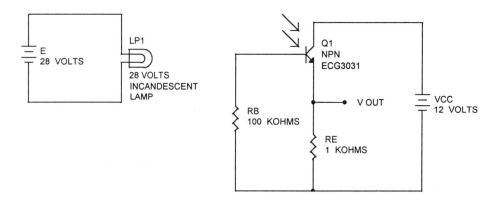

Figure 24.10: Test circuit to determine the effect that addition of a base resistor has on the operation of a phototransistor.

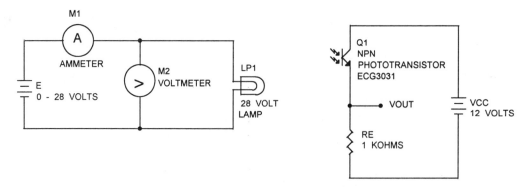

Figure 24.11: Test circuit to obtain calibration data for the phototransistor as a function of lamp voltage and lamp current.

Table 24.4: Phototransistor output voltage as a function of lamp voltage and lamp current with a fixed distance between the lamp and the phototransistor.

LAMP VOLTAGE AS A % OF RATED VOLTAGE	LAMP VOLTAGE, VOLTS	LAMP CURRENT, MILLIAMPS	PHOTOTRANSISTOR OUTPUT VOLTAGE, VOLTS
100%			
90%			
80%			
70%			
60%			
50%			
40%			
30%			
20%			
10%			

PART 4: The Optoisolator
Background:
Control relays provide for electrical isolation between a low voltage supply to the relay's coil and a high-voltage supply providing current to some type of load. Relays have been, and will continue to be, used in may control applications. However, relays do have some disadvantages. For example, they cause electrical noise (RFI), are subject to contact erosion (pitting), and are subject to unwanted contact closure due to vibration. When it is necessary to provide electrical isolation between two circuits and overcome the disadvantages associated with the control relay, a variety of optoelectronic devices may be used. This particular family of optoelectronic devices is referred to as opotoisolators or optocouplers. Optoisolators are constructed from two major parts--an emitter and a detector. Quite often an infrared-emitting diode is used as the emitter and a photodiode, phototransistor, photoSCR, or phototriac, are used as the detectors. Of course, the light sensitivity of the detector must be matched to the wavelength of light given off by the emitter. PhotoSCRs and phototriacs are covered in Experiment 25. In this part of the experiment, you will work with a phototransistor-based optoisolator. The package outline for the SK2040 is shown in Figure 24.12. In this particular configuration, the emitter and detector are completely enclosed in the plastic DIP package. Both components are then protected from the environment. Not all optoisolators are encapsulated. Figure 24.13 shows a special type of optoisolator known as a photointerrupter. Like the optoisolator in Figure 24.12, there is both an emitter and a detector in this package, but there is a notch or space between the emitter and detector halves. This particular device is used in conjunction with some type of mechanism that passes in between the emitter and detector, blocking and unblocking the path of light. In the example shown in Figure 24.13, a notched disk rotates past the emitter-detector pair. This arrangement could be used to determine the speed of a rotating device such as the shaft of a motor. The output voltage pulses from the detector circuitry could be fed to some type of electronic counter and in turn decoded to a display showing motor RPM. The circuit that you are to wire is shown in Figure 24.14. Note that there is a base connection to the phototransistor. In this experiment, the base is left open.

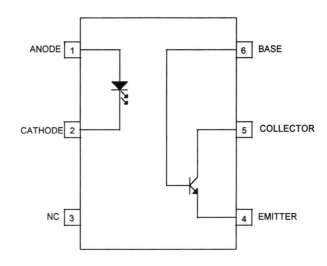

Figure 24.12: Package outline and pin identification for a phototransistor-based optoisolator.

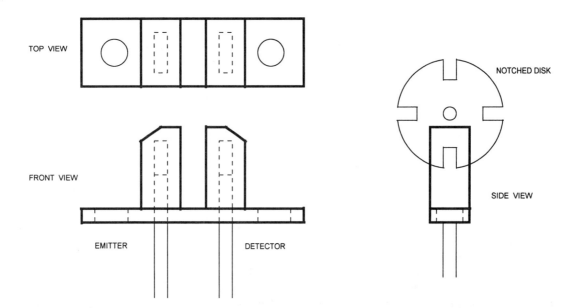

Figure 24.13: Top, front, and side views of a photointerrupter. The side view shows how a notched disk is used to block and unblock the path of light from the emitter (IRED) to the detector.

Procedure:

4.1 Obtain the parts required to construct the circuit shown in Figure 24.14. Before constructing the circuit, measure the resistance of R_S and R_L. Record the measured resistance values in Table 24.5.

Table 24.5: Nominal and measured resistance values for the resistors in Figure 24.14.

NOMINAL RESISTANCE, OHMS	330	1500
MEASURED RESISTANCE, OHMS		

4.2 Construct the circuit shown in Figure 24.14. Note the grounds for the emitter and detector circuits are isolated from each other. Do not connect the grounds together. For each of the emitter supply voltages, E1, shown in Table 24.6, measure and record the corresponding values for V_{RS}, V_{RL}, and V_{CE}.

4.3 Use Ohm's Law and the measured resistance values in Table 24.5 and the values for V_{RS} and V_{RL} to calculate the corresponding emitter and detector currents. Record the current values in Table 24.6.

4.4 Turn off the power supply.

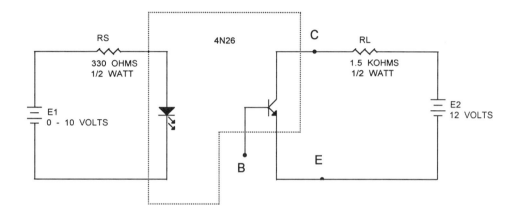

Figure 24.14: Optoisolator test circuit.

Table 24.6: Experimental data for the optoisolator circuit shown in Figure 24.14.

SUPPLY VOLTAGE, E1, VOLTS	V_{RS}, VOLTS	I_{RS}, mA	V_{RL}, VOLTS	I_{RL}, mA	V_{CE}, VOLTS
0					
1					
1.5					
2					
2.5					
3					
3.5					
4					
6					
8					
10					

PART 5: Questions

5.1 Using the data shown in Table 24.1, construct a graph of LED current as a function of LED voltage drop for each of the LEDs. Plot each set of data on the same X-Y coordinate system. Attach a copy of your graph to your report. Label both vertical and horizontal axes on your graph. Label your graph with the title: <u>Graph 24.1: LED I-V curves</u>.

5.2 Referring to the plots of your data from Table 24.1, approximately at what LED voltage does the I-V curve for each of the LEDs start to break over and go vertical?

5.3 What pins on your seven-segment display would you ground to produce the number 5?

5.4 How did the addition of the base resistor affect the operation of the phototransistor in Figure 24.10? What do you think would happen if the value of R_B were made smaller?

5.5 Using the data in Table 24.3, construct a graph of phototransistor output voltage, V_{OUT}, as a function of distance from the lamp for each set of data. Remember from algebra that $Y = f(X)$. In this case, $V_{OUT} = f(D)$, where V_{OUT} is the voltage across emitter resistor R_E and D is the distance of the phototransistor from the lamp. Plot each set of data on the same X-Y coordinate system. Attach a copy of your graph to your report. Label both vertical and horizontal axes on your graph. Label your graph with the title: <u>Graph 24.2: Phototransistor output voltage as a function of distance from the lamp</u>.

5.6 Using the data in Table 24.4, construct a graph of phototransistor output voltage, V_{OUT}, as a function of lamp current. Remember from algebra that $Y = f(X)$. In this case, $V_{OUT} = f(I)$, where V_{OUT} is the voltage across emitter resistor R_E and I is the current through the lamp. Attach a copy of your graph to your report. Label both vertical and horizontal axes on your graph. Label your graph with the title: <u>Graph 24.3: Phototransistor output voltage as a function of lamp current</u>.

5.7 Using the data in Table 24.4, construct a graph of phototransistor output voltage, V_{OUT}, as a function of lamp voltage. Remember from algebra that $Y = f(X)$. In this case $V_{OUT} = f(E)$ where V_{OUT} is the voltage across emitter resistor R_E and E is the voltage applied to the lamp. Attach a copy of your graph to your report. Label both vertical and horizontal axes on your graph. Label your graph with the title: <u>Graph 24.4: Phototransistor output voltage as a function of lamp voltage</u>.

5.8 Using the data in Table 24.6, construct a graph of load voltage, V_{RL}, as a function of LED current, I_{RS}. Attach a copy of your graph to your report. Label both vertical and horizontal axes on your graph. Label your graph with the title: <u>Graph 24.5: Phototransistor load voltage as a function of LED current</u>.

5.9 Based on the data from Tables 24.3 and 24.4, does the collector-emitter resistance of the phototransistor increase or decrease with increasing levels of light? Referring to your data, explain how you arrived at your answer.

5.10 Which of your graphs, if any, seemed to depict a linear response?

5.11 Under what conditions will the phototransistor in Figure 24.7 operate in cutoff? Bright light or darkness?

5.12 Referring to the data in Table 24.6, what value of emitter current causes the transistor to just go into saturation?

PART 6: Troubleshooting

6.1 You are troubleshooting a faulty seven-segment display like the one shown in Figure 24.4. You have been told that one of the segments common to numbers 0, 2, 3, 5, 7, 8, and 9 will not light. You remove the seven-segment display from its socket and proceed to test it by constructing the circuit shown in Figure 24.6. You ground the cathode of segment A through a 330 Ohms resistor as shown. The segment does not light. You then measure the voltage across the 330 Ohms resistor. You measure +5.0 volts. Has the LED failed as an open or short? Explain how you arrived at your answer.

6.2 In the circuit shown in Figure 24.7, what voltage would you expect to measure across the phototransistor if it failed as a short? What voltage would you expect to measure across emitter resistor, R_E? Explain how you arrived at your answer.

6.3 In the circuit shown in Figure 24.7, what voltage would you expect to measure across the phototransistor if it failed as an open? What voltage would you expect to measure across the emitter resistor, R_E? Explain how you arrived at your answer.

6.4 If the resistance of the current-limiting resistor R_S in Figure 24.14 gradually increased over time, what effect would there be on the operation of the output stage?

6.5 Assume for the moment that you are troubleshooting the circuit shown in Figure 24.14. You measure the voltage across the load resistor and find it to be 0.0 volt. Name at least three different possible causes for this phenomenon.

PART 7: Circuit Design

7.1 For the circuit shown in Figure 24.15, select a value for R_B that will turn on the 2N3905 (or 2N3906) transistor, which will turn on a control relay that is used to operate a 120 VAC light bulb. If your instructor wishes, construct the circuit and demonstrate it. Show all calculations in the space provided below.

Calculations:

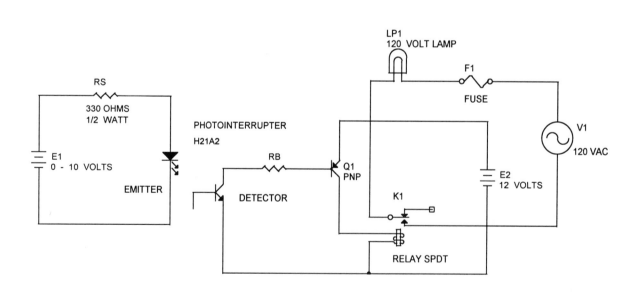

Figure 24.15: Schematic diagram for circuit design problem.

Light-Activated Thyristors: 25

INTRODUCTION:
In Experiments 19, 20, and 21, you studied the operational characteristics of two popular thyristors--the SCR and triac. In Experiment 24, you were introduced to a number of optoelectronic devices. In this experiment we will combine the two disciplines and study the characteristics of two more optoelectronic devices—the light-activated SCR (LASCR) and the light-activated triac (LAT). The two devices that you will work with in this experiment are also examples of photocouplers or optoisolators. These terms were also applied to the phototransistor studied in Experiment 24. Other terms applied to these devices include photoSCR, phototriac, or more generally, photothyristor. Just like the conventional SCR and triac studied earlier, the primary purpose of these two devices is to control or switch on and off the supply of current to some load. The LASCR and LAT act as switches; they are either on or off. The phototransistor has an active region in between the on and off states. The LASCR and LAT are low-power devices, switching on and off low-power loads. Applications may even include triggering the gate circuit of higher-power SCRs and triacs. The remaining difference between the LASCR and LAT and their high-power counterparts is that the gates of each device are composed of a light-sensitive semiconductor material that will cause the LASCR or LAT to be triggered into conduction when exposed to a light source rather than by a current source. As with the phototransistor, the light source is typically infrared (IRED). The light source and light-sensitive gate are typically enclosed in a plastic case. The package outlines for the LASCR and LAT that you will use in this experiment are shown in Figures 25.1 a) and b) respectively. Like the control relay, LASCRs and LATs are used where it is necessary to electrically isolate a low input or control voltage from the higher-voltage load that is being controlled.

SAFETY NOTE:
While both the LASCR and LAT can accept a wide range of input voltages (typically 5 to 15 volts), each one is still a low power device. It is important to select a current-limiting resistor that will provide enough current to the input IRED to produce light of sufficient amplitude to trigger the SCR and triac into conduction. But as with any LED, the forward-bias current must be limited to an amount that will not cause excessive power dissipation of the IRED. Likewise, the designer must ensure that the power dissipation of the output stage of each device is limited to the maximum amount recommended by the manufacturer. LEDs and IREDs have relatively low reverse breakdown voltages. It is therefore important to never inadvertently reverse bias the anode-cathode junction of either of these devices. Some designers will specify that a rectifier diode be placed in series with an LED or IRED to prevent accidental component failure. Another way to prevent LED or IRED failure due to reverse-biasing is to install a reverse-biased rectifier diode in parallel with the LED or IRED. Therefore, be sure to read and understand the manufacturer's data sheet for this device before using it in any future applications. Also, apply an appropriate factor of safety when specifying any electronic device. As always, wear safety glasses, and remove all jewelry from your hands and fingers prior to the start of the experiment.

OBJECT:
Upon successful completion of this experiment and all reading assignments, the student should be able to:
- construct an operational light-activated SCR circuit
- construct an operational light-activated triac circuit

REFERENCES:
Chapter 10 of Maloney's <u>Modern Industrial Electronics</u>

MATERIALS:
- 1 - 10 Kohms resistor, 1/4 watt
- 1 - 1.5 Kohms resistor, 1/2 watt
- miscellaneous lead wires and connectors
- 1 - optoisolator (light-activated SCR) H11C3 or SK4929
- 1 - 3.3 Kohms resistor, 1/4 watt
- 1 - 470 Ohms resistor, 1 watt
- 1 - solderless breadboard
- 1 - optoisolator (light-activated triac) MOC3021 or SK2048

EQUIPMENT:
- 1 - dual-outlet, adjustable D.C. power supply
- 1 - digital multimeter
- 1 - function generator
- 1 - oscilloscope
- 2 - 10X oscilloscope probes

PART 1: The Light-Activated SCR
Background:
The package outline and the pinouts for the SK2046 are shown in Figure 25.1 a). Note that there is an external gate connection even though this device is light activated. In this experiment, we will connect a resistor between the gate and cathode of the SCR to control the sensitivity of the gate. Make sure that you properly identify the anode and cathode of the IRED. As with any LED, the IRED is capable of withstanding only a small reverse voltage.

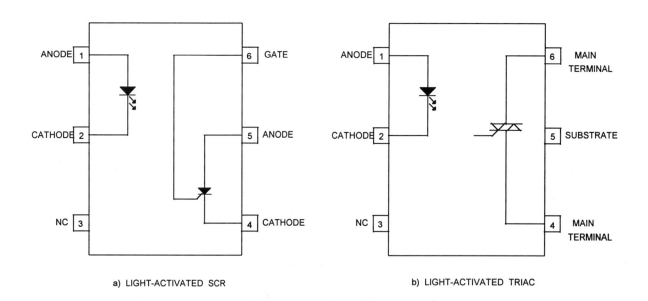

a) LIGHT-ACTIVATED SCR
b) LIGHT-ACTIVATED TRIAC

Figure 25.1: Package outlines and pin identification for a light-activated SCR and a light-activated triac.

Procedure:
1.1 Obtain all materials and equipment required to complete this experiment.
1.2 Use your multimeter to measure the resistance of each of your resistors. If any are out of tolerance, inform your instructor and obtain replacements. Record the measured value of the resistors in Table 25.1.

Table 25.1: Nominal and measured component values.

NOMINAL OR RATED COMPONENT VALUES	MEASURED COMPONENT VALUES
3300 OHMS	
470 OHMS	
10 KOHMS	
1.5 KOHMS	

1.3 Set one outlet of your D.C. power supply for 10 volts. This represents voltage source E2, which will be used to power the output stage of the SCR. Adjust the other outlet of your power supply for 0.0 volt. Turn off the power supply.

1.4 Construct the circuit shown in Figure 25.2. Figure 25.1 identifies the pins of the SK2046. If you are using a different optoisolator, consult the manufacturer's data manual to correctly identify the LED and SCR leads.

1.5 Please note that the input and output stages are isolated from each other. They do not have common grounds.

1.6 Turn on the power supply(ies).

1.7 Take a moment to review the circuit parameters specified in Table 25.2. Make a plan of action, then proceed to use your digital voltmeter to make the required measurements.

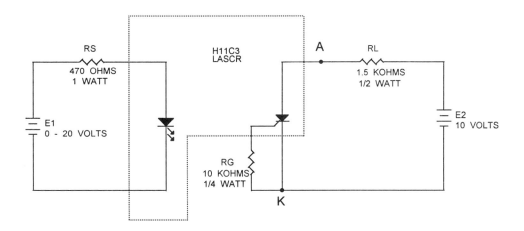

Figure 25.2: LASCR circuit with a D.C. source applied to the load.

1.8 With the SCR turned on, reduce E1 to 0.0 volt. What effect did this have on I_{RL}? Using your knowledge of SCRs acquired to date, explain this phenomenon.

___

___

___

Table 25.2: Circuit parameters for Figure 25.2 with a 10 Kohms gate resistor.

V_{RL} JUST BEFORE SCR TURN ON =	
I_{RL} JUST BEFORE SCR TURN ON = V_{RL}/R_L =	
V_{AK} JUST BEFORE SCR TURN ON =	
E_1 WITH THE SCR TURNED ON =	
V_{RS} WITH THE SCR TURNED ON =	
I_{RS} THAT JUST CAUSES SCR TURN ON = V_{RS}/R_S =	
V_{RL} WITH THE SCR TURNED ON =	
I_{RL} WITH THE SCR TURNED ON = V_{RL}/R_L =	
V_{AK} WITH THE SCR TURNED ON =	

1.9 How can you shut off the SCR in this type of circuit? Discuss at least two ways.

___

___

___

1.10 With E1 still set for 0.0 volt, slowly reduce E2 until the SCR turns off. Use your voltmeter to measure the voltage across R_L. Note the voltage across R_L just at the point the SCR turns off. Use this value to calculate the SCR's holding current, I_H.

$$I_H = V_{RL(\text{JUST BEFORE SCR TURN OFF})}/R_{L(\text{MEASURED})} =$$

1.11 Increase E2 to 10 volts. E1 should still be set for 0.0 volt. Turn off the power supply(ies).
1.12 Remove R_G from the circuit. The gate circuit should be left open for the next phase of the experiment. An open gate is equivalent to an $R_G = \infty$.
1.13 Take a moment to review the circuit parameters specified in Table 25.3. Make a plan of action, then proceed to use your digital voltmeter to make the required measurements.
1.14 Reduce E1 to 0.0 volt. Turn off the power supply(ies).

Table 25.3: Circuit parameters for Figure 25.2 with an open gate.

V_{RL} JUST BEFORE SCR TURN ON =	
I_{RL} JUST BEFORE SCR TURN ON = V_{RL}/R_L =	
V_{AK} JUST BEFORE SCR TURN ON =	
E_1 WITH THE SCR TURNED ON =	
V_{RS} WITH THE SCR TURNED ON =	
I_{RS} THAT JUST CAUSES SCR TURN ON = V_{RS}/R_S =	
V_{RL} WITH THE SCR TURNED ON =	
I_{RL} WITH THE SCR TURNED ON = V_{RL}/R_L =	
V_{AK} WITH THE SCR TURNED ON =	

1.15 Did the operational characteristics of the circuit change? If so, how? Explain, in detail, why these changes occurred.

_____

_____

_____

_____

_____

1.16 Install the 3.3 Kohms resistor in the gate circuit.
1.17 Take a moment to review the circuit parameters specified in Table 25.4. Make a plan of action, then proceed to use your digital voltmeter to make the required measurements.
1.18 Turn off the power supply(ies).
1.19 Based on your results (Tables 25.2, 25.3, 25.4), what is the relationship between the value of the gate resistance and the *sensitivity* of the LASCR's gate?

_____

_____

_____

Table 25.4: Circuit parameters for Figure 25.2 with a 3.3 Kohms gate resistor.

V_{RL} JUST BEFORE SCR TURN ON =	
I_{RL} JUST BEFORE SCR TURN ON = V_{RL}/R_L =	
V_{AK} JUST BEFORE SCR TURN ON =	
E_1 WITH THE SCR TURNED ON =	
V_{RS} WITH THE SCR TURNED ON =	
I_{RS} THAT JUST CAUSES SCR TURN ON = V_{RS}/R_S =	
V_{RL} WITH THE SCR TURNED ON =	
I_{RL} WITH THE SCR TURNED ON = V_{RL}/R_L =	
V_{AK} WITH THE SCR TURNED ON =	

1.20 Now replace E2 with a 12 volt p-p, 300 Hz sine wave as shown in Figure 25.3. Record the source waveform in the space provided on Graph 25.1.

1.21 With E1 set for +8 volts, use your scope probe to display the voltage across the SCR's anode-to-cathode, V_{AK}. With E1 set for +8 volts, the SCR should be fully turned on. If it is not, check your work. Your oscilloscope should be set for D.C. coupling. Record the V_{AK} waveform on Graph 25.2.

1.22 Next use the oscilloscope to display the voltage across the load, V_{RL}. Record the V_{RL} waveform on Graph 25.3.

1.23 Use a voltmeter to measure the D.C. voltage across the anode-cathode junction of the SCR, V_{AK}, and across the load resistor, V_{RL}. Record your measurements in Table 25.5.

1.24 Using the peak value of the load voltage waveform drawn in Graph 25.3, calculate the average value of the load waveform. Record your result in Table 25.5.

$$V_{RL} = V_{PEAK}/\pi =$$

1.25 Use the oscilloscope to monitor the voltage across the SCR. Slowly reduce the value of E1 until the SCR just fires--has a 90° firing delay angle. Record the SCR voltage waveform on Graph 25.4.

1.26 Use the oscilloscope to display the corresponding load voltage waveform, V_{RL}. Record the load voltage waveform on Graph 25.5.

1.27 Use a voltmeter to measure the voltage across the anode-cathode junction of the SCR, V_{AK}, and across the load resistor, V_{RL}. Record your measurements in Table 25.6.

1.28 Using the peak value of the load voltage waveform drawn in Graph 25.5, calculate the average value of the load waveform. Record your result in Table 25.6.

$$V_{RL} = V_{PEAK}/(2\pi) =$$

1.29 Turn off the power supplies.

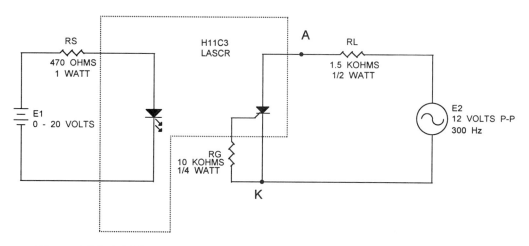

Figure 25.3: LASCR circuit with an A.C. source applied to the load.

Graph 25.1: Source voltage for the circuit shown in Figure 25.3.

E_2: _____ volts/div   _____ sec/div

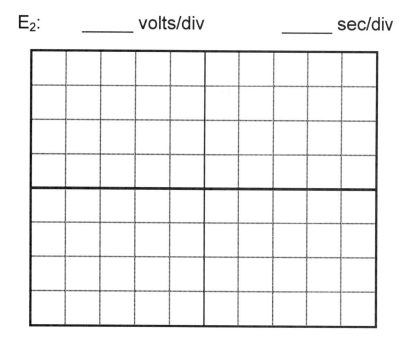

Graph 25.2: V_{AK} voltage waveform for the circuit shown in Figure 25.3 with the SCR fully turned on.

V_{AK}: _____ volts/div _____ sec/div

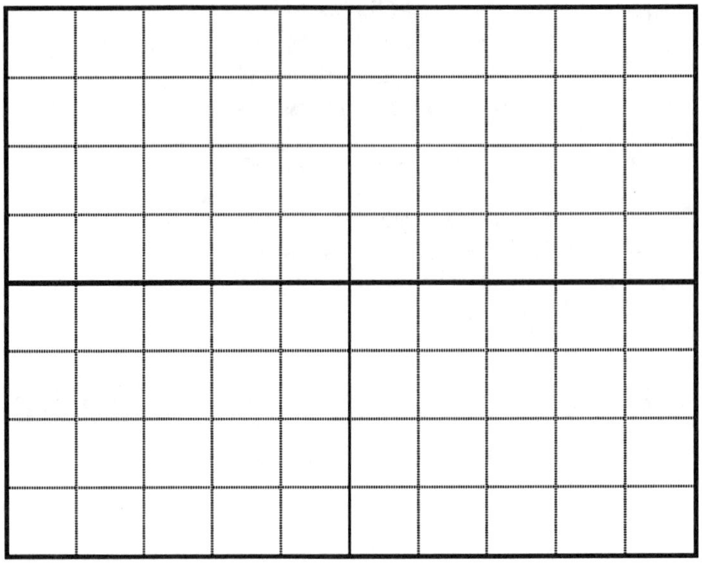

Graph 25.3: V_{RL} voltage waveform for the circuit shown in Figure 25.3 with the SCR fully turned on.

V_{RL}: _____ volts/div _____ sec/div

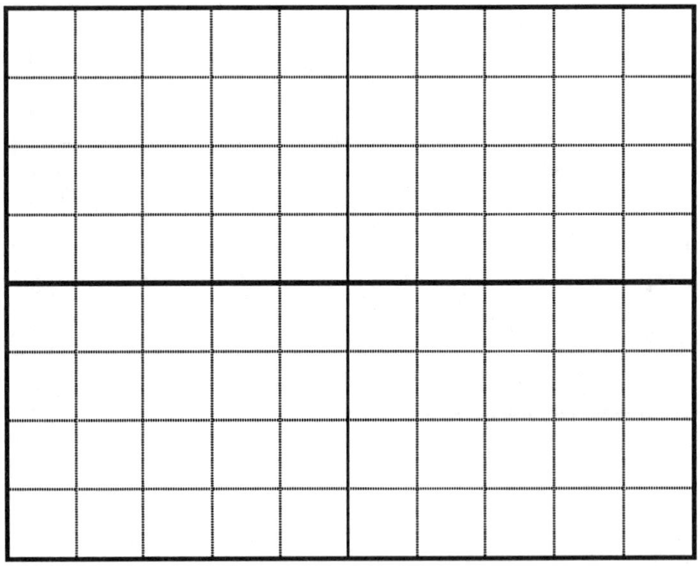

Table 25.5: Measured and calculated voltages (D.C. volts) for the output stage of the LASCR circuit shown in Figure 25.3 with the SCR fully turned on.

$V_{AK(MEASURED)}$	
$V_{RL(MEASURED)}$	
$V_{RL(CALCULATED)}$	

Graph 25.4: V_{AK} voltage waveform for the circuit shown in Figure 25.3 with the SCR just turned on.

V_{AK}: _____ volts/div _____ sec/div

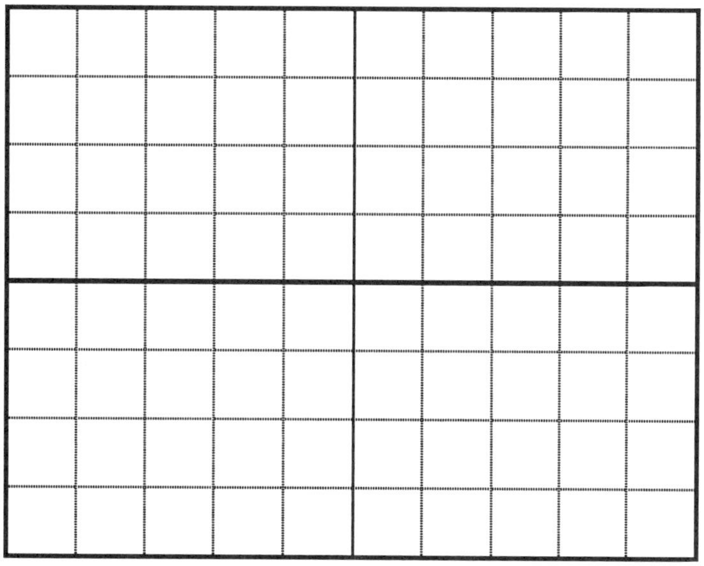

Table 25.6: Measured and calculated voltages (D.C. volts) for the output stage of the LASCR circuit shown in Figure 25.3 with the SCR just turned on.

$V_{AK(MEASURED)}$	
$V_{RL(MEASURED)}$	
$V_{RL(CALCULATED)}$	

Graph 25.5: V_{RL} voltage waveform for the circuit shown in Figure 25.3 with the SCR just turned on.

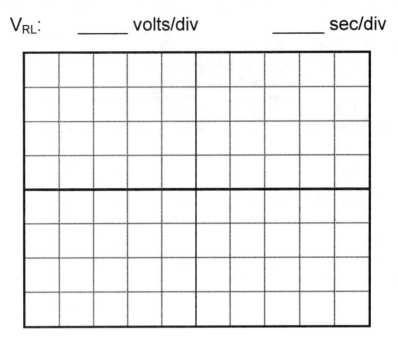

V_{RL}: _____ volts/div   _____ sec/div

PART 2: The Light-Activated TRIAC
Background:
The package outline and the pinouts for the SK2048 are shown in Figure 25.1 b). Note that this particular LAT does not have an external gate connection. As with the LASCR, make sure that you properly identify the anode and cathode of the IRED.

Procedure:
    2.1 Construct the circuit shown in Figure 25.4. E1 should be set initially for 0.0 volt. E2 should be a 12 volts p-p, 300 Hz sine wave.
    2.2 Use the oscilloscope to monitor the voltage across the triac. With both power supplies turned on, slowly increase the value of E1 until the triac is fully turned on.
    2.3 Display the corresponding waveform across the load. Record the load waveform on Graph 25.6.
    2.4 Use an A.C. voltmeter to measure the true RMS voltage across both the load and the triac. Record in Table 25.7.
    2.5 Slowly decrease the value of E1 until the firing delay angle for the triac is 90°. Record the resulting triac waveform on Graph 25.7.
    2.6 Record the corresponding load voltage waveform on Graph 25.8.
    2.7 Use an A.C. voltmeter to measure the true RMS voltage across both the load and the triac. Record in Table 25.8.
    2.8 Turn off the power supplies.

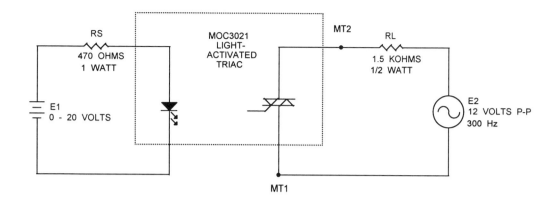

Figure 25.4: Light-activated triac circuit.

Graph 25.6: V_{RL} voltage waveform for the circuit shown in Figure 25.4 with the triac fully turned on.

V_{RL}: _____ volts/div _____ sec/div

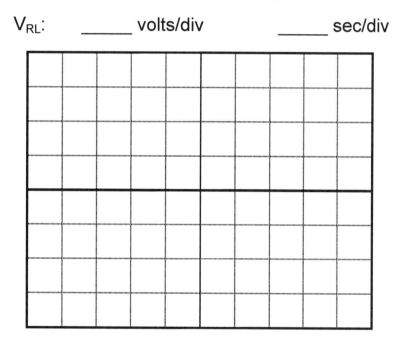

Table 25.7: Measured load and triac voltages (RMS units) for the output stage of the light-activated triac circuit shown in Figure 25.4 with the triac fully turned on.

$V_{TRIAC(MEASURED)}$	
$V_{LOAD(MEASURED)}$	

Graph 25.7: $V_{MT2-MT1}$ voltage waveform for the circuit shown in Figure 25.4 with a 90° firing delay angle.

$V_{MT2-MT1}$: _____ volts/div   _____ sec/div

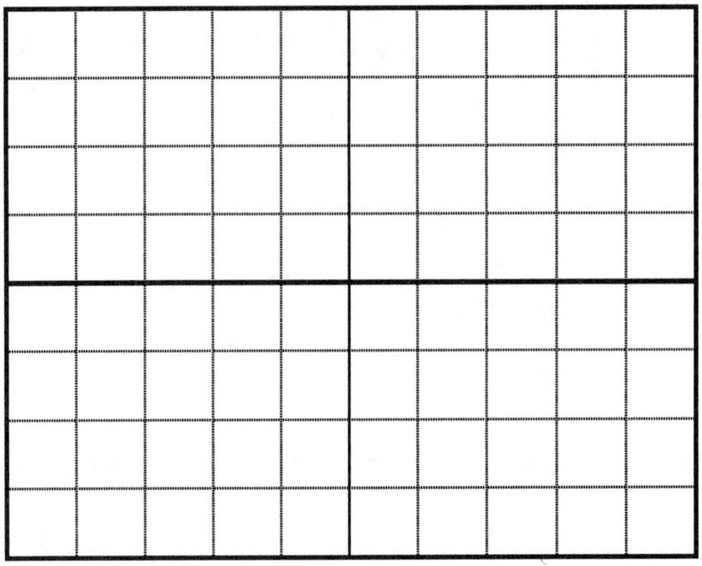

Graph 25.8: V_{RL} voltage waveform for the circuit shown in Figure 25.4 with a 90° firing delay angle.

V_{RL}: _____ volts/div   _____ sec/div

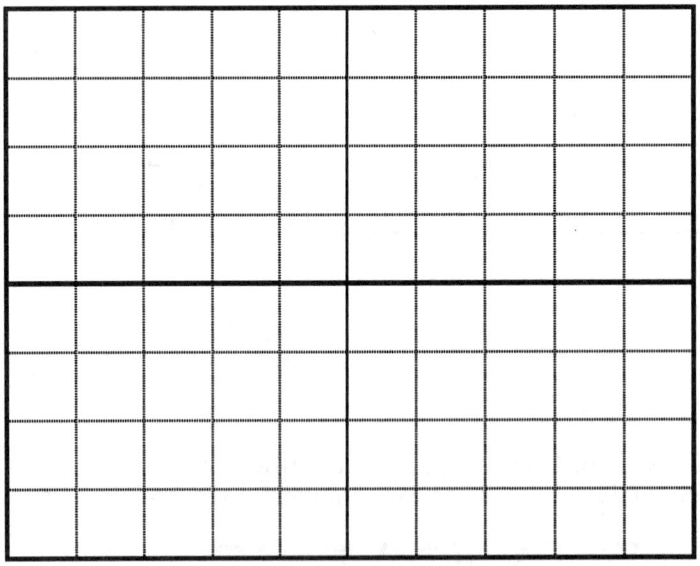

Table 25.8: Measured load and triac voltages (RMS units) for the output stage of the light-activated triac circuit shown in Figure 25.4 with a 90° firing delay angle.

V_{TRIAC(MEASURED)}	
V_{LOAD(MEASURED)}	

PART 3: Questions

3.1 Referring to the data in Table 25.5, what is the relationship between the measured values for V_{RL} and V_{AK}? Explain this relationship using Kirchhoff's Voltage Law.

3.2 Based on your experimental data, what effect does increasing LED current have on the D.C. voltage <u>across the load resistor</u> in the LASCR circuit?

3.3 If you completed Experiments 18, 19, and 20, compare the operation of the optoisolators (LASCR and LAT) to the operation of a conventional SCR and triac.

3.4 Based on your experimental data, what effect does increasing LED current have on the RMS voltage <u>across the triac</u> in the light-activated triac circuit?

3.5 If you completed Experiment 24, compare the operation of the LASCR and LAT to a phototransistor.

PART 4: Troubleshooting

4.1 If the LED in Figure 25.2 failed as an open, what voltage would you expect to measure across the 470 Ohms current-limiting resistor, R_S? What voltage would you expect to measure across the SCR?

4.2 In the circuit shown in Figure 25.2, what voltage would you expect to measure across the SCR if the SCR had failed as a short? What voltage would you expect to measure across the load resistor, R_L?

4.3 In the circuit shown in Figure 25.4, what voltage would you expect to measure across the triac if the triac had failed as an open? What voltage would you expect to measure across the load resistor, R_L?

4.4 Assume for the moment that you are troubleshooting the circuit shown in Figure 25.3. You measure the voltage across the load resistor and find it to be 0.0 volts. Name at least three different possible causes for this phenomenon.

PART 5: Circuit Design

5.1 For the circuit shown in Figure 25.5, select a value for R_B that will turn on the 2N3905 (or 2N3906) transistor, which will turn on a control relay that is used to operate a 120 VAC light bulb. If your instructor wishes, construct the circuit and demonstrate it. Show all calculations in the space provided below.

Calculations:

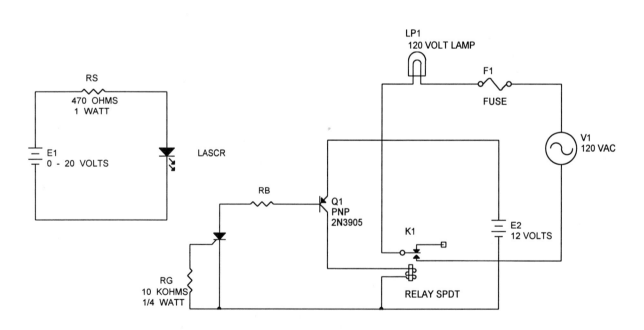

Figure 25.5: Schematic diagram for circuit design problem.

The Hall-Effect Switch: 26

INTRODUCTION:

In Experiment 1, you were introduced to a number of switching devices that are mechanically operated. In Experiment 3, you were introduced to the control relay, which is a type of switch that is operated by an electromagnetic field. In Experiments 24 and 25, you were introduced to optoelectronic devices that are capable of switching on and off current to a load by blocking or unblocking a light source, usually an IRED. The switching devices in Experiments 24 and 25 required no physical contact to cause the output to change states. Switches based on this type of non-contact operation are also referred to as *proximity* switches. This name was chosen because the input or stimulus that causes this type of switch to change states does not touch the sensor or switching device; rather, it comes in close proximity to it. In this experiment, you will study another type of proximity sensor--the Hall-effect switch. Hall-effect switches are based upon a phenomenon observed by physicist E. H. Hall in the late 19th century. The Hall effect principle is illustrated in Figure 26.1. In this figure, a current is passed through a semiconductor from an external power supply. If a magnetic field is brought into close proximity to the semiconductor material, the charges passing through the semiconductor are effectively reoriented or redistributed such that a small voltage (potential difference) appears across the semiconductor transverse or perpendicular to the flow of current through the semiconductor. The magnitude of the Hall voltage, E_H, is directly proportional to the strength of the magnetic field. In other words, the greater the flux density, Φ, the greater the value of E_H. If the Hall voltage is then applied to a Schmitt-triggered op-amp circuit similar to the ones studied in Experiment 17, the result will be an amplified square-wave output voltage. Hall-effect switches can be used in applications where it is not possible for the object or equipment being controlled to come into contact with a limit switch or other type of mechanically-operated switch. Proximity switches can be used as tachometers, in counting operations, and in other presence-sensing applications.

SAFETY NOTE:

The Hall-effect switch is a low-power device. Familiarize yourself with the manufacturer's ratings for your Hall-effect switch before constructing any circuits. As always, wear safety glasses, and remove all jewelry from your hands and fingers before the start of the experiment.

OBJECT:

Upon successful completion of this experiment and all reading assignments, the student should be able to:
- construct an operational Hall-effect switch circuit
- experimentally determine if a Hall-effect sensor is unipolar or bipolar

REFERENCES:

Chapter 10 of Maloney's <u>Modern Industrial Electronics</u>

MATERIALS:

- 1 - Hall-effect switch such as Micro Switch's® 8SS7 or 65SS4 or the equivalent
- 1 - small bar magnet (approximately 0.25" in diameter and no more than 0.5" in length)
- 1 - solderless breadboard
  - miscellaneous resistors
  - miscellaneous lead wires and connectors

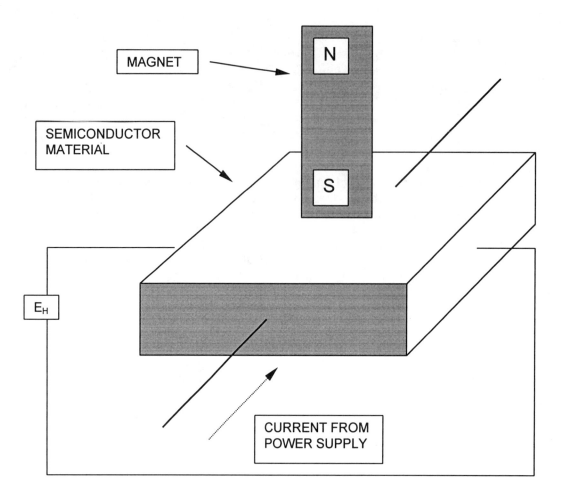

Figure 26.1: Model demonstrating the Hall-effect principle in a semiconductor. An external power supply provides a source of current through the semiconductor material. The presence of the external magnetic field "deflects" an excess of electrons to one side of the semiconductor material resulting in a potential difference, E_H.

EQUIPMENT:
    1 - adjustable D.C. power supply
    1 - digital multimeter
    1 - oscilloscope
    1 - oscilloscope probe
    1 - variable-speed, handheld electric drill

INSTRUCTOR/STUDENT NOTE: It is not necessary to use either of the Hall-effect switches included in the preceding parts list. Any similar device will be sufficient. However, choose the externally-connected pull-up resistor so that the maximum sink-current rating of your particular Hall-effect switch will not be exceeded.

PART 1: The Hall-Effect Switch
Background:
Hall-effect switches are available in a variety of configurations. Some are *unipolar* and some are *bipolar*. Unipolar Hall-effect sensors respond only to one magnetic pole of a magnet, either the north or the south, but not both. Bipolar Hall-effect sensors will respond to the presence of either the north or the south pole of a magnet. Some Hall-effect sensors are designed to sink current and others source it. Most Hall sensors have in common an input terminal (to which the positive lead of the supply voltage is applied), a ground terminal, and an output terminal. Some Hall sensors have a single output terminal and some multiple outputs. Figure 26.2 shows how Micro Switch's® Hall-effect sensor 8SS7 can be configured to produce a high output when no magnetic field is present and a low output when a magnetic field is present. The sensor has a small, circular, raised surface that is the target for the externally applied magnetic field. This particular sensor is rated for a supply voltage in the range of 4½ to 5½ volts and can sink up to 20 mA. Therefore, when the output voltage, V_{OUT}, of the circuit shown goes low, the sensor will sink at most 5V/470Ω = 10.6 mA which is well within the limits specified by the manufacturer. Figure 26.3 shows a similar circuit in which Micro Switch's® 65SS4 sensor is being used as a switch. This particular sensor is produced in the DIP configuration. Note that it has two outputs, pins 2 and 3. This sensor has the same supply voltage rating as the 8SS7 but can sink only 4 mA per output.

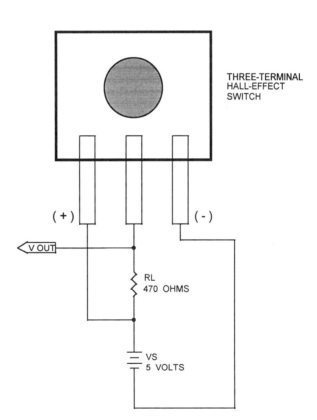

Figure 26.2: Hall-effect switch with single output.

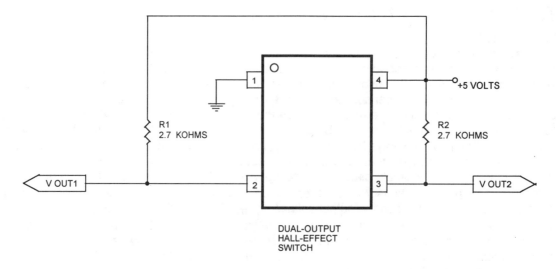

Figure 26.3: Hall-effect switch with dual outputs in a DIP configuration.

Procedure:

1.1 Obtain all materials and equipment required to complete this experiment. Adjust your power supply to provide the output voltage specified for your Hall-effect sensor. Turn off your power supply.

1.2 Based on the supply voltage and amount of current that your sensor can sink, use Ohm's Law to determine the value of the current-limiting, pull-up resistor.

1.3 Construct a circuit like that shown in either Figure 26.2 or 26.3. Turn on your power supply.

1.4 Use a digital voltmeter to measure the voltage V_{OUT} relative to ground. Record this value in Table 26.1.

1.5 Place the north pole of your magnet in close proximity to the sensor. Again measure the value for V_{OUT} and record in Table 26.1.

1.6 Now place the south pole of your magnet in close proximity to the sensor. Again measure the value for V_{OUT} and record in Table 26.1.

1.7 Hold your magnet about 1" from the sensor and slowly move the magnet directly toward it. Monitor V_{OUT}. Determine the distance between the magnet and the sensor at which the sensor output switches from high to low. Measure this distance in millimeters, and record your measurement in the space immediately following.

MAGNET-TO-SENSOR SWITCHING DISTANCE:_____ mm

Table 26.1: V_{OUT} for various operating conditions for the Hall-effect switch circuit.

CIRCUIT OPERATING CONDITION	V_{OUT}, VOLTS
NO MAGNETIC FIELD PRESENT	
NORTH POLE APPLIED TO SENSOR	
SOUTH POLE APPLIED TO SENSOR	

PART 2: The Hall-Effect Tachometer

Background:
As mentioned previously, Hall-effect sensors can be used in tachometer circuits. In this phase of your experiment, you will use your Hall-effect switch circuit and your oscilloscope to design a tachometer circuit.

Procedure:
2.1 For safety reasons, your instructor should directly supervise all of the remaining steps of this procedure. You must be wearing safety glasses!

2.2 Obtain a round wooden block 3" to 4" in diameter and about 0.75" thick. Drill four small holes in it to mount your magnets as shown in Figure 26.4. The diameter of the holes should be slightly smaller than the diameter of the magnets so that an interference fit will result. The fit of the magnets in the holes must be sufficiently tight so that they do not fly out when the wooden disk is rotated at high speed. Drill a hole in the center of the wooden block so that it will accept a 1/4" X 28 or 1/4" X 20 machine screw. Insert the machine screw through the hole, placing washers on both sides of the wooden disk and a lock nut as shown in Figure 26.4. The machine screw should protrude about 1" past the lock nut.

2.3 Insert the machine screw into the chuck of an electric hand drill. Securely tighten the chuck.

2.4 Use your oscilloscope to monitor the output voltage of the Hall-effect circuit.

2.5 Place the disk-drill assembly in close proximity to the Hall-effect sensor. <u>Slowly</u> increase the speed of the drill. If you have precisely prepared your wooden disk, you should see a square wave output on the oscilloscope. If not, call your instructor for assistance. Once your circuit is operating correctly, plot the resulting waveform in the space provided on Graph 26.1.

2.6 Vary the speed of the drill, noting the effect on the frequency of the voltage waveform.

2.7 Turn off the electric drill and the circuit's power supply.

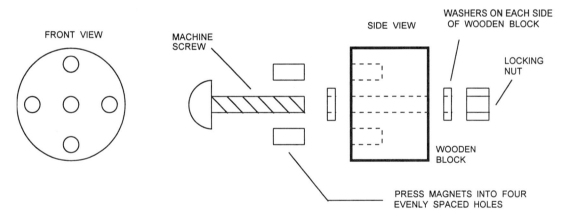

Figure 26.4: Magnet-block assembly for tachometer circuit.

2.8 Using the period of the voltage waveform drawn on Graph 26.1 and the fact that your disk has four magnets mounted on it, determine the drill's RPM. Show your calculations in the space provided below.

Calculations:

DRILL SPEED:_____ **RPM**

Graph 26.1: V_{OUT} for the Hall-effect tachometer circuit.

V_{OUT}: _____ volts/div   _____ sec/div

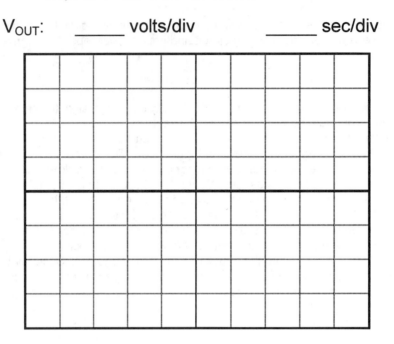

PART 3: Questions

3.1 Based on you experimental results, is your Hall-effect sensor bipolar or unipolar? Explain how you arrived at your answer.

3.2 Why didn't the voltage waveform drawn on Graph 26.1 have a 50% duty cycle? In other words, why didn't the time low equal the time high?

PART 4: Circuit Design

4.1 Modify your Hall-effect circuit so that the output current will be amplified by a transistor that will turn on and off a control relay that has a coil rated for 12 volts D.C. and a nominal resistance of 120 Ohms. Include a 120 VAC light bulb as a load. Draw and completely label a schematic diagram of this circuit showing all components in their normal or deactivated state. If a computer-aided drafting software package is available at your institution, use it to produce your drawing. Include your schematic with your report.

INSTRUCTOR NOTE: If time permits, have your students build and demonstrate this circuit. As an option, replace the control relay with a solid-state relay.

Thermistors: 27

INTRODUCTION:
When it is necessary to monitor and/or regulate the temperature of an industrial process, several different transducers are available to sense temperature change. These include thermistors, thermocouples, resistive temperature devices (RTDs), and integrated circuit (I.C.) sensors. The choice, of course, depends primarily on the application, as each device exhibits different operational characteristics. For example, the RTD is very linear in its response. The thermistor, as you will see in this lab, is very nonlinear. Most thermistors are constructed from semiconductor material. As you may recall from your earlier studies in an introductory electronic devices class, most semiconductor materials become better conductors as they are heated, because more electrons are available as conductors due to the increased thermal energy. Thermistors fall into two major categories--negative temperature coefficient (NTC) and positive temperature coefficient (PTC). As noted, thermistors are very nonlinear. Specifically, the thermistor's resistance as a function of applied temperature can be modeled mathematically as a series expansion of the natural logarithm. Because electronic technicians do not need to be exposed to this level of mathematical rigor, we will take a more practical approach and experimentally generate calibration data for a number of different thermistors. It is this type of data that will prove useful in future engineering applications. The calibration data that you will be generating today is usually available from the manufacturer upon request. At the minimum, most manufacturers publish the thermistor's type (NTC or PTC) and the resistance at 25°C. In this lab, you will generate calibration data for an NTC and a PTC thermistor by heating each of them over a range of temperatures and recording the resulting thermistor resistance. You will then plot this data, showing the resistance as a function of temperature. The resulting curves should look something like those shown in Figures 27.1 and 27.2. Figure 27.1 shows the general response of an NTC thermistor. Figure 27.2 shows the response of a particular PTC thermistor.

SAFETY NOTE:
In this experiment, you will be using some type of high temperature heat source such as an electric hot plate or a hot-air gun. If you use a hot air gun to heat your thermistor, do not bring the hot-air gun too close to the thermistor as a high temperature hot-air gun can melt the solder holding the leads to a thermistor and melt the plastic of a solderless breadboard. This observation is based on personal experience! In either case, take appropriate measures to avoid exposure of the skin to any heated surfaces. Allow all components to cool before cleaning and storing them. As always, wear safety glasses, and remove all jewelry from your hands and fingers before the start of the experiment.

OBJECT:
Upon successful completion of this experiment and all reading assignments, the student should be able to:
- generate and plot calibration data for an NTC and a PTC thermistor
- design an on/off temperature control circuit using a thermistor as the sensor

REFERENCES:
Chapter 10 of Maloney's Modern Industrial Electronics
Omega Engineering's® Temperature Handbook
www.omega.com

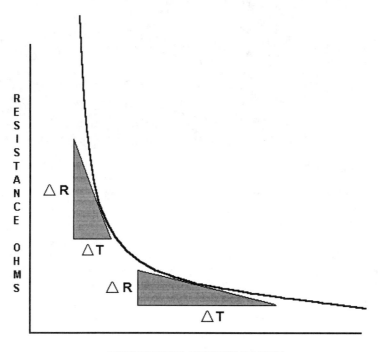

Figure 27.1: Generalizable plot of resistance as a function of temperature for an NTC thermistor. Tangent lines have been drawn at two different points to show different sensitivities or different rates of change of resistance with temperature.

MATERIALS:
  2 - NTC thermistors
  1 - PTC thermistor
  1 - test tube
  1 - 1000 mL or 2000 mL beaker
   - miscellaneous lead wires and connectors

EQUIPMENT:
  1 - digital ohmmeter
  1 - electric hot plate
  1 - thermometer or electronic temperature probe

PART 1: Thermistor Calibration Data
Background:
Figure 27.3 shows two typical schematic symbols used to represent the thermistor. While some engineering applications require the linearity of an RTD or I.C. sensor, the *sensitivity* of the thermistor is one of its advantages. If operated within a certain temperature range, the thermistor's change in resistance for a given change in input temperature is very great. For example, consider the NTC curve shown in Figure 27.1. A line has been drawn tangent to two possible operating points on the curve. Recall from your basic algebra course that the slope of a line is $\Delta Y/\Delta X$, or in this case $\Delta R/\Delta T$. Take a moment to compare the steepness of each curve. If operated at temperatures near the leftmost side of the graph, the change in resistance, ΔR, is very large for small changes in temperature, ΔT. Whereas, if the operating point on the thermistor curve is chosen near the rightmost side of the graph, the ΔR is very

small even for a large ΔT. If we wanted to take advantage of the thermistor's sensitivity, we would want to select a thermistor such that the desired operating temperature would be at a point on the curve where its tangent would be as steep as possible. Because the process of calibration involves applying a known input to a transducer then observing or measuring the transducer's response, you will need a heat source whose temperature can be controlled and a way to monitor the temperature of the heat source. Also, the response characteristic of the measuring device must be able to respond as quickly or, better yet, much faster than the device being measured. Because we can expect the thermistor's resistance to change rapidly with a small change in temperature, we should expose the thermistor to a heat source that will change slowly. One way to accomplish this is to place the thermistor in some type of enclosure of large mass and slowly heat the mass of the object. An object of large mass will have a slow thermal response. A temperature probe should then be installed immediately next to the thermistor to determine its temperature. A precision thermometer or electronic temperature probe would be an appropriate choice for monitoring the temperature of the thermistor. One way to accomplish this in a college laboratory setting is to place the thermometer and thermistor in a test tube, and in turn, place the test tube in a large beaker containing water as shown in Figure 27.4. The test tube acts as a *thermowell* insulating the thermistor both thermally and electrically from the water. The water serves as a sufficiently large mass whose temperature will change slowly. The water can then be heated with an electric hot plate (readily available in most chemistry labs). The resistance of the thermistor can be measured with a digital ohmmeter. If these resources are not available, you can then use a hot-air gun (readily available in most electronic labs) as a source of heat, but you must find a way to heat the thermistor very slowly.

INSTRUCTOR'S NOTE: Because the students will be obtaining calibration data for more than one thermistor, I suggest that you assign one thermistor per group then have the groups share their data upon completion of the lab. This will result in efficient use of lab time. Also, the students can alternate between collecting data and working on Part 4 of this lab. Lastly, make sure that the thermistors chosen for this experiment are capable of withstanding a temperature range of approximately 0°C to 100°C.

Procedure:

1.1 Obtain all materials and equipment required to complete this experiment.
1.2 Attach test leads from your digital ohmmeter to the NTC thermistor. Place the thermistor in a test tube. Insert a thermometer or temperature probe in the test tube next to the thermistor. Immerse the test tube in an ice bath as shown in Figure 27.4. Allow several minutes for the temperature of the test tube to stabilize.
1.3 Turn on the electric hot plate.
1.4 Take resistance measurements for each of the temperature values shown in Table 27.1.
1.5 Repeat steps 2 through 4 for a second NTC thermistor. Record your resistance values in Table 27.1.
1.6 Repeat steps 2 through 4 for a PTC thermistor. Record your resistance values in Table 27.1.

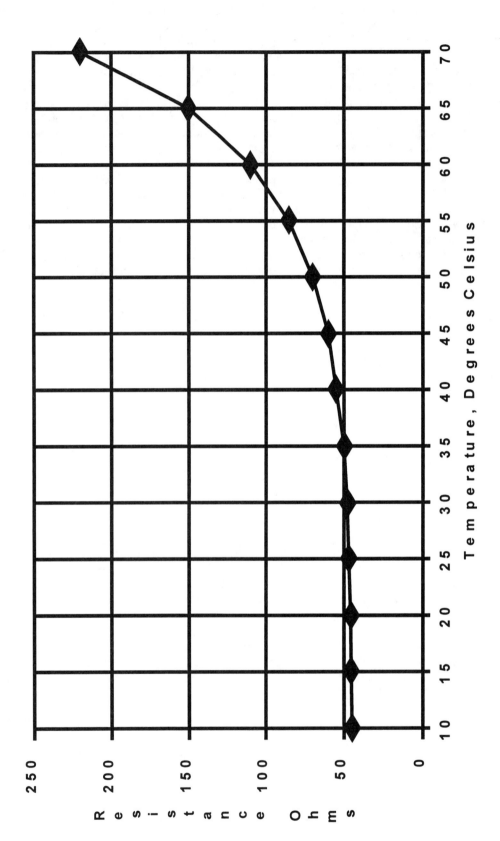

Figure 27.2: Plot of calibration data for a PTC thermistor.

Figure 27.3: Typical schematic symbols for a thermistor.

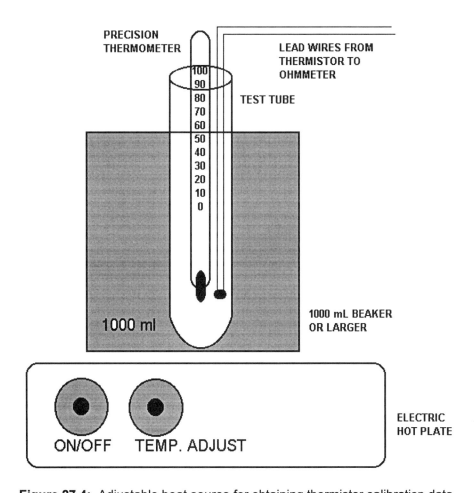

Figure 27.4: Adjustable heat source for obtaining thermistor calibration data.

Table 27.1: Thermistor resistance values for different temperatures.

TEMPERATURE, DEGREES CELSIUS	THERMISTOR #1 RESISTANCE, OHMS	THERMISTOR #2 RESISTANCE, OHMS	THERMISTOR #3 RESISTANCE, OHMS
10			
15			
20			
25			
30			
35			
40			
45			
50			
55			
60			
65			
70			

PART 2: Questions

2.1 Make a graph of resistance as a function of temperature for each set of data in Table 27.1. Try to use the same scale for each graph so that you can visually compare the different response of each thermistor. Attach your graphs to your report.

2.2 Which thermistor type, NTC or PTC, exhibited the greatest change in resistance over the entire range of temperatures? What are the implications of this difference when choosing a thermistor for a particular application?

2.3 Referring to your graph for thermistor #1, what is its resistance at a temperature of 52°C? Show your work on your graph. Write your answer in the space provided below.

2.4 Referring to your graph for thermistor #1, determine the slope of a line drawn tangent to the curve at 20°C. Show your work on the graph, and perform your calculations in the space provided below.

2.5 Referring to your graph for thermistor #1, determine the slope of a line drawn tangent to the curve at 60°C. Show your work on the graph, and perform your calculations in the space provided below.

2.6 At which point on the graph, 20°C or 60°C, is thermistor #1 the most sensitive? Explain your answer while referring to the slope of the tangent lines to these two points.

PART 3: Practice Problems

3.1 Figure 27.5 is a simple voltage divider network with a PTC thermistor whose resistance-temperature data is shown in Figure 27.2. What is the output voltage, V_{OUT}, at a temperature of 50°C?

Figure 27.5: Voltage-divider circuit. The value of the output voltage is dependent upon the temperature of the PTC thermistor.

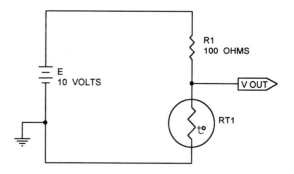

3.2 Figure 27.6 is a simple voltage divider network with an NTC thermistor whose resistance-temperature data is shown in Figure 27.7. What is the output voltage, V_{OUT}, at a temperature of 40°C?

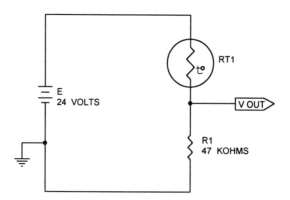

Figure 27.6: Voltage-divider circuit. The value of the output voltage is dependent upon the temperature of the NTC thermistor.

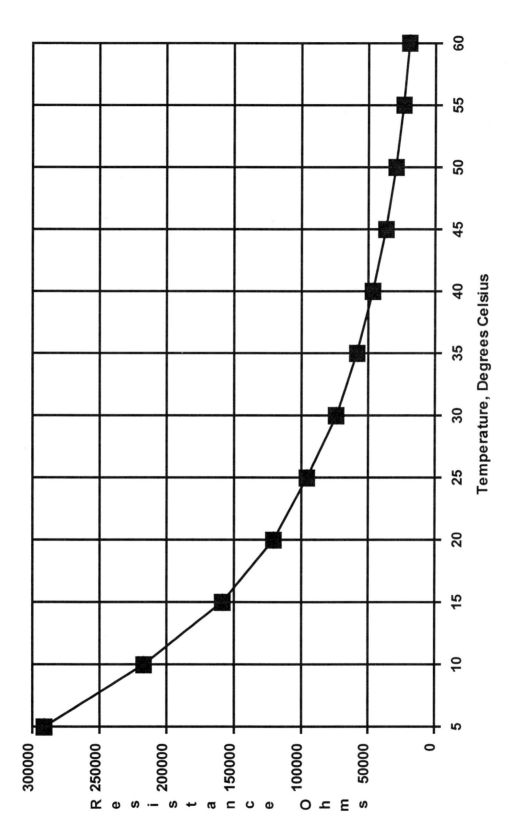

Figure 27.7: Plot of calibration data for an NTC thermistor.

3.3 If the temperature of the thermistor referred to in Practice Problem 3.1 decreases, what will happen to the output voltage, V_{OUT}? Explain how you arrived at your answer by making reference to the thermistor's resistance-temperature data and to the basic laws of D.C. circuit analysis.

3.4 If the temperature of the thermistor referred to in Practice Problem 3.2 increases, what will happen to the output voltage, V_{OUT}? Explain how you arrived at your answer by making reference to the thermistor's resistance-temperature data and to the basic laws of D.C. circuit analysis.

3.5 Figure 27.8 is a Wheatstone bridge with a thermistor whose resistance-temperature data is shown in Figure 27.2. What is the value of the differential output voltage, V_{DIFF}, at a temperature of 40°C?

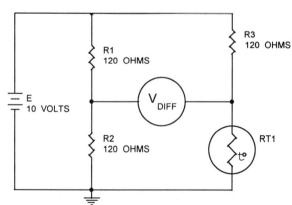

Figure 27.8: Wheatstone bridge circuit. The value of the differential voltage is dependent upon the temperature of the thermistor.

PART 4: Circuit Design

4.1 Using the data from any of the NTC thermistors that you worked with today, design an op-amp based comparator circuit like the one shown in Figure 27.9 that will switch from a positive output voltage, V_{OUT}, to a negative output voltage as the temperature reaches, then exceeds 50°C. At this point, the transistor should go into cutoff, de-energizing the relay coil, turning off the heating element. Show all of your calculations in the space provided below. If time permits, build this circuit and demonstrate it to your instructor.

Calculations:

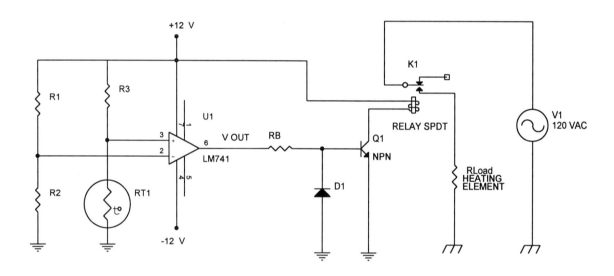

Figure 27.9: Op-amp comparator circuit connected to a Wheatstone bridge circuit. The output of the op-amp will switch from a positive output voltage to a negative output voltage when the temperature of the thermistor exceeds a certain set point.

I.C. Temperature Sensors: 28

INTRODUCTION:

While the thermistor offers the advantages of high sensitivity and fast response, its major drawback is that it is the most nonlinear of the common temperature sensors (RTD, thermocouple, I.C. sensor, and thermistor). For temperature-sensing applications requiring a high degree of linearity, the temperature transducer that best meets this requirement is the I.C. sensor. Another advantage of the I.C. sensor is its relatively low cost. Its major drawback is that it is limited to temperatures of less than 200°C. The thermocouple is much more durable than the I.C. sensor and is capable of operating over a wide range of temperatures. I.C. sensors are sometimes used in conjunction with thermocouples for cold junction compensation. Refer to National Semiconductor's® Data Acquisition Databook for example circuits. You should also consult this or a similar reference for the manufacturer's recommended mounting techniques because the accuracy of the I.C. sensor (as with any transducer) is highly dependent on the quality of the sensor's installation. The operation of the I.C. sensor is based on the characteristic that most semiconductors will produce more thermally-generated electron-hole pairs at elevated temperatures and fewer at lower temperatures. In other words, an I.C. sensor will become a better conductor at higher temperatures. With the addition of an external voltage source and discrete components, the I.C. sensor can be configured to produce an output voltage. This voltage can then be amplified (if necessary) and used as an input to an industrial control circuit or applied to an A/D converter that could drive a digital temperature display.

SAFETY NOTE:

In this experiment, you will be using some type of high-temperature heat source such as an electric hot plate or a hot-air gun. If heating your I.C. sensor with a hot-air gun, do not bring the hot-air gun too close to the sensor as the temperature near the outlet of a hot-air gun can rise to a very high level in just seconds. Take appropriate measures to avoid exposure of the skin to any heated surfaces. Allow all components to cool before handling and storing them. As always, wear safety glasses, and remove all jewelry from your hands and fingers before the start of the experiment.

OBJECT:

Upon successful completion of this experiment and all reading assignments, the student should be able to:
- generate and plot calibration data for an I.C. temperature sensor

REFERENCES:

Chapter 10 of Maloney's Modern Industrial Electronics
Omega Engineering's® Temperature Handbook
National Semiconductor's® Data Acquisition Databook
www.national.com

MATERIALS:

- 1 - temperature sensor, LM35CZ or the equivalent
- 1 - temperature sensor, LM335Z or the equivalent
- 1 - 10 Kohms, ten-turn potentiometer
- 1 - 1000 mL or 2000 mL beaker
- 1 - 15 Kohms, 1/4 watt resistor
- 1 - 12 Kohms, 1/4 watt resistor
- 2 - 1N914 diodes
- 1 - test tube

- miscellaneous lead wires and connectors   1 - solderless breadboard

EQUIPMENT:
1 - digital voltmeter calibrated for millivolts
1 - electric hot plate
1 - thermometer or electronic temperature probe

PART 1: I.C. Sensor Calibration Data
Background:
Figure 28.1 shows the bottom view and identifies the terminals of the LM35CZ I.C. temperature sensor. In addition to the plastic package depicted in Figure 28.1, this particular I.C. is also available in surface mount and metal can packages. The voltage output of this I.C. is linearly proportional to temperature in degrees Celsius. The manufacturer specifies the scale factor or sensor gain of this device to be 10.0 mV/°C. While thermocouples are active transducers capable of producing a small output voltage when heated, the I.C. sensor is a passive transducer like the thermistor. Hence, it requires an external power supply. The supply voltage is applied to the terminal labeled V_S. The GND pin is either connected to ground or a point in the temperature circuit at some reference voltage lower than that of the supply voltage. The remaining terminal is V_{OUT}. As shown in Figure 28.2, the V_{OUT} terminal is often connected to a resistor to establish an output voltage relative to the GND pin. Your second circuit will be constructed using the LM335 temperature sensor. Figure 28.4 shows the package outline and identifies the terminals of the LM335M surface mount package. Note the schematic symbol of the zener diode shown in the center of the package. As you may recall from your first electronic devices class, zener diodes are designed to operate in reverse breakdown. The breakdown voltage of this temperature sensor is directly proportional to temperature in degrees Kelvin. The scale factor for this particular device is 10 mV/°K. The second circuit you will construct is shown in Figure 28.5.

INSTRUCTOR'S NOTE: Because the students will be obtaining calibration data for more than one I.C. sensor, I suggest that you assign one I.C. sensor per group, then have the groups share their data upon completion of the lab. This will result in efficient use of lab time. Lastly, make sure that the students are warned not to increase the temperature of the I.C. sensors beyond 100°C. Temperatures in excess of 100°C will permanently damage both of these I.Cs.

Figure 28.1: Bottom view and pin identification of the LM35CZ Celsius (Centigrade) temperature sensor.

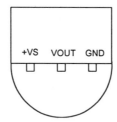

Procedure:
1.1 Obtain all materials and equipment required to complete this experiment. Make sure that you have properly identified the leads of the LM35CZ (or equivalent) I.C. before wiring the circuit.
1.2 Extend the leads of the I.C. so that the sensor may be placed inside a test tube as shown in Figure 28.3. You may wish to place heat shrink tubing over the sensor's leads so that they do not short out. Place the test tube containing the I.C. in a beaker containing ice water as shown in Figure 28.3. Insert a thermometer or temperature probe in the test tube next to the I.C. Allow several minutes for the temperature of the test tube to stabilize.
1.3 While the sensor cools, construct the circuit shown in Figure 28.2. Turn on the power supply.
1.4 Turn on the electric hot plate.
1.5 Measure the differential voltage, V_{OUT}, for each of the temperature values shown in Table

28.1.
1.6 Turn off the power supply. Turn off the hot plate.

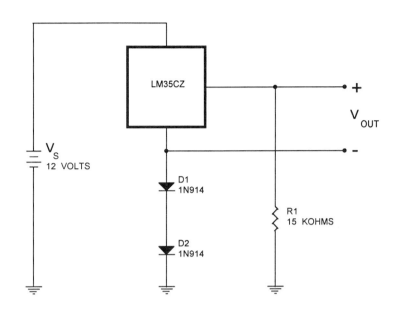

Figure 28.2: Temperature sensor circuit using the LM35CZ.

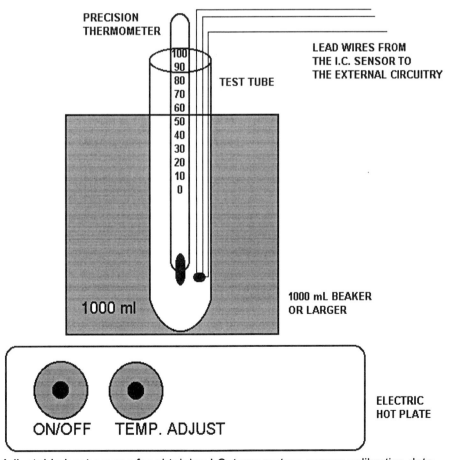

Figure 28.3: Adjustable heat source for obtaining I.C. temperature sensor calibration data.

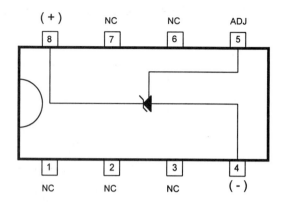

Figure 28.4: Top view of the LM335M surface mount package including pin identification.

Table 28.1: Temperature calibration data for the LM35CZ temperature sensor.

TEMPERATURE, DEGREES CELSIUS	DIFFERENTIAL OUTPUT VOLTAGE, VOUT, VOLTS
10	
15	
20	
25	
30	
35	
40	
45	
50	
55	
60	
65	
70	

1.7 Obtain the components required to construct the circuit shown in Figure 28.5. Make sure that you correctly identify the leads of the LM335 (or equivalent) I.C. before wiring the circuit.

1.8 Construct the circuit shown in Figure 28.5. Turn on the power supply. Before inserting the sensor in the ice bath, calibrate the sensor by adjusting the resistance of the 10 Kohms potentiometer so that the output voltage, V_{OUT}, is 2.982 volts at a temperature of 25°C (77°F). Since the temperature of your lab is already close to 77°F, warm the sensor slightly by <u>momentarily</u> touching it with your hand then adjusting the potentiometer to produce the desired output voltage. Once this phase is completed you should turn off the power supply and prepare the sensor to be inserted in the test tube.

1.9 Place the test tube containing the I.C. in a beaker containing ice water as shown in Figure 28.3. Insert a thermometer or temperature probe in the test tube next to the I.C. Allow several minutes for the temperature of the test tube to stabilize.

1.10 Turn on the electric hot plate. Turn on the power supply.

1.11 Measure the output voltage, V_{OUT}, for each of the temperature values shown in Table 28.2.

1.12 Turn off the power supply. Turn off the hot plate.

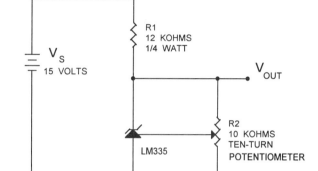

Figure 28.5: Temperature sensor circuit using the LM335Z.

Table 28.2: Temperature calibration data for the LM335 temperature sensor.

TEMPERATURE, DEGREES CELSIUS	SENSOR OUTPUT VOLTAGE, VOUT, VOLTS
10	
15	
20	
25	
30	
35	
40	
45	
50	
55	
60	
65	
70	

PART 2: Questions

2.1 Make a graph of output voltage as a function of temperature for each set of calibration data. Attach your graphs to your report.

2.2 Referring to your graphs from Question 2.1, which circuit, Figure 28.2 or Figure 28.5, is more sensitive? Yes, this is a trick question!

2.3 Determine the slope of the voltage versus temperature curve for the LM35CZ. This represents the calibration constant for the sensor.

$$\text{slope} = \Delta V/\Delta T$$

$$= (V_f - V_i)/(T_f - T_i) =$$

2.4 Using the calibration constant calculated in Question 2.3, what output voltage would you expect if the temperature of the I.C. was 75°C?

2.5 <u>Assume</u> for the moment that the output voltage of the circuit shown in Figure 28.5 is 3.030 volts at a temperature of 30°C. A temperature of 30°C is equal to 303°K (where °K = °C + 273). The calibration constant for this sensor (LM335) would then be found as follows:

$$\text{calibration constant} = V_{OUT}/°K$$

$$= 3.030 \text{ volts}/303°K$$

$$= 0.01 \text{ V}/°K$$

$$= 10 \text{ mV}/°K$$

Now use your data to calculate the calibration constant for the LM335 temperature sensor. Pick several points, and calculate the calibration constant for each point to ensure consistent results. Express your answer in mV/°K.

2.6 How does the calibration constant you calculated in Question 2.5 compare to the calibration constant specified by the manufacturer (10 mV/°K)?

2.7 If the calibration constant for the circuit shown in Figure 28.5 is 10 mV/°K, what is the temperature of the sensor if the output voltage is 3.16 volts? Express your answer in degrees Celsius and in degrees Kelvin.

PART 3: Circuit Design

3.1 Add an instrumentation amplifier to the output of the circuit shown in Figure 28.2 as shown in Figure 28.6 below. Connect the output of the instrumentation amplifier to the (+) input of the op-amp comparator circuit shown in Figure 28.7. Select values for R_G, R1, R2, and R_B that will cause the heater element to turn on and off at a temperature specified by your instructor. Show all of your calculations in the space provided below.

Calculations:

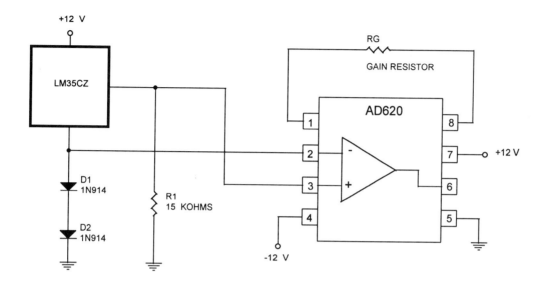

Figure 28.6: LM35CZ temperature sensor with instrumentation amplifier.

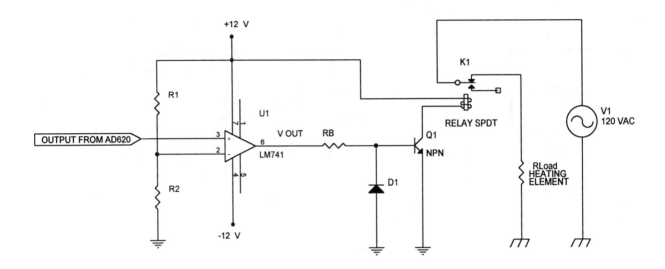

Figure 28.7: Op-amp comparator circuit. The output of the op-amp will switch from a negative output voltage to a positive output voltage when the temperature of the I.C. sensor exceeds a certain set point.

Thermocouples: 29

INTRODUCTION:

When designing a circuit to monitor the temperature of a product or manufacturing process, there are a wide number of transducers that the engineer has to choose from. This includes the thermocouple, RTD, thermistor, and I.C. sensor. The operating characteristics of each of these was briefly discussed in the introduction to Experiment 27. However, the most popular choice of temperature sensor for industrial process control is the thermocouple. The thermocouple is less expensive than RTDs, more linear than the thermistor, and able to withstand harsh operating environments. The thermocouple can operate over a wide range of temperatures. Thermocouples are an example of what is referred to as an *active transducer*. Active transducers produce a voltage when an external stimulus of some kind is applied. For example, when light is applied to a photovoltaic cell, it produces a small voltage. In a similar fashion, the thermocouple will produce a millivolt output when heated. This small voltage can then be amplified to a larger value for control purposes. The operation of the thermocouple is based on the principle that when two wires composed of dissimilar metals are joined, the resulting junction will produce a small electromotive force (EMF) when heated. This phenomenon is referred to as the Seebeck Effect, after Thomas Seebeck, who made this discovery. The schematic symbol for a thermocouple is shown in Figure 29.1.

As pointed out a moment ago, thermocouples are very popular in industrial process control. This is because there are several different types of thermocouples that can operate over a wide range of temperatures with differing levels of sensitivity, and they can operate in a wide variety of environments. Amongst the more popular thermocouples are the Type J, K, E, T, S, R, B, and N. Each of these thermocouples is constructed by joining together two different metal conductors. For example, the Type J thermocouple is formed by joining an iron wire to a wire made from a copper-nickel alloy as shown in Figure 29.2. An alloy is a mixture of two or more elements. For example, steel is made from iron and carbon. This is not to be confused with a compound such as salt, which results from a chemical reaction between the elements sodium and chlorine. In this experiment, you will generate calibration data for a number of different thermocouples. Table 29.1 shows the voltages produced by a number of different thermocouples as a function of temperature.

Figure 29.1: Schematic symbol for a thermocouple.

SAFETY NOTE:

In this experiment, you will be using some type of high-temperature heat source such as an electric hot plate or a hot-air gun. Therefore, take appropriate measures to avoid exposure of the skin to any heated surfaces. Allow all components to cool before cleaning and storing them. As always, wear safety glasses, and remove all jewelry from your hands and fingers before the start of the experiment.

Figure 29.2: Type J thermocouple. Thermocouples are made from two different metals. The point where the two metals are joined is referred to as the *junction*.

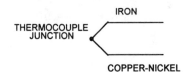

OBJECT:
Upon successful completion of this experiment and all reading assignments, the student should be able to:

- generate and plot calibration data for a thermocouple

REFERENCES:
Chapter 10 of Maloney's Modern Industrial Electronics
Omega Engineering's® Temperature Handbook
www.omega.com
www.analogdevices.com

MATERIALS:
- miscellaneous thermocouples
- 2 - 1000 mL or 2000 mL beakers
- miscellaneous lead wires and connectors

EQUIPMENT:
- 1 - digital voltmeter calibrated for millivolts
- 1 - electric hot plate
- 1 - thermometer or electronic temperature probe

Table 29.1: Thermocouple data for Types J, K, and E thermocouples.

TEMPERATURE, DEGREES CELSIUS	TYPE J mV OUTPUT	TYPE K mV OUTPUT	TYPE E mV OUTPUT
-200	-7.89	-5.891	-8.824
-100	-6.499	-4.912	-5.237
-50	-4.632	-3.553	-2.787
0	0	0	0
50	2.585	2.022	3.047
100	5.268	4.095	6.317
150	8.008	6.137	9.787
200	10.777	8.137	13.419
250	13.553	10.151	17.178
300	16.325	12.207	21.033

PART 1: Thermocouple Calibration Data Without Ice-Point Compensation

Background:

In this experiment, you will be obtaining calibration data for a variety of thermocouples by heating them in a controlled environment and measuring the voltage produced by the thermocouple using a precision voltmeter. While at first appearance this may seem rather straightforward, it is not. First consider the process just described by examining Figure 29.3. At first glance this appears to be a rather simple circuit. Figure 29.4 shows the equivalent circuit for Figure 29.3 as we have replaced the thermocouple by a small voltage source. So, what could be more simple to an advanced electronics student such as yourself than a circuit in which all you have to do is measure a simple voltage source? Well, let's consider in more detail just what we are dealing with by looking at Figure 29.5. Don't forget that no matter what type of circuit we connect our thermocouple to, we will have to attach the thermocouple leads to some other metallic conductor. In your case, this will most likely be copper test leads. In Figure 29.5, the copper test leads from the voltmeter are shown attached to the thermocouple. You might be asking, "So what's the problem?" Well, let's examine Figure 29.6, which is the electrical equivalent of Figure 29.5. Not only do we have a thermocouple junction produced by the iron and copper-nickel, we have also created two new thermocouple junctions--iron/copper and copper-nickel/copper. Because there are three different combinations, there will be three different EMFs produced by each of the junctions at a given temperature. As a result, we can now describe the circuit of Figure 29.6 by its equivalent as shown in Figure 29.7. The consequences are that the voltmeter will now read 27 mV rather than the true thermocouple output of 30 mV. This will lead to an incorrect temperature reading that is below the actual temperature.

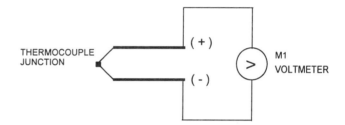

Figure 29.3: When heated, a thermocouple acts as a low-voltage (millivolt) source.

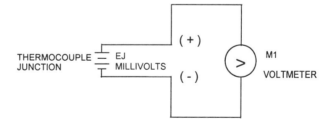

Figure 29.4: Equivalent circuit for the circuit shown in Figure 29.3.

To overcome this problem a number of compensation techniques can be applied. Consult Omega Engineering's® Temperature Handbook for methods of electronically compensating for this source of error. An early method used to overcome this problem was to configure the thermocouple wire like that shown in Figure 29.8. Take a moment to examine the relationship of the thermocouple lead wires. By creating two thermocouple junctions--one a *reference junction* and the other our *active junction*--we now have the copper test leads from the voltmeter connected to the iron leads of the thermocouple. This creates two iron/copper junctions that produce equal and opposite junction voltages, hence canceling each other out. By placing the reference junction in an ice bath, the resulting millivolt output from the reference junction at 0°C is 0.0 mV. As a result, the 30 mV produced at the active junction is now read at the voltmeter. Electronic instrumentation is available to simulate the circuit shown in Figure 29.8 without the inconvenience of keeping a fresh supply of ice around. In order to obtain as precise calibration data as possible, you should expose the thermocouple to a heat source that will change slowly. One way to accomplish this is to place the thermocouple in some type of enclosure having a large mass and slowly heat the mass of the object. An object of large mass will have a slow thermal response. A temperature probe should then be installed immediately next to the thermocouple to determine its temperature. A precision thermometer or electronic temperature probe would be an appropriate choice for monitoring the temperature of the thermocouple. A way to accomplish this in a college laboratory setting is to place the thermometer and thermocouple in a large beaker containing ice water as shown in Figure 29.9. The water serves as a sufficiently large mass whose temperature will change slowly. The water can then be heated with an electric hot plate (readily available in most chemistry labs). The millivolt output of the thermocouple can be measured with a precision digital voltmeter. If these resources are not available, you can then use a hot-air gun (readily available in most electronic labs) as a source of heat, but you must find a way to heat the thermocouple very slowly.

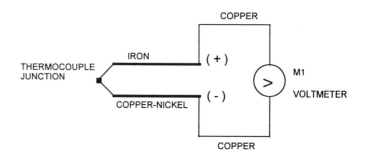

Figure 29.5: The circuit of Figure 29.3 redrawn showing the additional junctions created by connecting copper test leads from the voltmeter to the thermocouple.

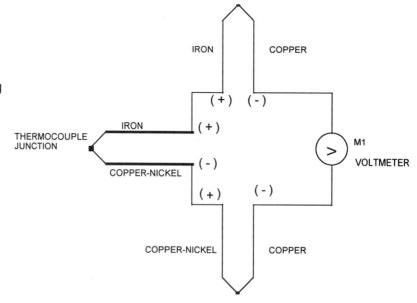

Figure 29.6: The circuit of Figure 29.5 redrawn showing three active thermocouple junctions.

Figure 29.7: Equivalent circuit for the circuit shown in Figure 29.6.

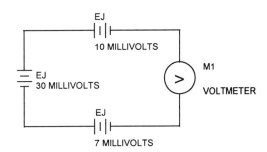

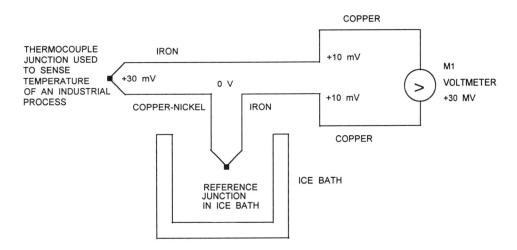

Figure 29.8: Thermocouple reference junction at ice point produces no voltage. The iron/copper thermocouple junctions produce an equal and opposite voltage that cancels out, leaving only the 30 mV from the measuring junction applied to the voltmeter.

INSTRUCTOR'S NOTE: Because the students will be obtaining calibration data for more than one thermocouple, I suggest you assign one thermocouple per group, then have the groups share their data upon completion of the lab. This will result in efficient use of lab time.

Procedure:
1.1 Obtain all materials and equipment required to complete this experiment.
1.2 Place a thermocouple in a beaker of ice water as shown in Figure 29.9. Attach the free ends of the thermocouple to a voltmeter like that shown in Figure 29.3. Insert a thermometer or temperature probe into the ice water in close proximity to the thermocouple. Allow five minutes for the temperature of the thermocouple and thermometer to stabilize.
1.3 Turn on the electric hot plate.
1.4 Take voltage measurements for each of the temperature values shown in Table 29.2.
1.5 Repeat steps 2 through 4 for a second thermocouple. Record your measurements in Table 29.2.
1.6 Repeat steps 2 through 4 for a third thermocouple. Record your measurements in Table 29.2.

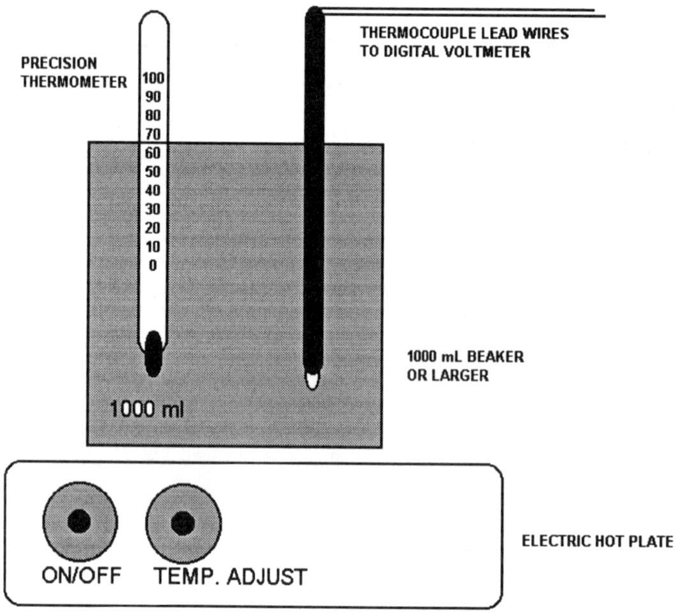

Figure 29.9: Adjustable heat source for obtaining thermocouple calibration data.

Table 29.2: Thermocouple output voltages for different temperatures without ice-point compensation.

TEMPERATURE, DEGREES CELSIUS	THERMOCOUPLE #1 MILLIVOLT OUTPUT	THERMOCOUPLE #2 MILLIVOLT OUTPUT	THERMOCOUPLE #3 MILLIVOLT OUTPUT
10			
20			
30			
40			
50			
60			
70			
75			
80			
85			
90			
95			
100			

PART 2: Thermocouple Calibration Data with Ice-Point Compensation

Background:
The first phase of this experiment involved obtaining thermocouple data without compensating for the additional thermocouple junctions created by the test leads. This part of the experiment will involve using a simple technique to eliminate the error caused by the two additional thermocouple junctions. This technique was described in the background information of Part 1 of this experiment, and the circuitry is shown in Figure 29.8. A variety of electronic-compensation techniques can be used, including an I. C. temperature sensor as the primary compensating device. For more information on the circuits that can be used for this purpose, consult either Omega Engineering's® Temperature Handbook or the temperature sensor section of National Semiconductor's® Data Acquisition Databook.

Procedure:
2.1 Obtain two thermocouple wires of the same type. Connect them as shown in Figure 29.8, and prepare two separate ice baths. Place one thermocouple in one beaker of ice water and the second thermocouple in another beaker of ice water. One of these thermocouples will serve as the reference junction, and the other will be the active junction to be heated as shown in Figure 29.9.

2.2 Attach the free ends of the thermocouple wires to a voltmeter like that shown in Figure 29.8. Insert a thermometer or temperature probe into the ice bath of the active thermocouple in close proximity to it. Allow five minutes for the temperature of the thermocouple and thermometer to stabilize.

2.3 Turn on the electric hot plate.

2.4 Take voltage measurements for each of the temperature values shown in Table 29.3.

2.5 Repeat steps 1 through 4 for two additional types of thermocouples. Record your measurements in Table 29.3.

Table 29.3: Thermocouple output voltages for different temperatures with ice-point compensation.

TEMPERATURE, DEGREES CELSIUS	THERMOCOUPLE #1 MILLIVOLT OUTPUT	THERMOCOUPLE #2 MILLIVOLT OUTPUT	THERMOCOUPLE #3 MILLIVOLT OUTPUT
10			
20			
30			
40			
50			
60			
70			
75			
80			
85			
90			
95			
100			

PART 3: Questions

3.1 Make a graph of millivolt output as a function of temperature for each set of data in Table 29.3. Try to use the same scale for each graph so that you can visually compare the different response of each thermocouple. Attach your graphs to your report.

3.2 Referring to your graphs of millivolts versus temperature, do the thermocouples seem to have a linear response?

3.3 Referring to your graphs of millivolts versus temperature, which of the thermocouples is the most sensitive? In other words, which one had the greatest change in millivolt output for a given change in temperature?

3.4 How does your data recorded in Tables 29.2 and 29.3 compare to the published values for these thermocouples? Which table of data seems to be more accurate?

3.5 For each of the thermocouple types shown in Table 29.4, record the two types of metal used in their construction.

Table 29.4: Thermocouple composition for Types T, K, and E thermocouples.

THERMOCOUPLE TYPE	THERMOCOUPLE COMPOSITION
T	
K	
E	

3.6 Thermocouple wire has a unique color code for each type of thermocouple, much like resistors have a color code. For each of the thermocouple types shown in Table 29.5, record the colors used in the color identification scheme for each.

Table 29.5: Thermocouple color code for Types T, K, and E thermocouples.

THERMOCOUPLE TYPE	THERMOCOUPLE COLOR CODE
T	
K	
E	

PART 4: Practice Problems

4.1 A Type E thermocouple is installed in a circuit like that shown in Figure 29.8. The reference junction is maintained at a temperature of 0°C. The voltmeter reads 2.544 mV. What is the temperature of the active junction?

4.2 A Type J thermocouple is installed in a circuit like that shown in Figure 29.8. The reference junction is maintained at a temperature of 0°C. The temperature of the active junction is 453°C. What is the reading on the digital voltmeter?

4.3 A Type J thermocouple is installed in a circuit like that shown in Figure 29.8. The reference junction is maintained at a temperature of 50°C. The temperature of the active junction is 453°C. What is the reading on the digital voltmeter?

4.4 A Type E thermocouple is installed in a circuit like that shown in Figure 29.8. The reference junction is maintained at a temperature of 50°C. The voltmeter reads 2.544 mV. What is the temperature of the active junction? Express your answer to the nearest degree.

4.5 You have recently been employed by a company that pasteurizes, then packages, a variety of drink products that are sold in grocery stores. The maintenance supervisor has just come to you about a problem with one of the pasteurization machines. The digital readout of the machine indicates that the required temperature of 150°C is being reached, but subsequent tests show that the bacteria count is too high, indicating that in reality the liquid is not getting hot enough. This machine uses a Type K thermocouple. The maintenance supervisor pointed out that the thermocouple had recently been replaced by the plant electrician. Upon closer examination of the color coding of the thermocouple leads, you find out that the electrician incorrectly replaced the Type K thermocouple with a Type J. You obtain a Type K and install it. The next day the maintenance supervisor informs you that the machine is working properly. Explain how installing the wrong thermocouple resulted in what seemed to be the right temperature according to the digital readout but that actually was too low to properly pasteurize the liquid drink. Hint: Start by referring to the millivolt output for each thermocouple at a temperature of 150°C.

PART 5: Circuit Design

5.1 Design an instrumentation amplifier circuit to amplify the millivolt signal from one of your thermocouples. You are to build a circuit like that shown in Figure 28.6, but use a thermocouple as the temperature sensor instead of the I.C. sensor. Connect the output of the instrumentation amplifier to the (+) input of the op-amp comparator circuit shown in Figure 28.7. Select values for R_G, R1, R2, and R_B that will cause the heater element to turn on and off at a temperature specified by your instructor. Show all of your calculations in the space provided below. Refer to the web site www.analogdevices.com for the AD620 manufacturer's data sheets. Here you will find an example amplifier circuit with thermocouple input.

Calculations:

The D/A Converter: 30

INTRODUCTION:
There are many applications in which a computer must control an analog device or machine. The digital-to-analog (D/A) converter is an I. C. that can be used as an interface between a microcomputer and some analog device we wish to control. In this experiment, you will work with an 8-bit digital-to-analog converter (DAC). Because the output of the DAC acts a current source, rather than a voltage source, it will also be necessary to use an op-amp to convert the DAC's output current into a voltage. The DAC can be configured in three common arrangements--the *unipolar* configuration, the *symmetrical offset binary mode of operation*, and the *bipolar* configuration.

SAFETY NOTE:
Be sure to observe proper polarity when wiring the DAC and the op-amp since you will have to use a dual-outlet power supply set for +15 volts and a -15 volts with a common ground. Improper wiring techniques could result in either a shorted power supply or "blowing the top off of" your DAC and/or op-amp(s). As always, wear safety glasses, and remove all jewelry from your hands and fingers before the start of the experiment.

OBJECT:
Upon successful completion of this experiment and all reading assignments, the student should be able to:
- construct an operational unipolar DAC circuit
- construct an operational DAC circuit configured in the symmetrical offset binary mode of operation
- calculate the reference current in a DAC circuit
- calculate the full-scale output current
- calculate I_{OUT} and $I_{OUT}/$ for a given binary input to a DAC operating in either the uniploar or the symmetrical offset binary configuration
- calculate the output voltage, V_{OUT}, for a given binary input to a DAC operating in either the uniploar or the symmetrical offset binary configuration

REFERENCES:
National Semiconductor's® Data Acquisition Databook
www.national.com

MATERIALS:
- 8 - 330 Ohms resistors or resistor pack, 1/4 watt
- 8 - LEDs or 1 ten-segment bar display
- 1 - dip switch pack with eight SPST switches
- 1 - op-amps, LM741 or the equivalent
- miscellaneous lead wires and connectors
- 1 - transistor, 2N3904
- 1 - 220 Ohms, 1 watt resistor
- 1 - 10 Kohms, ten-turn potentiometer
- 2 - 5 Kohms, ten-turn potentiometers
- 4 - 5 Kohms, 1/4 watt resistors
- 2 - 1.2 Kohms, 1/4 watt resistors
- 2 - capacitors, 0.1 µF
- 1 - capacitor, 0.01 µF
- 1 - transistor, TIP29
- 1 - 100 Kohms, 1/4 watt resistor
- 1 - solderless breadboard

1 - digital-to-analog converter, DAC0800
or the equivalent

1 - incandescent lamp, 24 volt

EQUIPMENT:
1 - dual-outlet D.C. power supply, 0 - 20 volts
1 - 5 volt D.C. power supply
1 - digital multimeter
1 - muffin-style fan, 24 VDC

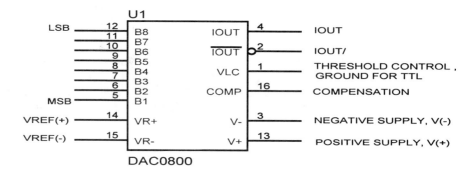

Figure 30.1: Package outline and pin assignment for the DAC0800.

PART 1: The Unipolar DAC
Background:
The package outline and the pinouts for a DAC0800 are shown in Figure 30.1. A typical unipolar DAC circuit is shown in Figure 30.8. Since the DAC has eight input pins (B1 through B8) to which TTL compatible logic levels can be applied, 256 different binary input combinations are possible. The number of possible input combinations can be found using the following formula:

$$\text{INPUT COMBINATIONS} = 2^n \qquad \textbf{30.1}$$

where n is the number of DAC input bits. Much like an op-amp, the DAC requires (+) and (-) supplies at pins 13 and 3 respectively. In this circuit, $V_{(+)}$ is +15 volts and $V_{(-)}$ is -15 volts. In some applications, the DAC may be configured where $V_{(-)}$ is 0.0 volt. In either case, as with the op-amp, it is important not to exceed the manufacturer's rated differential voltage, $E_{DIFF(MAX)}$, at these two pins. For example, if $E_{DIFF(MAX)}$ is 30 volts, we could apply +15 and -15 volts at pins 13 and 3, respectively, or apply +30 volts and 0.0 volt at pins 13 and 3. Pin 1, V_{LC} (Logic Control Voltage), is connected to ground when interfacing the DAC to TTL logic devices. If other logic families (such as CMOS) are used, a different logic level may be required. As always, consult the manufacturer's data manual or an applications engineer regarding your particular requirements. Pins 14 (V_{R+}) and 15 (V_{R-}) serve as the current or voltage reference that determines the maximum output current from pins 4 (I_{OUT}) and 2 ($I_{OUT}/$). These two currents will be converted to an output voltage, V_{OUT}, by the op-amp shown in Figure 30.8. To better understand the op-amp's role in this process, let's review a few common op-amp circuits. Figure 30.2 shows a typical inverting op-amp configuration. An input voltage of +5.0 volts is being applied to the circuit. This causes a 5 mA current to flow through the 1 Kohm resistor, R1. Remember from your study of op-amps that an op-amp has an extremely high input impedance. As a result, there is no current flow through the op-amp from the $V_{(-)}$ input to the $V_{(+)}$ input. Also, there is an extremely small differential voltage (E_{DIFF}) across these two inputs. This differential voltage is so small that for all intents and purposes we can consider it to be 0.0 volt. Because the $V_{(+)}$ input is at ground potential, and keeping in mind that there is no voltage drop from the $V_{(+)}$ input to the $V_{(-)}$ input, the $V_{(-)}$ input is also at ground potential or, more correctly, at what we refer to as *virtual* ground. Since the 5 mA of current must then flow through R_F, it will generate a potential across R_F of 10 volts. Because of the direction of the current (conventional current flow), the resulting polarity at the output relative to ground is negative. Hence, V_{OUT} is -10 volts.

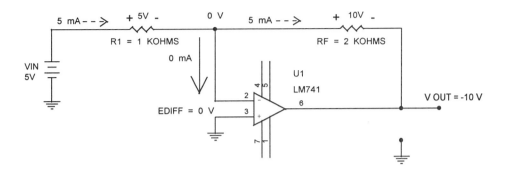

Figure 30.2: Inverting op-amp circuit showing the relationship of voltage magnitudes, current flow, and voltage polarity.

NOTE: As we go through this analytical process, all currents shown will be conventional current. When conventional current flows into a resistor, that side of the resistor is the positive terminal. Conventional current then comes out the negative terminal of the device. If you are more familiar with electron flow, reverse the direction of all currents shown on the drawings. With electron flow, electrons flowing into a resistor go in the negative terminal and come out the positive terminal. This results in the same polarity regardless of the current convention that you adhere to. Please also keep in mind the direction of the currents I_{OUT} and $I_{OUT}/$ shown in Figures 30.8 and 30.9. This is the opposite of the current direction shown in Figures 30.2 and 30.4. You must master the following examples to truly understand the resulting output voltage's magnitude and polarity for the circuits shown in Figure 30.8 and 30.9.

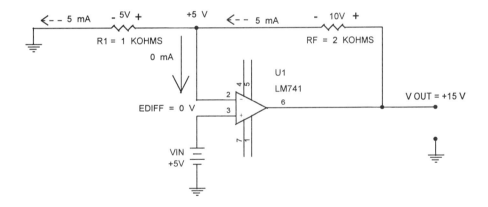

Figure 30.3: Non-inverting op-amp circuit showing the relationship of voltage magnitudes, current flow, and voltage polarity.

In Figure 30.3, a non-inverting op-amp circuit is shown. Here +5 volts is applied to the non-inverting input. Again because the differential voltage from $V_{(+)}$ to $V_{(-)}$ is 0.0 volt, we have +5.0 volts at the $V_{(-)}$ input of our op-amp. This voltage is applied to resistor R1, causing 5 mA of current to flow in the direction shown. This same 5 mA must also flow through R_F for the reason previously discussed. This results in a 10 volt drop across R_F. The resulting total voltage at the output is then +15 volts. Because the polarity of the input and output voltage to ground is the same, this circuit operates as a non-inverting amplifier. If we reverse the connections to V$_{IN}$, we will have the circuit as shown in Figure 30.4. The resulting output is now -15 volts.

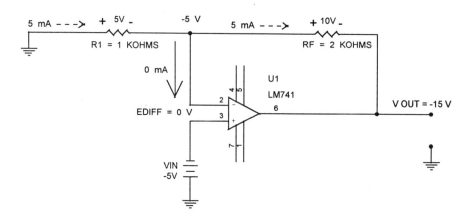

Figure 30.4: Non-inverting op-amp circuit showing the relationship of voltage magnitudes, current flow, and voltage polarity. The polarity of V$_{IN}$ has been changed from that shown in Figure 30.3.

Another inverting op-amp circuit is shown in Figure 30.5. I will leave it up to you to confirm the magnitude and direction of the currents shown as well as the magnitude and polarity of the output voltage.

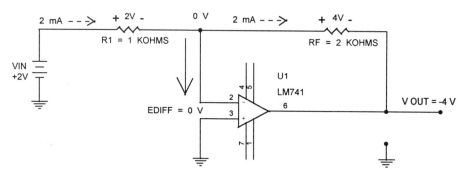

Figure 30.5: Inverting op-amp circuit showing the relationship of voltage magnitudes, current flow, and voltage polarity.

In Figure 30.6, we effectively have what is a combination of the circuits shown in Figures 30.3 and 30.5. In other words, we have both an inverting and a non-inverting circuit in one. To understand how this circuit operates, apply the principle of superposition from your earlier D.C. or A.C. circuit analysis class(es). First analyze the circuit as if it was an inverting circuit. Then analyze it as if it was a non-inverting circuit. You should note that the resulting output of +11 volts is the sum of the outputs from Figures 30.3 and 30.5, that is, +15 V + (-4 V) = +11 V. It is these same current and voltage relationships in Figures 30.1 through 30.6 that help to explain how the op-amps in Figures 30.8 and 30.9 converts the output currents of the DAC into either positive or negative output voltages.

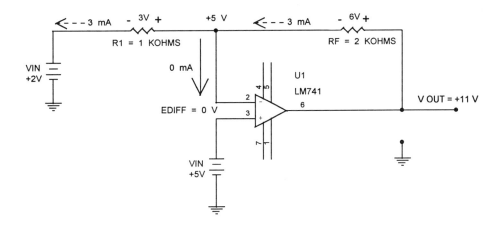

Figure 30.6: Combined inverting and non-inverting op-amp circuits showing the relationship of voltage magnitudes, current flow, and voltage polarity.

Continuing on with our explanation of how Figure 30.8 works, a fixed reference current, I_{REF}, is applied to the DAC via the reference voltage, V_{REF}, and the reference resistor, R_{REF}. For this circuit, I_{REF} can be found using Ohm' Law as follows:

$$I_{REF} = V_{REF}/R_{REF} \qquad\qquad 30.2$$

This results in a reference current, I_{REF}, of 2 mA. The maximum or *full-scale* output current, I_{FS}, that the DAC can *source* (or *sink* as shown in Figures 30.8 and 30.9) will occur when all of the DAC's binary inputs are at a logical high, that is, 1111 1111. This corresponds to a decimal 255. The formula for the full-scale output current is as follows:

$$I_{FS} = [(2^n - 1)/(2^n)] \cdot I_{REF} \qquad\qquad 30.3$$

As with Formula 30.1, *n* represents the number of binary bits. For our eight-bit DAC we would have

$$I_{FS} = [(2^8 - 1)/(2^8)] \cdot (2 \text{ mA})$$

$$= (255/256) \cdot (2 \text{ mA})$$

$$= 1.992 \text{ mA}$$

Again, this is the maximum possible value for I_{OUT}. You may wish to review Experiment 18 on the summing amplifier for an explanation of why I_{FS} is not 2 mA, the value for the reference current. If we now apply the op-amp principles reviewed earlier, we can develop the formula for the maximum or full-scale output voltage, V_{FS}, for the circuit shown in Figure 30.8. The formula becomes

$$V_{FS} = (I_{FS}) \cdot (R_F) \qquad\qquad 30.4$$

Applying Formula 30.4, we can now calculate the maximum or full-scale output voltage as follows:

$$V_{FS} = (1.992) \cdot (5 \text{ Kohms})$$

$$= 9.961 \text{ volts}$$

Substituting Formula 30.3 into Formula 30.4, we have

$$V_{FS} = [(2^n - 1)/(2^n)] \cdot I_{REF} \cdot R_F \qquad \textbf{30.5}$$

Since $R_F = R_{REF}$, we have

$$V_{FS} = [(2^n - 1)/(2^n)] \cdot I_{REF} \cdot R_{REF} \qquad \textbf{30.6}$$

Applying Ohm's Law, Formula 30.6 becomes

$$V_{FS} = [(2^n - 1)/(2^n)] \cdot V_{REF} \qquad \textbf{30.7}$$

For binary inputs less than decimal 255, we can calculate the output current using the following formula:

$$I_{OUT} = [(D_{EQ})/(2^n)] \cdot I_{REF} \qquad \textbf{30.8}$$

D_{EQ} represents the decimal equivalent of the binary input to the DAC. For example, if the binary input to the DAC was 1100 0010, we would first convert the binary pattern to its decimal equivalent. For this pattern, D_{EQ} is 194. Substituting into Formula 30.8, we have

$$I_{OUT} = (194/2^8) \cdot (2 \text{ mA})$$

$$= (0.7578) \cdot (2 \text{ mA})$$

$$= 1.5156 \text{ mA}$$

Pick some eight-bit binary patterns at random and apply Formula 30.8 for practice. Applying Ohm's Law, we can write a formula for V_{OUT} in terms of I_{OUT} and R_F as follows:

$$V_{OUT} = (I_{OUT}) \cdot (R_F) \qquad \textbf{30.9}$$

If we substitute Formula 30.8 for I_{OUT} into Formula 30.9 for V_{OUT} and substitute R_{REF} for R_F, we have

$$V_{OUT} = [(D_{EQ})/(2^n)] \cdot (I_{REF}) \cdot (R_{REF}) \qquad \textbf{30.10}$$

Since $V_{REF} = (I_{REF}) \cdot (R_{REF})$, we then have

$$V_{OUT} = [(D_{EQ})/(2^n)] \cdot (V_{REF}) \qquad \textbf{30.11}$$

For example, if the binary input is 0110 0000, then the decimal equivalent is 96 and the output voltage would be

$$V_{OUT} = (96/256) \cdot (10 \text{ volts})$$

$$= 3.75 \text{ volts}$$

As you may have surmised by now, the DAC is capable of producing only a finite number of different output voltages. The number of possible outputs depends on the number of binary inputs. This relationship is described mathematically by Formula 30.1. While Formula 30.1 represents the formula for the number of different binary combinations at the DAC's input, this formula also tells us that this is the maximum number of different analog outputs possible. A true analog device or power supply is capable of producing an infinite number of different current or voltage outputs over a given range of values. The output of our DAC will not be smooth but rather will look somewhat like a staircase as shown in Figure 30.7.

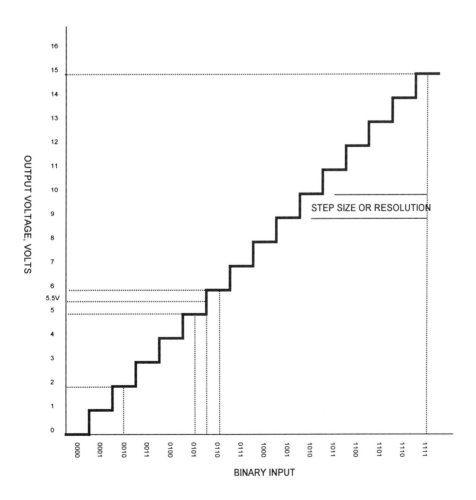

Figure 30.7: Plot of V_{OUT} as a function of the binary input pattern applied to a 4-bit DAC. The step size or resolution is 1.0 volt.

The height of each step, or the step size, is referred to as the *resolution* of the DAC. The resolution is the smallest change in the DAC's output voltage that will occur for an input change of one binary bit. For example, if the binary input is 0110 0001 (decimal 97), this represents a 1 bit increase from 0110 0000 (decimal 96). The output voltage from the DAC of Figure 30.8 for an input of 0110 0001 is

$$V_{OUT} = (97/256) \cdot (10 \text{ volts})$$

$$= 3.789 \text{ volts}$$

The difference in the DAC's output from decimal 96 to decimal 97 is the DAC's resolution. The change in output voltage, ΔV_{OUT} is then

$$\Delta V_{OUT} = 3.789 \text{ volts} - 3.75 \text{ volts}$$

$$= 0.039 \text{ volt}$$

$$= 39 \text{ mV}$$

This should also equal the value of the output of the DAC when a binary pattern of 0000 0001 (decimal 1) is applied to the DAC's input. This leads to a general formula for the step size or resolution of a DAC.

$$\text{RESOLUTION} = V_{REF}/2^n \qquad \textbf{30.12}$$

If we solve Formula 30.7 for V_{REF} and substitute into Formula 30.12, we have a formula for the resolution in terms of V_{FS}.

$$\text{RESOLUTION} = V_{FS}/(2^n - 1) \qquad \textbf{30.13}$$

Now let's assume for the moment that you are designing a circuit that will be interfaced to an eight-bit data bus of a microcomputer. You want an output of 3.8 volts for your particular application. You decide to use a circuit like that in Figure 30.8. What binary pattern must you write to the data bus to produce the required output voltage? If we solve Formula 30.11 for D_{EQ}, we will have

$$D_{EQ} = [(V_{OUT}/V_{REF})] \cdot (2^n) \qquad \textbf{30.14}$$

Substituting +10 volts for V_{REF} and +3.8 volts for V_{OUT}, we have

$$D_{EQ} = (3.8 \text{ V}/10 \text{ V}) \cdot (2^8)$$

$$= 97.28$$

The decimal number 97.28 rounded to the nearest whole number is 97. We must do this because there is no binary pattern on our DAC that can represent 97.28 as the weight of the least significant bit (LSB) is a 1. Therefore, the closest binary pattern that we can select is that corresponding to a decimal 97. That binary pattern then is 0110 0001. A moment ago we found that this particular pattern will produce an output voltage of 3.789 volts. As a result, we have a slight error or deviation from the desired output of 3.8 volts. This error may be referred to as the *quantizing error* or the *absolute error*. The formula for the quantizing error (Q.E.) is

$$Q.E. = V_{OUT(ACTUAL)} - V_{OUT(DESIRED)} \qquad 30.15$$

The quantizing error for this problem then is

$$Q.E. = 3.789 - 3.8$$
$$= -0.011 \text{ volt}$$
$$= -11 \text{ mV}$$

From this we can calculate the percent error to see if the design is within acceptable engineering specifications. The percent error can be found using the following formula:

$$\% \text{ ERROR} = [(\text{QUANTIZING ERROR})/(V_{OUT(DESIRED)})] \cdot (100\%) \qquad 30.16$$

For our example we have

$$\% \text{ ERROR} = (-0.011 \text{ V}/3.8 \text{ V}) \cdot (100\%)$$
$$= -0.289\%$$

Based on prior experience and/or engineering department specifications, a decision would then have to be made to determine if this level of error is acceptable. If it is not, then the designer has the choice of selecting a DAC with more bits or to reduce the full-scale output voltage by reducing V_{REF} or R_F. Implementing either one or both of these changes will improve the resolution (make it smaller) and in turn reduce the absolute and percent errors. In any case though, the maximum error is $\pm\frac{1}{2}$ LSB.

NOTE: Like most digital devices, DACs and op-amps are capable of supplying only a few milliamperes of current. To drive devices requiring much more than 10 milliamperes, it will be necessary to interface the output of the DAC or the op-amp (or whatever your final output device is) with some device that will act as a buffer between the DAC or op-amp and the load to be driven. You must find out what the current requirements of your final load are and determine the rated output current of the device that is supplying the current by referring to the manufacturer's data manual for that device.

Procedure:
1.1 Obtain all materials and equipment required to complete this experiment.
1.2 Use your multimeter to measure the resistance of each of your resistors, and record in Table 30.1.
1.3 If your lab has one, use a capacitance tester to check the capacitance values of your capacitors. If any of them are out of tolerance, inform your instructor and obtain replacements.
1.4 Construct the circuit shown in Figure 30.8.
1.5 In this circuit, closing one of the switches will cause a ground at the corresponding pin of the DAC and will turn on the corresponding LED. Therefore, a closed switch will result in a logic low. If a switch is open, the corresponding LED will be off and a logic high will be applied to the corresponding DAC input. Turn on all power supplies, and check the operational status of your LEDs. If they seem to be working properly, go on to the next step.

Table 30.1: Nominal and measured resistor values.

NOMINAL RESISTANCE	MEASURED RESISTANCE
R_{REF} = 5 Kohms	
R_1 = 5 Kohms	
R_F = 5 Kohms	
$R_F/$ = 5 Kohms	

1.6 For each of the binary input patterns shown in Table 30.2, measure and record the corresponding output voltage, V_{OUT}.

1.7 Using the formulas developed on the preceding pages, calculate the theoretical value of the output voltage for each of the binary input patterns shown in Table 30.3 for the circuit shown in Figure 30.8.

Table 30.2: Binary input pattern and measured output voltages for the unipolar DAC.

INPUT BINARY PATTERN	DECIMAL EQUIVALENT	MEASURED OUTPUT VOLTAGE, VOLTS
0000 0000	0	
0000 0001	1	
0000 0010	2	
0000 0011	3	
0000 0100	4	
0000 1000	8	
0001 0000	16	
0010 0000	32	
0100 0000	64	
0110 0000	96	
0111 0000	112	
1000 0000	128	
1010 0000	160	
1100 0000	192	
1110 0000	224	
1111 1101	253	
1111 1110	254	
1111 1111	255	

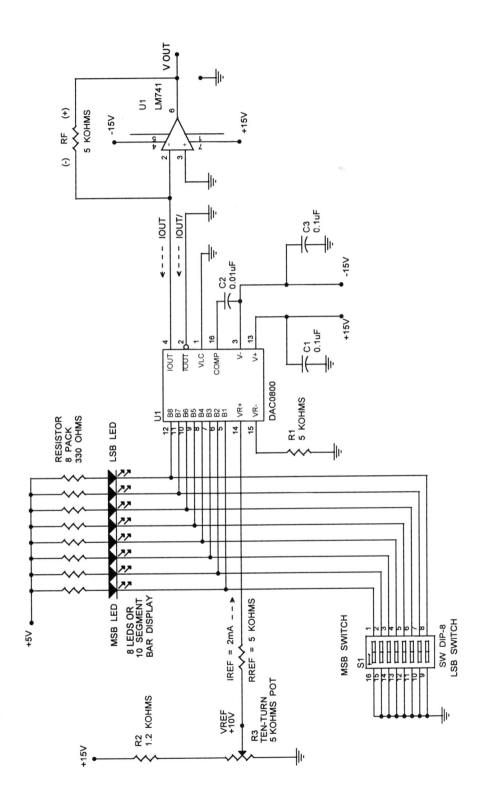

Figure 30.8: Digital-to-analog converter circuit. This is the unipolar configuration. The op-amp is being used as a current-to-voltage converter.

Table 30.3: Binary input pattern and calculated output voltages for the unipolar DAC.

INPUT BINARY PATTERN	DECIMAL EQUIVALENT	CALCULATED OUTPUT VOLTAGE, VOLTS
0000 0000	0	
0000 0001	1	
0000 0010	2	
0000 0011	3	
0000 0100	4	
0000 1000	8	
0001 0000	16	
0010 0000	32	
0100 0000	64	
0110 0000	96	
0111 0000	112	
1000 0000	128	
1010 0000	160	
1100 0000	192	
1110 0000	224	
1111 1101	253	
1111 1110	254	
1111 1111	255	

PART 2: Symmetrical Offset Binary Operation
Background:
Some applications require a negative output voltage as well as a positive output voltage. The symmetrical offset binary mode of operation will provide both positive and negative output voltages. The positive and negative output voltages will be symmetrical; for example, the magnitude of the maximum positive output voltage will be equal to the magnitude of the most negative output voltage. However, while the operation of the DAC in this configuration is symmetrical, the output will not go exactly to 0.0 volt. This is the major drawback of this configuration. Figure 30.9 shows a DAC configured to operate in the symmetrical offset binary mode.

At this point, we should probably discuss the relationship between the currents I_{OUT} and $I_{OUT}/$. As discussed previously, the maximum or full-scale output current from a DAC can be calculated according to Formula 30.3. For an eight-bit DAC such as the DAC0800, we would then have

$$I_{FS} = (255/256) \cdot (I_{REF})$$

where I_{REF} is found according to Formula 30.2. The formula relating I_{OUT} and $I_{OUT}/$ is as follows:

$$I_{FS} = I_{OUT} + I_{OUT}/ \qquad \textbf{30.17}$$

I_{OUT} can be found according to Formula 30.8. Rearranging Formula 30.17 and solving for $I_{OUT}/$, we have

$$I_{OUT}/ = I_{FS} - I_{OUT}$$

Substituting Formulas 30.3 and 30.8 into Formula 30.17 and solving for $I_{OUT}/$ we have

$$I_{OUT}/ = \{[(2^n - 1)/(2^n)] \cdot I_{REF}\} - \{[(D_{EQ})/(2^n)] \cdot I_{REF}\} \qquad \textbf{30.18}$$

The maximum possible value for either I_{OUT} or $I_{OUT}/$ is the full-scale output current, I_{FS}. I_{OUT} is maximum when $I_{OUT}/$ is 0.0 ampere, and $I_{OUT}/$ is maximum when I_{OUT} is 0.0 ampere. If we now examine Figure 30.9, we see that the flow of current I_{OUT} through R_F contributes a positive voltage at the output of the op-amp relative to ground. The flow of current through $R_F/$ is such that it produces a negative voltage at pin 3 of the op-amp relative to ground. Because there is no voltage drop from pin 3 of the op-amp to pin 2 of the op-amp, the voltage at pin 3 due to $I_{OUT}/$ appears at pin 2 of the op-amp. The sum of the voltages produced by I_{OUT} and $I_{OUT}/$ results in the value for V_{OUT}. In other words,

$$V_{OUT} = [(I_{OUT}) \cdot (R_F)] - [(I_{OUT}/) \cdot (R_F/)] \qquad \textbf{30.19}$$

Since $R_F = R_F/$, we can substitute R_F for $R_F/$ in Formula 30.19 and have

$$V_{OUT} = (I_{OUT} - I_{OUT}/) \cdot (R_F) \qquad \textbf{30.20}$$

Procedure:
2.1 Construct the circuit shown in Figure 30.9.
2.2 For each of the binary input patterns shown in Table 30.4, measure and record the corresponding output voltage, V_{OUT}.
2.3 Using the formulas developed on the preceding pages, calculate the theoretical value for V_{OUT} for the binary input patterns shown in Table 30.5 for the circuit shown in Figure 30.9.

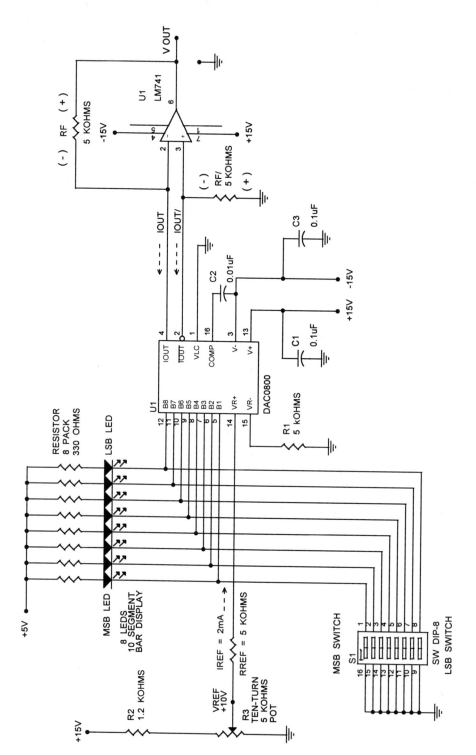

Figure 30.9: Digital-to-analog converter circuit. This is the symmetrical offset binary mode of operation. The op-amp is being used as a current-to-voltage converter.

Table 30.4: Binary input pattern and measured output voltages for the DAC when operated in the symmetrical offset binary mode.

INPUT BINARY PATTERN	DECIMAL EQUIVALENT	MEASURED OUTPUT VOLTAGE, VOLTS
0000 0000	0	
0000 0001	1	
0000 0010	2	
0000 0011	3	
0000 0100	4	
0000 1111	15	
0001 0000	16	
0001 1111	31	
0011 1111	63	
0100 0000	64	
0111 1111	127	
1000 0000	128	
1010 0000	160	
1100 0000	192	
1110 0000	224	
1111 0000	240	
1111 1110	254	
1111 1111	255	

Table 30.5: Binary input pattern and calculated output voltages for the DAC when operated in the symmetrical offset binary mode.

INPUT BINARY PATTERN	DECIMAL EQUIVALENT	CALCULATED OUTPUT VOLTAGE, VOLTS
0000 0000	0	
0000 0001	1	
0000 0010	2	
0000 0011	3	
0000 0100	4	
0000 1111	15	
0001 0000	16	
0001 1111	31	
0011 1111	63	
0100 0000	64	
0111 1111	127	
1000 0000	128	
1010 0000	160	
1100 0000	192	
1110 0000	224	
1111 0000	240	
1111 1110	254	
1111 1111	255	

PART 3: Application
Background:
This activity is intended to demonstrate an actual application of digital-to-analog conversion and signal conditioning. The D/A converter allows us to convert a digital pattern that might be created by a computer or possibly a programmable controller (refer to Experiment 36) to an analog signal. Computers are not only digital devices, but they are also low power devices. So, as part of the signal conditioning process (making one type of signal or voltage--such as digital--compatible with another--such as analog), the designer must also provide for current gain. The circuits shown in Figures 30.8 and 30.9 can be used to convert a digital pattern to a low-power analog output. The circuit shown in Figure 30.10 can be used to control a high-power consuming load (a fan in this case) with a low current device (the DAC).

Procedure:
    3.1 Construct the circuit shown in Figure 30.10. Adjust pot R3 for a resistance of 5 Kohms. You may use a 20-volt D.C. supply in place of the 24-volt source if you do not have enough power supplies.

3.2 Connect the output of the circuit shown in Figure 30.8 to the input of the circuit shown in Figure 30.10.

3.3 Use a voltmeter to measure the collector-to-emitter voltage, V_{CE}, of the TIP29.

3.4 Test your circuit by opening and closing the switches for the most significant bits and monitor V_{CE}. If you measure a large change in voltage, your circuit is working. If not, ask your instructor for help. Your circuit may be wired incorrectly, or you may need to change the resistance values in the interface circuit to achieve the desired current gain for your particular fan.

3.5 Vary the binary input to determine the range of operation. What binary pattern produces the greatest fan speed? What binary pattern produces the slowest fan speed without the fan turning off?

Binary pattern$_{FASTEST}$:_____   Binary pattern$_{SLOWEST}$:_____

3.6 Replace the fan with a 24-volt incandescent light bulb. You should be able to vary the light intensity of the lamp by changing the dip switch settings.

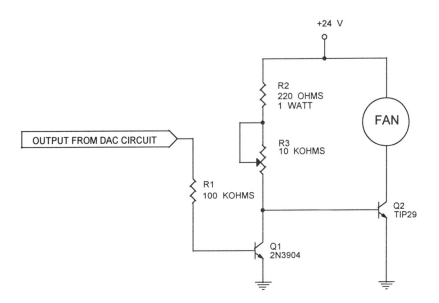

Figure 30.10: Fan circuit interface to the circuit shown in Figure 30.8.

PART 4: Questions

4.1 On a piece of graph paper plot the data from Table 30.2 showing the output voltage as a function of the input expressed as a decimal. In other words, plot the output voltage on the Y-axis and the decimal equivalent of the input on the X-axis. Attach your graph to your lab report.

4.2 Based on your graph from Question 4.1, what can you say about the output response or characteristic of your DAC?

4.3 For the circuit shown in Figure 30.8, how much did the analog output voltage change when the input binary pattern changed from 1111 1110 to 1111 1111? What is the relationship of this value to the remaining values recorded in Table 30.2? What should its relationship be? In other words, what does this voltage change represent?

4.4 Define the term *resolution* in your own words.

4.5 Does your DAC use the binary-weighted or the R/2R method of conversion?

4.6 How does the measured data in Table 30.2 compare to the calculated data in Table 30.3?

4.7 How does the measured data in Table 30.4 compare to the calculated data in Table 30.5?

4.8 How does the measured data in Table 30.6 compare to the calculated data in Table 30.7?

4.9 How would you modify the circuit shown in Figure 30.8 so that the output voltage would vary from 0.0 to +5.0 volts for binary inputs varying from 0000 0000 to 1111 1111? There is more than one possible solution.

4.10 If you applied a binary pattern of 1111 1111 to the DAC circuit shown in Figure 30.8, would it cause the fan to turn at a high speed, slow speed, or turn off? Explain your answer in detail. In your answer, include the expected DAC output voltage and what would happen to Q1, Q2, and the fan in the circuit shown in Figure 30.10 as a result.

PART 5: Practice Problems

5.1 Assume that you are designing a circuit like that shown in Figure 30.8 but you require a 10-bit DAC because of the need for a finer resolution. A V_{REF} of +5.0 volts has been specified. Both R_{REF} and R_F are to be 5 Kohms. Answer the following questions.

a) How many different binary input combinations are possible?

b) What is the value of the reference current, I_{REF}?

c) What is the value of the full-scale output current, I_{FS}?

d) What is the value of the full-scale output voltage, V_{FS}?

e) What is the value for I_{OUT} when the binary input is 10 0000 0000 (decimal 512)?

f) What is the value for V_{OUT} when the binary input is 10 0000 0000?

g) What is the resolution of this DAC?

h) What is the decimal equivalent, D_{EQ}, of the binary input that will produce an output voltage of 1.25 volts?

i) What is the decimal equivalent, D_{EQ}, of the binary input that will produce an output voltage of 1.256 volts?

j) What is the quantizing error or absolute error that will result for the answer to the previous question, part i)?

k) What is the percent error for the previous question, part j)?

l) What is the maximum possible quantizing error for this DAC? Express your answer as a voltage.

5.2 Now assume that you modify the circuit described in Problem 5.1 to operate in the symmetrical offset binary mode like the circuit shown in Figure 30.9. V_{REF} is still +5.0 volts, and $R_{REF} = R_F = R_F/ = 5$ Kohms. Answer the following questions.

a) What is the value for I_{OUT} when the binary input pattern is 11 0000 0000 (decimal 768)?

b) What is the value for $I_{OUT}/$ when the binary input pattern is 11 0000 0000 (decimal 768)?

c) What is the value for V_{OUT} when the binary input pattern is 11 0000 0000 (decimal 768)?

PART 6: DAC Formulas

Uniploar Configuration -

INPUT COMBINATIONS = 2^n	**30.1**
$I_{REF} = V_{REF}/R_{REF}$	**30.2**
$I_{FS} = [(2^n - 1)/(2^n)] \cdot I_{REF}$	**30.3**
$V_{FS} = (I_{FS}) \cdot (R_F)$	**30.4**
$V_{FS} = [(2^n - 1)/(2^n)] \cdot I_{REF} \cdot R_F$	**30.5**
$V_{FS} = [(2^n - 1)/(2^n)] \cdot I_{REF} \cdot R_{REF}$	**30.6**
$V_{FS} = [(2^n - 1)/(2^n)] \cdot V_{REF}$	**30.7**
$I_{OUT} = [(D_{EQ})/(2^n)] \cdot I_{REF}$	**30.8**
$V_{OUT} = (I_{OUT}) \cdot (R_F)$	**30.9**
$V_{OUT} = [(D_{EQ})/(2^n)] \cdot (I_{REF}) \cdot (R_{REF})$	**30.10**
$V_{OUT} = [(D_{EQ})/(2^n)] \cdot (V_{REF})$	**30.11**
RESOLUTION = $V_{REF}/2^n$	**30.12**
RESOLUTION = $V_{FS}/(2^n - 1)$	**30.13**
$D_{EQ} = [(V_{OUT}/V_{REF})] \cdot (2^n)$	**30.14**
Q.E. = $V_{OUT(ACTUAL)} - V_{OUT(DESIRED)}$	**30.15**
% ERROR = [(QUANTIZING ERROR)/($V_{OUT(DESIRED)}$)] $\cdot$ (100%)	**30.16**

Symmetrical offset binary mode of operation -

$I_{FS} = I_{OUT} + I_{OUT}/$ **30.17**

$I_{OUT}/ = \{[(2^n - 1)/(2^n)] \cdot I_{REF}\} - \{[(D_{EQ})/(2^n)] \cdot I_{REF}\}$ **30.18**

$V_{OUT} = [(I_{OUT}) \cdot (R_F)] - [(I_{OUT}/) \cdot (R_F/)]$ **30.19**

$V_{OUT} = (I_{OUT} - I_{OUT}/) \cdot (R_F)$ **30.20**

The A/D Converter: 31

INTRODUCTION:
In Experiment 30, you learned how to take a binary pattern and convert into an analog voltage using a digital-to-analog converter (DAC). When monitoring industrial processes it is usually necessary to convert an analog signal (whether it be from a thermistor, thermocouple, pressure transducer, flow meter, strain gage, optoelectronic device, or any other transducer) into a digital format that can be read from the data bus of a mainframe or microcomputer or possibly a dedicated microcontroller or programmable logic controller. The computer or controller can then use the digital information for decision-making purposes and adjust the industrial process accordingly. The analog-to-digital (A/D) converter is an I. C. that can be used as an interface between a microcomputer and some analog device whose output we wish to monitor. In this experiment you will work with an 8-bit successive approximation analog-to-digital converter (ADC).

SAFETY NOTE:
Be sure to read the manufacturer's data sheet before wiring the ADC so that you do not exceed any of its voltage, current, or power ratings. As always, wear safety glasses, and remove all jewelry from your hands and fingers before the start of the experiment.

OBJECT:
Upon successful completion of this experiment and all reading assignments, the student should be able to:
- construct an operational ADC circuit using a successive approximation ADC
- determine the output binary pattern from an ADC for a given analog input voltage

REFERENCES:
National Semiconductor's® Data Acquisition Databook
www.national.com

MATERIALS:
- 2 - 10 Kohms resistors, 1/4 watt
- 5 - 1 Kohms resistors, 1/4 watt
- 3 - 2 Kohms, ten- or twenty-turn potentiometers
- 8 - 330 Ohms resistors or eight pack
- 8 - LEDs or 1 ten-segment bar display
- 1 - solderless breadboard
  - miscellaneous lead wires and connectors
- 1 - analog-to-digital converter, ADC0803 or ADC0804 or the equivalent
- 1 - capacitor, 0.001 µF
- 1 - capacitor, 0.1 µF
- 1 - capacitor, 150 pF
- 1 - 1 or 10 µF tantalum capacitor
- 1 - 74LS08
- 1 - 74LS240
- 2 - switches, SPST

EQUIPMENT:
- 1 - dual-outlet D.C. power supply or two single-outlet power supplies
- 1 - digital multimeter
- 1 - oscilloscope

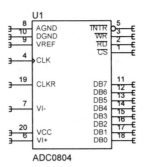

Figure 31.1: Package outline and pin identification for the ADC0804.

PART 1: The Unipolar ADC
Background:
The package outline and the pinouts for an ADC0804 are shown in Figure 31.1. A typical DAC circuit is shown in Figure 31.4. This particular ADC configuration is unipolar, because the range of digital outputs correspond to all positive input voltages. The ADC0801 through ADC0805 series of analog-to-digital converters accepts analog inputs from 0.0 to +5.0 volts. The ADC0800 accepts analog inputs in the range from -5.0 to +5.0 volts; hence it can be operated in the bipolar mode. The ADC has eight binary output pins (DB0 through DB7) that are compatible with the data bus of most microprocessors. The binary output pattern corresponds to the differential voltage applied to the $V_{IN(+)}$ and $V_{IN(-)}$ inputs. For example, if the voltage applied to the $V_{IN(+)}$ input is +3.0 volts and the voltage applied to the $V_{IN(-)}$ input is +1.0 volt, the binary pattern on the output will correspond to the differential voltage of +2.0 volts. If the $V_{IN(-)}$ input is grounded and +3.0 volts is applied to the $V_{IN(+)}$ input, then the binary output pattern will correspond to the differential voltage of +3.0 volts. This feature allows designers to apply input voltages that vary either from some voltage above ground to some maximum or from ground potential to some maximum. The maximum allowable differential voltage is specified in the manufacturer's data tables. Exceeding the maximum allowable input to V_{CC}, $V_{IN(+)}$, $V_{IN(-)}$, or to $V_{REF}/2$ will cause breakdown of the ADC's internal components. V_{CC} for the ADC0804 is typically +5.0 volts, but it may be as high as +6.5 volts. The span, or range, of the analog input voltage is determined either by the voltage applied to the V_{CC} pin or by the value applied to the $V_{REF}/2$ input (pin 9). If the $V_{REF}/2$ input is not connected, then the span of the analog input defaults to the value of V_{CC}. In other words, if V_{CC} is +5.0 volts and the $V_{IN(-)}$ input is at ground potential, then the range of analog voltage that may be applied to the $V_{IN(+)}$ input is from 0.0 to +5.0 volts. According to the National Semiconductor® <u>Data Acquisition Databook</u>, the maximum analog input voltage range for the ADC0801 through ADC0805 series of analog-to-digital converter is V_{CC} + 0.05 volts. Because $V_{CC(MAX)}$ for this series of devices is +6.5 volts, then the maximum range is +6.55 volts. If a range of voltages other than that established by V_{CC} is desired, then the $V_{REF}/2$ input may be used. If the voltage applied to the $V_{REF}/2$ pin is +2.0 volts, then the value for V_{REF} is $(2)\cdot(+2.0 \text{ volts})$, or +4.0 volts. As a formula, we could write this as

$$V_{REF} = (2) \cdot \text{(the voltage applied to the } V_{REF}/2 \text{ input)} \qquad \textbf{31.1}$$

V_{REF} then represents the *span* or *range* of analog input voltages. With a V_{REF} of +4.0 volts (+2.0 volts applied to the $V_{REF}/2$ pin) and +1.0 volt applied to the $V_{IN(-)}$ input, the valid range of analog input voltages that could then be applied to the $V_{IN(+)}$ input would be from +1.0 to +5.0 volts, where an input of +1.0 volts would correspond to a binary output pattern of 0000 0000. An analog input of +5.0 volts would then correspond to a binary output pattern of 1111 1111. Now this is important to remember, because anytime that you apply a voltage to the $V_{IN(-)}$ pin you must connect the V_{REF} input with a voltage equal to your expected range of analog input voltage divided by 2. Pin 8 is the ADC's common analog ground connection, while pin 10 is the ADC's common ground for digital signals. The successive approximation ADC requires a clock input at pin 4, CLK_{IN}. This may be an externally generated signal or, as shown in Figure 31.4, an external R-C network powered by the ADC's CLK_R output will provide a clock frequency of

$$f_{CLK} = 1/[(1.1)\cdot(R)\cdot(C)] \qquad \textbf{31.2}$$

The typical value of f_{CLK} for the ADC0801 through ADC0805 series of ADCs is 640 kHz, with a maximum of 1460 kHz.

The remaining four pins to be discussed include active-low inputs CS/, RD/, WR/, and active-low output INTR/. The DAC that was studied in Experiment 30 continuously converts its digital input to the corresponding analog output once power is applied. That is not the case with the ADC. The ADC must go through an initialization sequence to start the process of converting the analog input to the corresponding digital output. The CS/ (*Chip Select*) input has two functions. First, CS/ and WR/ (*Start Conversion*) must both be low at the same time to reset the ADC's internal shift register and successive-approximation latches. As long as both of these inputs are held low, the ADC will remain in its reset mode. CS/ must be brought low before or at the same instant that WR/ is brought low. The conversion process will be initialized when either CS/ or WR/ is brought high. During the conversion process, INTR/ (*End-of-Conversion*) is high. The INTR/ output will signal the end of conversion by a high-to-low transition. This signal may be connected to a microprocessor to signal that valid digital output data is available from the ADC, or this signal may be connected to the WR/ input to initialize another conversion cycle. Remember that I said that the CS/ input has two functions. In addition to acting in conjunction with WR/ to reset the ADC and initialize the conversion process, the CS/ input also acts in concert with the RD/ (*Output Enable*) to enable the Tri-State® output latches. A typical microcomputer interface to the ADC would involve connecting the ADC's WR/ and RD/ inputs to the microprocessors WR/ and RD/ lines. Refer to Figure 31.2 as a simplified model of an interface of the ADC to a microcomputer's bus.

A program could then be written to output a particular address on the microcomputer's address lines via the address bus. This address would be then be latched and decoded by some form of combinational logic circuitry, PAL, or other logic device that would in turn generate an active-low output chip-select signal that would be applied to the CS/ input of the ADC. At the same time, or a few microseconds (or nanoseconds) later, the WR/ line would go low, indicating that the microprocessor is writing data to the data bus. As I mentioned earlier, taking both the CS/ and WR/ lines low would then cause the ADC's internal registers to be cleared and the conversion process to start when the WR/ line returned high upon the microprocessor's completion of writing data to the data bus. If the INTR/ line from the ADC were connected to the microcomputer, then handshaking could occur. The ADC would, in effect, be able to inform the microcomputer that it had completed the conversion process by taking the INTR/ line low. This would then cause the microcomputer to initiate a read cycle, during which the ADC's data would be transferred from one of the microcomputers I/O ports (from the ADC) to the microprocessor's data bus. The I/O port address to be read from will again be decoded to cause the CS/ line to go low, and upon initiation of the read cycle, the RD/ line would also go low, enabling the ADC's three-state output latches putting the ADC's output data on the data bus. Because this is an 8-bit ADC, the total number of different output combinations is $2^8 = 256$ (0000 0000 through 1111 1111). The general formula for the number of different output binary patterns for an n-bit ADC is as follows:

$$\text{OUTPUT COMBINATIONS} = 2^n \qquad \textbf{31.3}$$

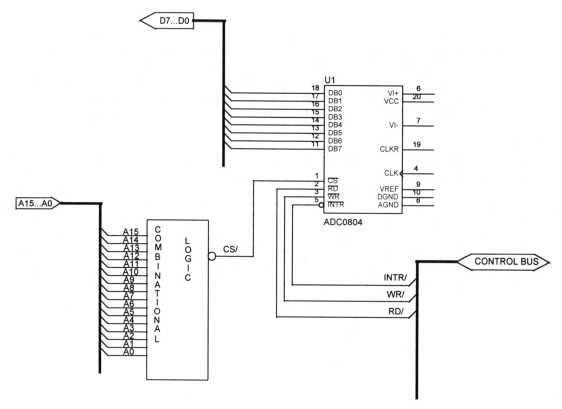

Figure 31.2: Simplified circuit for interfacing an ADC to a microprocessor.

As with the DAC, the relationship between the reference voltage, V_{REF}, and the full-scale input voltage (the voltage that will produce a binary output pattern of 1111 1111), V_{FS}, is as follows:

$$V_{FS} = [(2^n - 1)/(2^n)] \cdot V_{REF} \qquad \textbf{31.4}$$

Referring to the ADC circuit shown in Figure 31.4, the voltage applied to the $V_{REF}/2$ pin is +2.50 volts. The resulting value for V_{REF} then is (2)·(2.50 V), or +5.00 volts. The value for V_{FS} then becomes

$$V_{FS} = [(2^8 - 1)/(2^8)] \cdot (5.00 \text{ volts})$$

$$= +4.98 \text{ volts}$$

With the $V_{IN(-)}$ input at ground potential as shown in Figure 31.4 and +4.98 volts applied to the $V_{IN(+)}$ input, a binary pattern of 1111 1111 should be produced at the output. The mathematical relationship between the reference voltage, the differential analog input voltage, and the decimal value, D_{EQ}, of the resulting binary output pattern can be expressed as follows:

$$V_{DIFF(IN)} = [(D_{EQ})/(2^n)] \cdot (V_{REF}) \qquad \textbf{31.5}$$

As a reminder, the formula for $V_{DIFF(IN)}$ is

$$V_{DIFF(IN)} = V_{IN(+)} - V_{IN(-)} \qquad \textbf{31.6}$$

Rearranging Formula 31.5 and solving for D_{EQ}, we have

$$D_{EQ} = [(V_{DIFF(IN)}/V_{REF})] \cdot (2^n) \qquad \textbf{31.7}$$

For example, an analog input voltage of +4.004 volts at the $V_{IN(+)}$ input of the ADC shown in Figure 31.4 will result in a $V_{DIFF(IN)}$ of $V_{IN(+)} - V_{IN(-)}$ = +4.004 volts - 0.0 volt = +4.004 volts. The corresponding decimal equivalent can then be found as follows:

$$D_{EQ} = (4.004 \text{ volts}/5.00 \text{ volts}) \cdot (2^8)$$

$$= 205$$

The corresponding binary output pattern is 1100 1101 (CD_{16}). As you learned in Experiment 30, the DAC is capable of producing only a finite number of different analog output voltages (256 for an 8-bit DAC). Similarly, an 8-bit ADC can produce *only* 256 different binary (digital) output patterns. I say *only*, because there are an infinite number of different analog voltages possible that can be applied to the input, even with a range of voltages as small as 5 volts. As a result, a small degree of inaccuracy is inherent with our ADC. The output of our ADC will look somewhat like a staircase as shown in Figure 31.3. The height of each step, or the step size, is referred to as the *resolution* of our ADC. In Figure 31.3, the resolution is 1 volt since a 1-bit change in the binary output corresponds to a change of 1 volt at the analog input (a change in V_{DIFF} of 1 volt). By definition, the resolution is the smallest change in analog input voltage that will result in a change of 1 LSB in the binary output. For example, if the binary output is 0000 0101 and changes to 0000 0110, this represents a 1-bit increase. In Figure 31.3, this represents a change from +5.0 to +6.0 volts, but what if the input voltage is somewhere between +5.0 and +6.0 volts? In other words, what output binary pattern would result if we had an input voltage of +5.5 volts? Well, as it turns out, any voltage from +5.0 volts to just below +6.0 volts (say, +5.99 volts) will result in a binary pattern of 0000 0101. As a result, we could have an output pattern of 0000 0101 (05_{16}) for an input of +5.99 volts, which would represent an error of 0.99 volt (5.99 V - 5.00 V), or an error of 1 LSB. In other words, since 5.99 volts is approximately 6.0 volts, we probably would expect a binary output pattern of 0000 0110. The reality is that we are 1 LSB below this for an error of -1 LSB. So, the overall possible range of our error can be from 0 to -1 LSB. Now let's consider the resolution of the circuit shown in Figure 31.4. Because the resolution represents a change of 1 LSB, let's determine the analog input voltage that will produce a binary output from our ADC of 0000 0001. Applying Formula 31.5 for a binary 0000 0001 (decimal 1), we have

$$V_{DIFF(IN)} = (1/2^8) \cdot (5.00 \text{ volts})$$

$$= 0.0195 \text{ volt}$$

$$= 19.5 \text{ mV}$$

For binary 0000 0010 (decimal 2), we have

$$V_{DIFF(IN)} = (2/2^8) \cdot (5.00 \text{ volts})$$

$$= 0.039 \text{ volt}$$

$$= 39 \text{ mV}$$

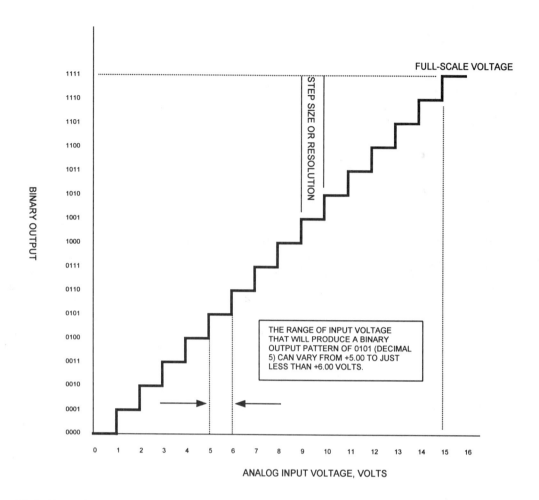

Figure 31.3: Plot of the binary output pattern as a function of the analog input voltage applied to a 4-bit ADC. The step size or resolution is 1.0 volt.

For the circuit shown in Figure 31.4, a differential input of 19.5 mV will produce a binary output pattern of 0000 0001 and a differential input of 39 mV will produce an output pattern of 0000 0010. This also represents a change of 1 bit and an input voltage change of 19.5 mV (39 mV - 19.5 mV). What about the voltages between 19.5 and 39 mV. Well, this 19.5 mV range (39 mV - 19.5 mV) from one bit pattern to the next higher one represents the *uncertainty* or *resolution* of our ADC. Differential input voltages in the range of 19.5 through 38 mV will produce an output pattern of 0000 0001. As noted previously, this represents an infinite number of different possibilities. However, the ADC is limited to a finite number of output combinations. Because the binary output range, or change, over which the 19.5 mV input change occurs is 1 binary bit, the uncertainty or quantizing error is 1 bit (1 LSB). This leads to a general formula for the resolution of our ADC.

$$\text{RESOLUTION} = V_{REF}/2^n \qquad \textbf{31.8}$$

If we solve Formula 31.4 for V_{REF} and substitute into Formula 31.8, we have a formula for the resolution in terms of V_{FS}.

$$\text{RESOLUTION} = V_{FS}/(2^n - 1) \qquad \textbf{31.9}$$

NOTE: Like most digital devices, ADCs are capable of supplying only a few milliamperes of current. To drive devices requiring much more than 10 milliamperes per output, it will be necessary to install a buffer between the ADC's digital outputs and the load(s) to be driven. You must find out what the current requirements of your final load are and determine the rated output current capabilities of the ADC by referring to the manufacturer's data sheets. Also keep in mind that there is usually a difference between the amount of current that an ADC can source versus what it can sink.

Procedure:

1.1 Obtain all materials and equipment required to complete this experiment.

1.2 Use your multimeter to measure the resistance of each of your resistors. If any of them are out of tolerance, inform your instructor and obtain replacements.

1.3 If your lab has one, use a capacitance tester to check the capacitance values of your capacitors. If any of them are out of tolerance, inform your instructor and obtain replacements.

1.4 Examine the circuit shown in Figure 31.4. If you look closely, you will see that different ground symbols have been used at various points on the circuit. This is an attempt to simulate as closely as possible the manufacturer's recommendations to establish both digital and analog grounds when interfacing an ADC to a microcomputer (which is what we are trying to "simulate" with this circuit). Note the different schematic symbols for the digital and analog grounds shown on your schematic. Because the resolution of this ADC circuit is so small, it is susceptible to error due to low-level noise. To get results that are as accurate as possible, you should try to establish digital and analog grounds that are as solidly at ground potential as possible; otherwise, you may not be able to get a true 0000 0000 output from your ADC when the $V_{IN(+)}$ sees just a few millivolts of noise. It has been my experience that this can happen when using long test leads from power supplies and voltmeters and solderless breadboards that are either cheaply constructed or have seen a lot of "wear and tear." I suggest that you use one common strip, or bus, on your breadboard for the digital grounds and another for the analog grounds. Once both of these grounds are established, connect them together with jumper wires at two different points. This should help to ensure that you have a well-grounded circuit. Refer to Figure 31.4 as a guide.

1.5 The 74LS240 is being used as a buffer to drive the LED segments rather than having the ADC driving the LEDs. The inverting output of the 74LS240 will also result in a lit LED when the output of the ADC is high and each LED will be off when the corresponding ADC output is low.

1.6 Before constructing the circuit, use your ohmmeter to adjust the resistance of each of the pots to its midrange.

1.7 Adjust the output of one of your power supplies to +5.00 volts. This voltage and all remaining voltage adjustments must be made as precisely as possible. You should select a voltmeter range that will allow you to make millivolt readings. Adjust the output of the other power supply outlet (or the other supply if your are using two different supplies) to +10.0 volts.

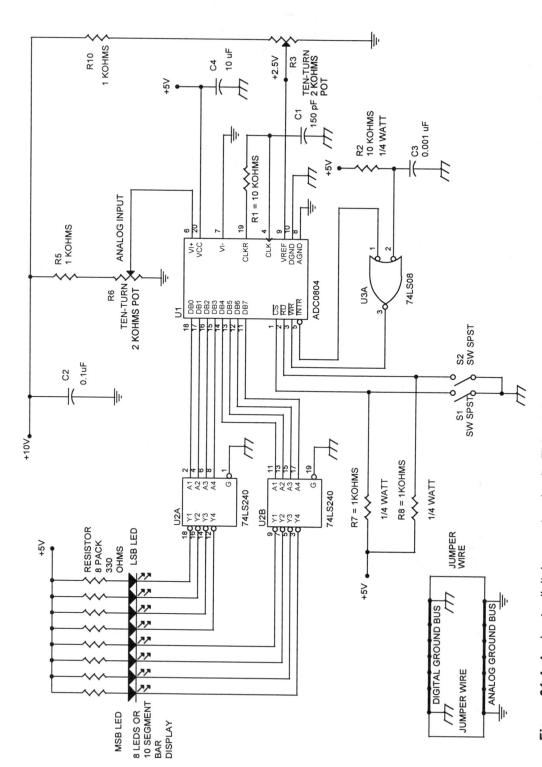

Figure 31.4: Analog-to-digital converter circuit. This ADC operates in the unipolar mode. VI(−) is at ground potential.

1.8 Turn off the power supplies and construct the circuit shown in Figure 31.4.

1.9 Close both switches S1 and S2.

1.10 When you turn on your power supply, the voltage at C3 will momentarily be at ground potential. This will act as a power-on start-up signal to the ADC, initializing the analog-to-digital conversion process. Turn on your power supply(ies).

1.11 Adjust potentiometer R3 until the voltage applied to pin 9 ($V_{REF}/2$) is +2.50 volts. Refer to National Semiconductor's® Data Acquisition Databook for a more detailed description of the adjustment procedure for R3 by reading the section describing full-scale adjustment.

1.12 The voltage from R6 is the simulated analog input voltage from a transducer.

1.13 For each of the binary output patterns shown in Table 31.1, measure and record the corresponding input voltage, $V_{IN(+)}$. Adjust pot R6 until the specified pattern just starts to light.

1.14 Using the formulas developed on the preceding pages, calculate the theoretical value for $V_{IN(+)}$ for the binary input patterns shown in Table 31.2 for the circuit shown in Figure 31.4.

1.15 Now adjust the analog input voltage until it is +2.0 volts. Make a mental note of the binary pattern. Open switch S2. What happened to the LED display? Record your observation in the space provided below.

Observation:_____

_____

1.16 With S2 still open, adjust the analog input voltage until it is +3.0 volts. What, if anything, happened to the LED display? Now close switch S2. Did the LED display change? If so, how? Record your observations and responses in the space provided below.

Observation:_____

_____

_____

_____

1.17 Now adjust the analog input voltage until it is +2.0 volts. Make a mental note of the binary pattern. Open switch S1. What happened to the LED display? Record your observation in the space provided below.

Observation:_____

_____

1.18 With S1 still open, adjust the analog input voltage until it is +3.0 volts. What, if anything, happened to the LED display? Now close switch S1. Did the LED display change? If so, how? Record your observations and responses in the space provided below.

Observation:_____

_____

_____

_____

Table 31.1: Binary output pattern and corresponding differential input voltage for the unipolar ADC shown in Figure 31.4.

DIFFERENTIAL INPUT VOLTAGE, VOLTS	BINARY OUTPUT PATTERN	DECIMAL EQUIVALENT
	0000 0000	
	0000 0001	
	0000 0010	
	0000 0011	
	0000 0100	
	0000 1000	
	0001 0000	
	0010 0000	
	0011 0000	
	0100 0000	
	0110 0000	
	0111 0000	
	1000 0000	
	1100 0000	
	1110 0000	
	1111 0000	
	1111 1110	
	1111 1111	

Table 31.2: Binary output pattern and calculated differential input voltage pattern and decimal equivalent for the unipolar ADC for the circuit shown in Figure 31.4.

DIFFERENTIAL INPUT VOLTAGE, VOLTS	BINARY OUTPUT PATTERN	DECIMAL EQUIVALENT
	0000 0000	
	0000 0001	
	0000 0010	
	0000 0011	
	0000 0100	
	0000 1000	
	0001 0000	
	0010 0000	
	0011 0000	
	0100 0000	
	0110 0000	
	0111 0000	
	1000 0000	
	1100 0000	
	1110 0000	
	1111 0000	
	1111 1110	
	1111 1111	

1.19 Use your oscilloscope to display the waveform at pins 4 and 5 in proper time phase. Draw these two waveforms in the space provided on Graph 31.1. What is the relationship between these two waveforms? Specifically, what is the relationship between the frequency of the $CLK_{(IN)}$ signal and the frequency of the INTR/ output.

Answer:_____

_____

_____

Using the period of the $CLK_{(IN)}$ waveform from you oscilloscope, calculate its frequency (Remember: $f = 1/T$ where T is the period of the waveform). Also use Formula 31.2 to calculate the theoretical value of this waveform's frequency.

Calculations:

$f_{CLK} = 1/T =$

$f_{CLK} = 1/[(1.1) \cdot (R) \cdot (C)] =$

How do the two values compare? Explain any differences.

Answer: _____

_____

_____

1.20 While monitoring the INTR/ signal on your oscilloscope, open switch S1. What happens to INTR/ when CS/ is brought high?

Answer: _____

_____

_____

1.21 Close switch S1. While monitoring the INTR/ signal on your oscilloscope, open switch S2. What, if anything, happens to INTR/ when RD/ is brought high?

Answer: _____

_____

_____

Graph 31.1: CLK$_{(IN)}$ and INTR/ in proper time phase.

CLK$_{(IN)}$: _____ volts/div _____ sec/div
INTR/: _____ volts/div _____ sec/div

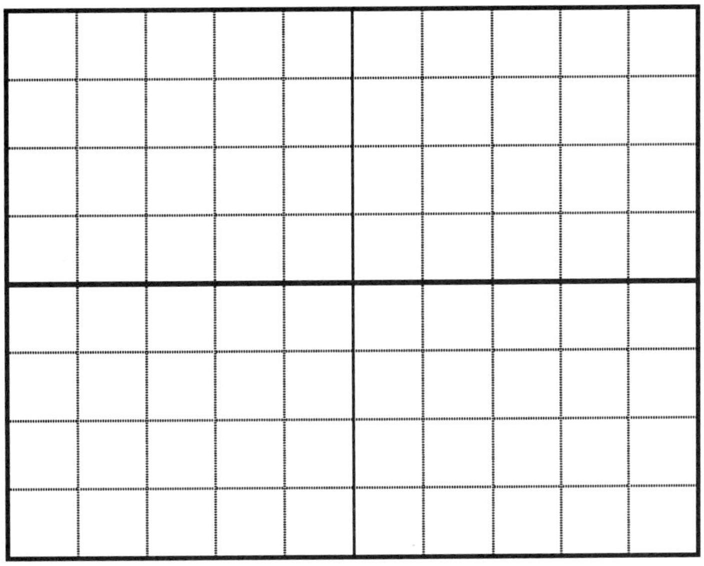

1.22 Turn off all power. Construct the circuit shown in Figure 31.5. Note that the V$_{REF/2}$ input is no longer +2.50 volts but should be adjusted to +1.50 volts. Also note that V$_{IN(-)}$ is no longer grounded. Adjust the voltage applied to V$_{IN(-)}$ to +1.0 volt.
1.23 Switches S1 and S2 should be closed. Turn on the power supply(ies).
1.24 For each of the binary output patterns shown in Table 31.3, measure and record the corresponding input voltage, V$_{IN(+)}$. Adjust pot R6 until the specified pattern just starts to light.
1.25 Using the formulas developed on the preceding pages, calculate the theoretical value for V$_{IN(+)}$ for the binary input patterns shown in Table 31.4 for the circuit shown in Figure 31.5.

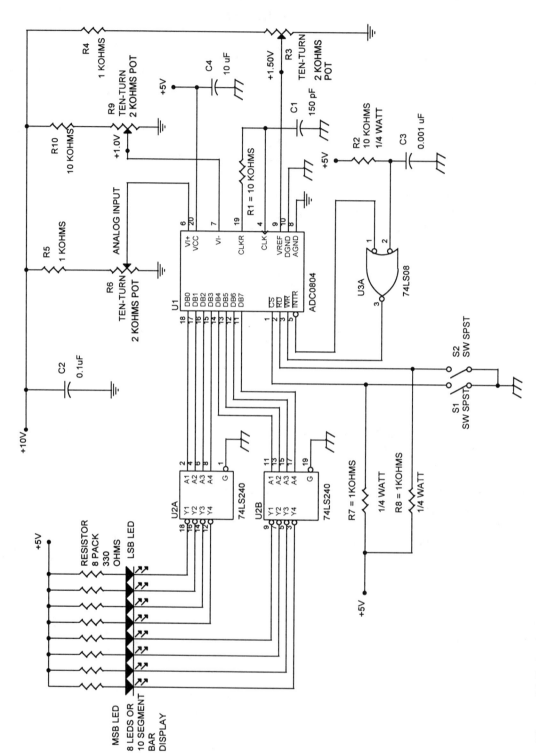

Figure 31.5: Analog-to-digital converter circuit. This ADC operates in the unipolar mode. VI(-) has an input of +1.0 volt.

Table 31.3: Binary output pattern and corresponding differential input voltage for the unipolar ADC shown in Figure 31.5.

DIFFERENTIAL INPUT VOLTAGE, VOLTS	BINARY OUTPUT PATTERN	DECIMAL EQUIVALENT
	0000 0000	
	0000 0001	
	0000 0010	
	0000 0011	
	0000 0100	
	0000 1000	
	0001 0000	
	0010 0000	
	0011 0000	
	0100 0000	
	0110 0000	
	0111 0000	
	1000 0000	
	1100 0000	
	1110 0000	
	1111 0000	
	1111 1110	
	1111 1111	

Table 31.4: Binary output pattern and calculated differential input voltage pattern and decimal equivalent for the unipolar ADC for the circuit shown in Figure 31.5.

DIFFERENTIAL INPUT VOLTAGE, VOLTS	BINARY OUTPUT PATTERN	DECIMAL EQUIVALENT
	0000 0000	
	0000 0001	
	0000 0010	
	0000 0011	
	0000 0100	
	0000 1000	
	0001 0000	
	0010 0000	
	0011 0000	
	0100 0000	
	0110 0000	
	0111 0000	
	1000 0000	
	1100 0000	
	1110 0000	
	1111 0000	
	1111 1110	
	1111 1111	

PART 2: Questions

2.1 On a piece of graph paper plot the data from Table 31.1 showing the binary output pattern (expressed as a decimal) as a function of the analog input voltage. In other words, plot the decimal equivalent of the binary output pattern on the Y-axis and the analog input voltage on the X-axis. Attach your graph to your lab report.

2.2 Based on your graph from Question 2.1, what can you say about the output response or characteristic of your ADC?

2.3 Based on your results from step 1.20, what affect does the logic level of the CS/ input have on the analog-to-digital conversion process?

2.4 Based on your results from step 1.21, what affect does the logic level of the RD/ input have on the analog-to-digital conversion process?

2.5 Referring to the data in Table 31.1, how much did the input voltage change from binary 1111 1110 to binary 1111 1111? What is the relationship of this value to the remaining values recorded in Table 31.1? What should its relationship be? Another way to ask this might be, what does this voltage change represent?

2.6 Based on your experimental results, what is the resolution of the ADC circuit shown in Figure 31.4?

2.7 Based on your experimental results, what is the resolution of the ADC circuit shown in Figure 31.5?

2.8 Explain why the answers to Questions 2.6 and 2.7 differ.

2.9 In terms of a real-world, future application, how might you take advantage of the fact that the resolution of the circuit shown in Figure 31.4 differs from that shown in Figure 31.5?

2.10 What determined the *range* of valid analog input voltages for the circuit shown in Figure 31.5? Refer to your experimental data and the circuit voltages.

2.11 What determined the analog input voltage that produced a binary output pattern of 0000 0000 for the circuit shown in Figure 31.5? Refer to your experimental data and the circuit voltages.

2.12 What binary output pattern would you expect the circuit shown in Figure 31.5 to produce if the analog input voltage to $V_{IN(+)}$ was +0.5 volt?

PART 3: Practice Problems

3.1 Assume that you are designing a circuit like that shown in Figure 31.4 but you require a 10-bit ADC because of the need for a finer resolution. V_{CC} is +5.0 volts, and the V_{REF} input is not connected. $V_{IN(-)}$ is connected to ground. Answer the following questions.

   a) How many different binary output combinations are possible?

b) What is the value of the full-scale input voltage that will produce an output binary pattern of 11 1111 1111?

c) What is the value of the analog input voltage that will produce a binary output pattern of 10 0000 0000?

d) What binary output pattern will result when the analog input voltage is +2.0 volts?

e) What is the resolution of this ADC?

3.2 Now assume that you modify the circuit described in Problem 3.1 so that there is +2.0 volts applied to the V_{REF} input and +0.5 volt applied to the $V_{IN(-)}$ input. Answer the following questions.

a) What is the value of the analog input voltage that will produce a binary output pattern of 00 0000 0000?

b) What is the value of the full-scale input voltage that will produce an output binary pattern of 11 1111 1111?

c) Now what is the resolution of this ADC configuration?

PART 4: Unipolar ADC Formulas

$V_{REF} = (2) \cdot$ (the voltage applied to the $V_{REF}/2$ input)   **31.1**

$f_{CLK} = 1/[(1.1) \cdot (R) \cdot (C)]$ **31.2**

OUTPUT COMBINATIONS $= 2^n$ **31.3**

$V_{FS} = [(2^n - 1)/(2^n)] \cdot V_{REF}$ **31.4**

$V_{DIFF(IN)} = [(D_{EQ})/(2^n)] \cdot (V_{REF})$ **31.5**

$V_{DIFF(IN)} = V_{IN(+)} - V_{IN(-)}$ **31.6**

$D_{EQ} = [(V_{DIFF(IN)}/V_{REF})] \cdot (2^n)$ **31.7**

RESOLUTION $= V_{REF}/2^n$ **31.8**

RESOLUTION $= V_{FS}/(2^n - 1)$ **31.9**

The 555 Timer: 32

INTRODUCTION:
The 555 timer is a versatile, low-power I. C. capable of performing a variety of timing functions. In this experiment, you will construct two common 555 timer circuits. One is the *astable multivibrator*, and the other is the *monostable multivibrator*. A monostable multivibrator (or *one-shot* as it is sometimes called) as the name suggests is stable in only one state. As you will see when completing the second part of this experiment, the output of the 555 timer, when configured as a one-shot, will momentarily go high and then return low. The low state represents its stable state. A bi-stable multivibrator such as a flip-flop is stable in either of two states. The output of a flip-flop can be either high or low and stay in either state for an indefinite period of time. The 555 timer, when configured as an astable multivibrator, has no stable state. In this configuration, the 555 timer acts as an oscillator with its output continuously toggling between a high and a low state. This mode of operation is also referred to as *free-running*.

SAFETY NOTE:
While the 555 timer can accept a wide range of input voltages (typically 5 to 15 volts), it is still a low-power device. Therefore, be sure to read and understand the manufacturer's data sheet for this device before using it in any future applications. As always, wear safety glasses, and remove all jewelry from your hands and fingers before the start of the experiment.

OBJECT:
Upon successful completion of this experiment and all reading assignments, the student should be able to:
- use the 555 timer to construct an operational astable multivibrator circuit
- use the 555 timer to construct an operational monostable multivibrator circuit
- calculate the output time high and low for a given 555 timer astable multivibrator circuit
- calculate the output time high for a given 555 timer monostable multivibrator circuit

REFERENCES:
National Semiconductor's® Linear Application Specific IC's Databook
www.national.com

MATERIALS:
- 1 - LM555 or the equivalent
- 1 - 12 Kohms resistors, 1/4 watt
- 1 - 15 Kohms resistor, 1/4 watt
- 1 - 10 Kohms resistor, 1/4 watt
- 2 - 1 Kohms resistors, 1/4 watt
- 1 - 270 Kohms resistor, 1/4 watt
- 1 - transistor, 2N3904 or the equivalent
- 1 - LED
- 1 - ceramic or tantalum capacitor, 0.01 µF
- 1 - ceramic or tantalum capacitor, 0.1 µF
- 1 - capacitor, 10 µF
- 1 - diode, 1N914
- 1 - normally open pushbutton switch
- 1 - solderless breadboard
- miscellaneous lead wires and connectors

EQUIPMENT:
1 - D.C. power supply
1 - digital multimeter
1 - oscilloscope
1 - oscilloscope probe
1 - stop watch
1 - function generator

Figure 32.1: Schematic symbol for the 555 timer.

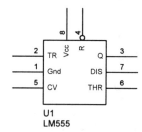

PART 1: The Astable Multivibrator
Background:
The package outline and the pinouts for the LM555 are shown in Figure 32.1. Consult the manufacturer's data manual or an electronic devices book for a detailed description of the function of each pin. A typical 555 astable multivibrator circuit is shown in Figure 32.2. V_{CC} is applied to pin 8 and ground to pin 1. V_{CC} and ground set up two internal reference voltages, which are applied to an op-amp comparator. If a voltage level other than that established by the internal resistor network is desired, it is possible to apply a different control voltage at pin 5, C_V. In this configuration, C2, a tantalum or ceramic capacitor, is connected to pin 5 to act as a decoupling capacitor to pass any noise to ground. The circuit will work without C2 but it is not recommended by the manufacturer. In Figure 32.2, resistors R_A and R_B and capacitor C1 establish an R-C network that will continuously charge up the capacitor through R_A and R_B, then discharge it through R_B into the 555 timer at pin 7, DIS. Pin 3, Q, is where the timer's oscillating output voltage will appear. The time that the output voltage waveform is high, T_{HIGH}, can be found from the following formula:

$$T_{HIGH} = (R_A + R_B) \cdot (C_1) \cdot \text{Ln}(2) \qquad \textbf{32.1}$$

where Ln (2) is the natural logarithm of 2. Substituting the value for Ln (2) in Formula 32.1, we now have

$$T_{HIGH} = 0.693(R_A + R_B) \cdot (C_1) \qquad \textbf{32.2}$$

The time that the output voltage waveform is low, T_{LOW}, can be found from the following formula:

$$T_{LOW} = (R_B) \cdot (C_1) \cdot \text{Ln}(2) \qquad \textbf{32.3}$$

Substituting the value for Ln (2) in Formula 32.3, we now have

$$T_{LOW} = 0.693(R_B) \cdot (C_1) \qquad \textbf{32.4}$$

When working with oscillator circuits like the astable multivibrator circuit, it is sometime desirable to calculate the *duty cycle* of the output voltage. This is important, because we will often use the output of the timer to drive other devices that are limited by the amount of time that they can be high or on. As mentioned in the relay lab, solenoids and many other devices are often rated for intermittent duty while some are rated for continuous duty. A device that is rated for intermittent duty can only be on for a certain amount of time, then must be off so that it may cool down. If an electrical or electronic device that is rated for intermittent duty is kept on continuously, it will eventually fail due to overheating. The duty cycle is defined as a percent of the time that a waveform or a device is high or on. For our 555 timer circuit, the duty cycle can be found as follows:

$$\text{DUTY CYCLE} = [(T_{HIGH})/(T_{HGH} + T_{LOW})] \cdot 100\% \qquad \textbf{32.5}$$

Procedure:

1.1 Obtain all materials and equipment required to complete this experiment.
1.2 Use your multimeter to measure the resistance of each of your resistors. If any of them are out of tolerance, inform your instructor and obtain replacements. Record the measured value of the resistors in Table 32.1.
1.3 If your lab has one, use a capacitance tester to check the capacitance values of your capacitors. If any of them are out of tolerance, inform your instructor and obtain replacements. Record the measured capacitance values in Table 32.1.

Table 32.1: Nominal and measured component values.

NOMINAL OR RATED COMPONENT VALUES	MEASURED COMPONENT VALUES
R1 = 1 KOHMS	
RA = 12 KOHMS	
RB = 15 KOHMS	
RA = 270 KOHMS	
R2 = 10 KOHMS	
C1 = 0.1 uF	
C2 = 0.01 uF	
C1 = 10 uF	

1.4 Construct the circuit shown in Figure 32.2.
1.5 Turn on the power supply.

Figure 32.2: 555 timer configured as an astable multivibrator.

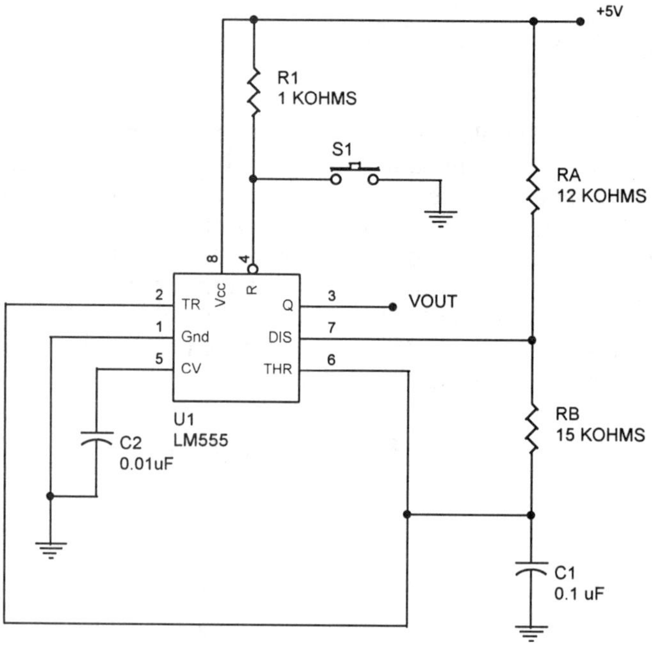

1.6 Use an oscilloscope to display the voltage waveform at pin 3, V_{OUT}. Record the resulting waveform on Graph 32.1.

1.7 Press and hold S1. What happens to the voltage waveform, V_{OUT}? Record your observation in the space provided below.

Observation:_____

_____

1.8 Turn off the power supply. Use the measured values for R_A, R_B, and C1 from Table 32.1 and Formulas 32.2 and 32.4 to <u>calculate</u> the <u>theoretical</u> time high and time low for the output voltage waveform for the circuit shown in Figure 32.2.

Calculations:

$T_{HIGH} =$

$T_{LOW} =$

1.9 Compare the calculated times high and low to the <u>measured</u> times high and low shown on Graph 32.1. If there is a significant difference between the calculated values and the measured values, check your calculations, then repeat your oscilloscope measurements.

432

Graph 32.1: V_{OUT} for the astable multivibrator circuit shown in Figure 32.2.

V_{OUT}: _____ volts/div   _____ sec/div

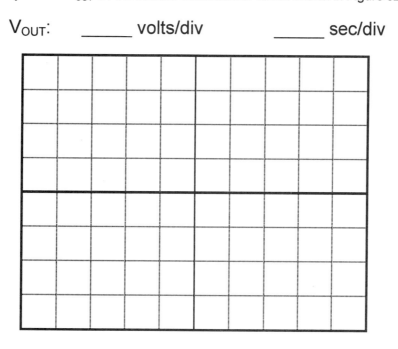

1.10 Revise your circuit so that it appears as in Figure 32.3.
1.11 Again display the voltage waveform V_{OUT} on your oscilloscope. Record on Graph 32.2.

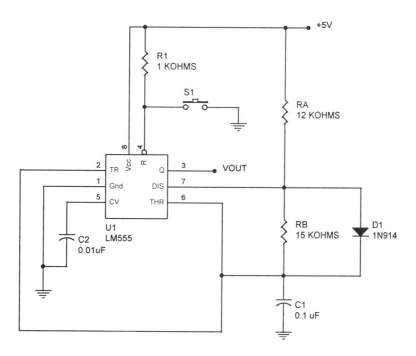

Figure 32.3: 555 timer configured as an astable multivibrator with a 50% duty cycle.

Graph 32.2: V_{OUT} for the astable multivibrator circuit shown in Figure 32.3.

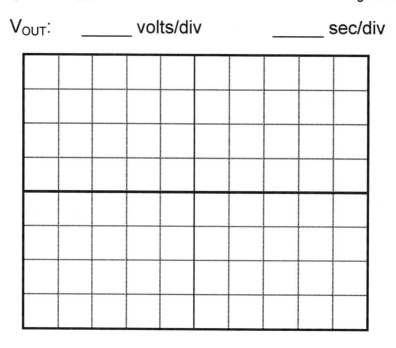

V_{OUT}: _____ volts/div   _____ sec/div

PART 2: The Monostable Multivibrator
Background:
A typical 555 monostable multivibrator circuit is shown in Figure 32.4. In Figure 32.4, resistor R_A and capacitor C1 establish an R-C network that will charge up the capacitor through R_A and then will discharge through the 555 timer at pin 7, DIS. To initiate the timer, a momentary low pulse must be applied to pin 2, TR, trigger. The time that the output voltage waveform is high, T_{HIGH}, can be found from the following formula:

$$T_{HIGH} = (R_A) \cdot (C_1) \cdot \text{Ln}(3) \qquad \textbf{32.6}$$

Substituting the value for Ln (3) in Formula 32.6, we now have

$$T_{HIGH} = 1.1(R_A) \cdot (C_1) \qquad \textbf{32.7}$$

Procedure:
2.1 Construct the circuit shown in Figure 32.4.
2.2 Turn on the power supply.
2.3 In this circuit, the LED will be used to indicate the time that the output of the 555 timer is high. The LED will be off when the output of the 555 timer is high. You will need a digital watch or stop watch to record the time high.

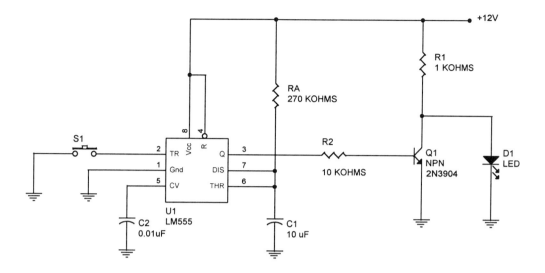

Figure 32.4: 555 timer configured as a monostable multivibrator. The output of the 555 timer is used to drive the base of a transistor switch.

2.4 Momentarily press and release switch S1. Record the time that the LED is off.
2.5 Do this a number of times to ensure the accuracy of your measurements. Record your final time high (LED off time) in the space provided below.

T_{HIGH} = _____

2.6 Turn off the power supply. Use the measured values for R_A and C1 from Table 32.1 and Formula 32.7 to calculate the theoretical time high for the 555 timer circuit shown in Figure 32.4.

Calculations:

T_{HIGH} =

2.7 Compare the <u>calculated</u> time high to the <u>measured</u> time high. If there is a significant difference between the calculated value and the measured value, check your calculations, then repeat your measurement.
2.8 Turn on the power supply. Press and hold switch S1. Do not release it until 30 seconds have passed. Record the <u>total time</u> that the LED is off (time that the Q output is high).

T_{HIGH} = _____

2.9 Release the switch. What happened when you released the switch? Record your observation in the space provided below.

Observation:_____

_____

PART 3: The Voltage-Controlled Oscillator
Background:
The 555 timer is a versatile I.C. In addition to the previous two circuits, it can also be configured as a voltage-controlled oscillator, VCO. The output frequency of the astable multivibrator is determined by the external resistors and capacitor. The output frequency of a VCO is controlled, or varied, by the amplitude of the input voltage.

Procedure:
3.1 Construct the circuit shown in Figure 32.5.
3.2 Set the function generator to produce a 5 V_{P-P}, 10 Hz sawtooth waveform having +2.5 volts D.C. offset.
3.3 Turn on the D.C. power supply. Connect the function generator to the circuit.
3.4 Use an oscilloscope to display V_{OUT} and the sawtooth waveform in proper time phase.
3.5 Draw the resulting waveforms on Graph 32.3. Label both axes.

Figure 32.5: 555 timer configured as a voltage-controlled oscillator.

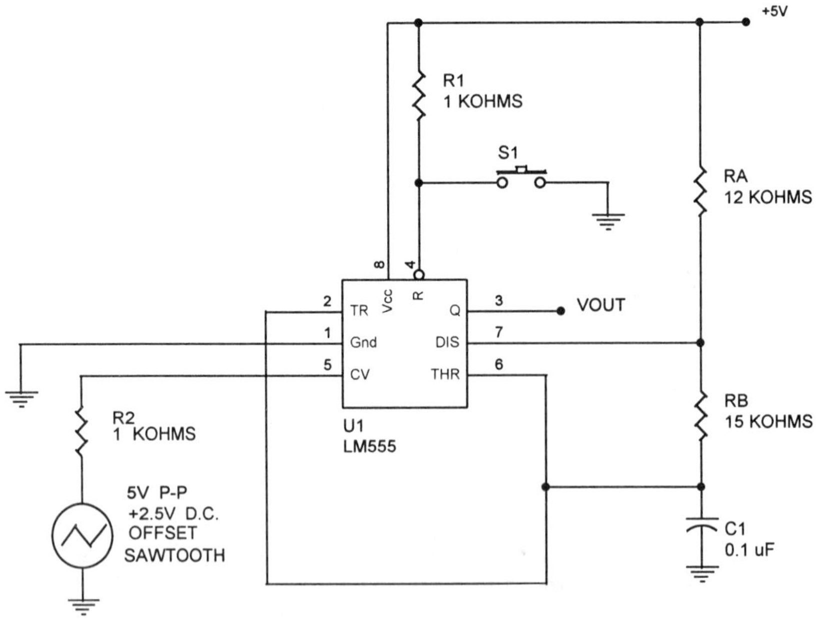

Graph 32.3: Input and output voltage waveforms for the circuit shown in Figure 32.5.

V_{OUT}: _____ volts/div _____ sec/div
V_{IN}: _____ volts/div _____ sec/div

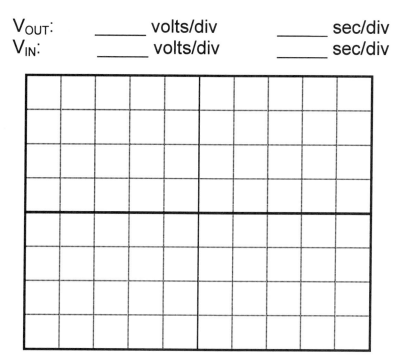

PART 4: Questions

4.1 How did the addition of the 1N914 diode to the circuit shown in Figure 32.2 change its operation?

4.2 Using your data from Graph 32.1, calculate the duty cycle of the circuit shown in Figure 32.2.

4.3 Using your data from Graph 32.2, calculate the duty cycle of the circuit shown in Figure 32.3.

4.4 What effect did pressing switch S1 have on the operation of the circuit shown in Figure 32.2?

4.5 How did pressing and holding the pushbutton for 30 seconds change the operation of the circuit shown in Figure 32.4?

4.6 What happened to the output frequency of the VCO shown in Figure 32.5 as the amplitude of the input voltage increased?

PART 5: Practice Problems

5.1 You have just built a circuit like that shown in Figure 32.2. Both R_A and R_B are 10 Kohms. $C1 = 1\ \mu F$. Answer the following questions.

a) What is the theoretical time high for the output voltage waveform?

b) What is the theoretical time low for the output voltage waveform?

c) What is the duty cycle of the output voltage waveform?

5.2 You have just built a circuit like that shown in Figure 32.4. R_A = 10 Kohms, and C1 = 1 μF. What is the theoretical time high of the output voltage pulse?

PART 6: 555 Timer Formulas

Astable Multivibrator -

$T_{HIGH} = (R_A + R_B) \cdot (C_1) \cdot \text{Ln}(2)$ **32.1**

$T_{HIGH} = 0.693(R_A + R_B) \cdot (C_1)$ **32.2**

$T_{LOW} = (R_B) \cdot (C_1) \cdot \text{Ln}(2)$ **32.3**

$T_{LOW} = 0.693(R_B) \cdot (C_1)$ **32.4**

$\text{DUTY CYCLE} = [(T_{HIGH})/(T_{HGH} + T_{LOW})] \cdot 100\%$ **32.5**

Monostable Multivibrator -

$T_{HIGH} = (R_A) \cdot (C_1) \cdot \text{Ln}(3)$ **32.6**

$T_{HIGH} = 1.1(R_A) \cdot (C_1)$ **32.7**

Frequency-to-Voltage Converters: 33

INTRODUCTION:

In Experiment 30, you used the D/A converter to convert a binary input pattern to an analog voltage. Conversely, in Experiment 31, you used an A/D converter to convert an analog input to a binary output. In Experiment 32, you used the 555 timer with an external R-C network to convert an input D.C. voltage to a square wave output. That was an example of voltage-to-frequency conversion. As is the case with the D/A and A/D converters, where it is necessary to convert from either digital to analog or vice versa, it is also sometimes necessary to convert from a D.C. voltage to an A.C. voltage or from an A.C. waveform to a D.C. voltage. The frequency-to-voltage converter (FVC) can be used to take sine wave, square wave, or sawtooth waveforms and convert them to a corresponding D.C. value. The magnitude of the D.C. value is dependent on the frequency of the input voltage. A simplified schematic of the LM2917 (SK9209) frequency-to-voltage converter is shown in Figure 33.1. For a more detailed schematic, refer to National Semiconductor's® Linear Applications Handbook. A schematic of a FVC circuit is shown in Figure 33.2. In this experiment, you will use a function generator to simulate an A.C. input of varying frequency that will be applied to pin 1 of the FVC. However, in industrial applications, the A.C. waveform will be generated by a sensor or transducer that is being used to monitor an industrial process. For example, consider Figure 33.3. A magnetic (variable reluctance) proximity pickup has been installed close to a steel gear. As the teeth pass by the pickup, the reluctance of the path of magnetism from the pickup to the gear will be constantly changing resulting in the output square wave shown. This square wave will in turn be applied to the frequency input (pin 1) of a FVC. If it was desired to keep the RPM of this gear constant, and hence the speed of the D.C. motor driving it constant, then the output of the FVC could be applied to an amplifier to control the excitation (and in turn motor speed) of the motor's armature windings. For example, if the motor started to slow down due to a sudden load, the frequency of the output waveform from the pickup would also decrease. In turn, the D.C. output of the FVC would decrease. This decreasing voltage could then be applied to an inverting current amplifier that in turn would increase the current to the motor's armature. This would then increase the motor's torque output and return its speed to the desired level. Some handheld tachometers use a notched disk like that shown in Figure 33.4 to block and unblock the light given off by some type of emitter (typically infrared) that will be received by some type of detector (photodiode in the circuit shown). The result will be a square wave output from the detector circuit as shown in Figure 33.4. The faster the notched disk rotates, the higher the frequency of the output from the detector circuit. This A.C. waveform could then be applied to a FVC to convert it to a D.C. voltage that in turn could be used for some type of process control.

SAFETY NOTE:

While the frequency-to-voltage converter I.C. can accept a wide range of supply voltages (V_{CC} typically 0 to 28 volts), it is still a low-power device. Therefore, be sure to read and understand the manufacturer's data sheet for this device before using it in any future applications. In particular, if using the LM2917 or and equivalent such as the SK9209, be sure to follow the manufacturer's recommendations when selecting power supply dropping resistors (resistor R3 in Figure 33.2). This is important to prevent excessive power dissipation of the I.C. at higher-power supply voltages (basically any V_{CC} over +5 volts). As always, wear safety glasses, and remove all jewelry from your hands and fingers before the start of the experiment.

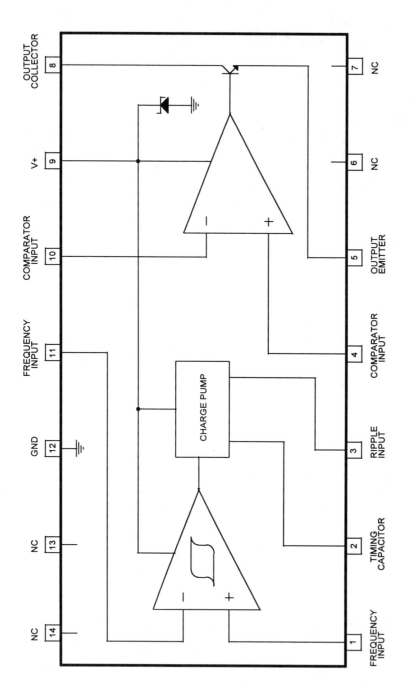

Figure 33.1: Simplified schematic diagram of a frequency-to-voltage converter showing pin identification.

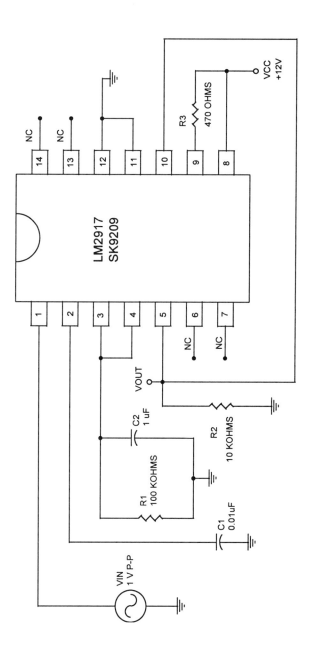

Figure 33.2: Frequency-to-voltage converter circuit.

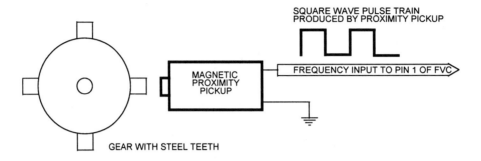

Figure 33.3: Square wave output from a proximity pickup sent to a FVC in a speed control circuit.

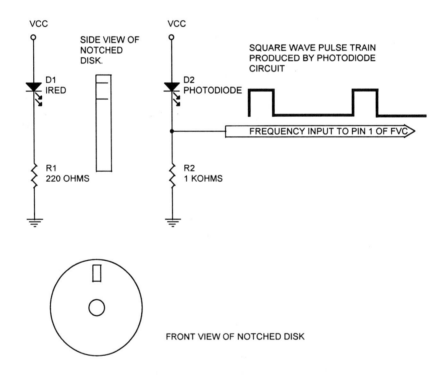

Figure 33.4: Infrared emitted by D1 is blocked and unblocked by the rotating notched disk. The result is a square wave pulse train from the photodetector circuit (photodiode) that is in turn supplied to the input of a FVC circuit.

OBJECT:
Upon successful completion of this experiment and all reading assignments, the student should be able to:
- use the LM2917, SK9209, or LM2907 to construct an operational frequency-to-voltage converter circuit
- explain the effect that changing power supply voltage has on the output of the FVC
- calculate the output voltage from a FVC circuit at a given input frequency
- explain the effect A.C. input waveform shape has on the output of the FVC
- explain the effect that D.C. offset on the input A.C. waveform has on the output of the FVC
- explain the relationship of the size (microfarad rating) of the external capacitors to the magnitude of the output of the FVC at a given frequency

REFERENCES:
National Semiconductor's® Linear Application Specific IC's Databook and Linear Applications Handbook
www.national.com

MATERIALS:
- 1 - LM2917, or SK9209, or the equivalent
- 1 - 100 Kohms resistor, 1/4 watt
- 1 - 470 Ohms resistor, 1/2 watt
- 1 - 10 Kohms resistor, 1/4 watt
- 1 - 220 Ohms, 1 watt resistor
- 1 - 10 Kohms, ten-turn potentiometer
- 1 - 100 Kohms, ten-turn potentiometer
- 1 - ceramic or tantalum capacitor, 0.01 µF
- 1 - ceramic or tantalum capacitor, 0.001 µF
- 1 - ceramic or tantalum capacitor, 1 µF
- 1 - solderless breadboard
- 1 - transistor, 2N3904
- 1 - transistor, TIP29
- miscellaneous lead wires and connectors

EQUIPMENT:
- 1 - D.C. power supply
- 1 - function generator
- 1 - digital multimeter
- 1 - oscilloscope
- 1 - oscilloscope probe
- 1 - muffin-style fan, 24 VDC

INSTRUCTOR'S NOTE: If your lab does not have any LM2917 or SK9209 FVCs, you may substitute the LM2907. Refer to National Semiconductor's® Linear Application Specific IC's Databook or the Linear Applications Handbook for the proper pin identification.

PART 1: The Frequency-to-Voltage Converter
The package outline and the pinouts for the LM2917 are shown in Figure 33.1. The input stage is shown by the Schmitt-triggered op-amp symbol on the left. Pins 1 and 11 are the frequency inputs. In this experiment, the input A.C. waveform will be applied to pin 1 while pin 11 will act as the comparison voltage level that will be at ground potential. The output of the Schmitt trigger is connected to the charge pump that converts the input frequency to a D.C. voltage. This is accomplished by connecting a timing capacitor (C1 in Figure 33.2) to pin 2 of the FVC and an integrating R-C network (R1 and C2 in Figure 33.2) at pin 3 of the FVC. The output of the charge pump is then connected to a unity (or nearly unity) gain voltage comparator. In Figure 33.2, this is accomplished by connecting pin 3 to pin 4. The output voltage from the op-amp is applied to the base of an NPN transistor, which acts as a current source. The reverse-biased zener diode establishes the supply voltage to the op-amps and, indirectly, the magnitude of the output voltage. The output current is converted to a voltage by connecting an external resistor (R2 in Figure 33.2) to pin 5 of the op-amp. The output voltage from pin 5 is then returned to the inverting input of the op-amp comparator. Output current is provided by V_{CC}, which is connected to pin 8. The output voltage, V_{OUT}, from pin 5 can be approximated by the following formula:

$$V_{OUT} = (V_{(+)}) \cdot (f_{IN}) \cdot (C1) \cdot (R1) \qquad \textbf{33.1}$$

where $V_{(+)}$ is the voltage at pin 9 of the FVC. For the LM2907, the formula for the output voltage is

$$V_{OUT} = (V_{CC}) \cdot (f_{IN}) \cdot (C1) \cdot (R1) \qquad \textbf{33.2}$$

Procedure:

1.1 Obtain all materials and equipment required to complete this experiment.
1.2 Use your multimeter to measure the resistance of each of your resistors. If any of them are out of tolerance, inform your instructor and obtain replacements. Record the measured value of the resistors in Table 33.1.
1.3 If your lab has one, use a capacitance tester to check the capacitance values of your capacitors. If any of them are out of tolerance, inform your instructor and obtain replacements. Record the measured capacitance values in Table 33.1.

Table 33.1: Nominal and measured component values.

NOMINAL OR RATED COMPONENT VALUES	MEASURED COMPONENT VALUES
R1 = 100 KOHMS	
R2 = 10 KOHMS	
R3 = 470 OHMS	
C1 = 0.01 uF	
C2 = 1 uF	
C1 = 0.001 uF	

1.4 Construct the circuit shown in Figure 33.2. Connect the output of your function generator to an oscilloscope. Adjust the function generator to provide a 1 volt peak-to-peak sine wave with a frequency of 50 Hz and no D.C. offset.
1.5 Turn on the D.C. power supply, and set it for an output of +12 volts D.C. Turn off your function generator and D.C. power supply. Connect the outputs of your D.C. power supply and function generator to your circuit.
1.6 Turn on your D.C. power supply and your function generator. Starting at a frequency of 50 Hz, measure V_{OUT} using a digital voltmeter. Repeat the voltage measurement for each of the frequencies shown in Table 33.2.
1.7 Set the function generator for a square wave output. The amplitude should still be 1 volt p-p. Measure and record V_{OUT} for the frequencies shown in Table 33.3.
1.8 Now set the function generator for a sawtooth wave output. The amplitude should still be 1 volt p-p. Measure and record V_{OUT} for the frequencies shown in Table 33.4.

Table 33.2: Output voltage, V_{OUT}, as a function of frequency with C1 = 0.01 uF and a sine wave input.

INPUT VOLTAGE VIN, FREQUENCY, HERTZ	OUTPUT VOLTAGE VOUT, VOLTS D.C.
50	
100	
200	
300	
400	
500	
600	
700	
800	
900	
1000	
2000	
3000	
5000	

1.9 Reset the function generator for a sine wave output, and return the frequency to 500 Hz. Record the resulting V_{OUT} in Table 33.5. Also measure and record the voltage at pin 9 of the FVC relative to ground. Record your measurement in Table 33.5.

1.10 Increase V_{CC} to +14 volts, and again measure the resulting V_{OUT} at a frequency of 500 Hz. Also measure the voltage at pin 9 of the FVC relative to ground. Record your measurements in Table 33.5.

1.11 Reduce V_{CC} to +10 volts, and again measure the resulting V_{OUT} at a frequency of 500 Hz. Also measure the voltage at pin 9 of the FVC relative to ground. Record your measurements in Table 33.5.

1.12 Return V_{CC} to +12 volts.

1.13 With V_{CC} set to +12 volts and a sinusoidal waveform with a frequency of 500 Hz applied to pin 1 of the FVC, measure and record V_{OUT} for the different sine wave amplitudes shown in Table 33.6. Also measure and record the voltage at pin 9 of the FVC relative to ground.

Table 33.3: Output voltage, V_{OUT}, as a function of frequency with C1 = 0.01 uF and a square wave input.

INPUT VOLTAGE VIN, FREQUENCY, HERTZ	OUTPUT VOLTAGE VOUT, VOLTS D.C.
100	
300	
500	
700	

Table 33.4: Output voltage, V_{OUT}, as a function of frequency with C1 = 0.01 uF and a sawtooth wave input.

INPUT VOLTAGE VIN, FREQUENCY, HERTZ	OUTPUT VOLTAGE VOUT, VOLTS D.C.
100	
300	
500	
700	

Table 33.5: Output voltage, V_{OUT}, and the voltage at pin 9 with different values of V_{CC} and a 1 volt p-p sine wave input with a frequency of 500 Hz.

SUPPLY VOLTAGE VCC, VOLTS D.C.	OUTPUT VOLTAGE VOUT, VOLTS D.C.	VOLTAGE AT PIN 9, VOLTS D.C.
12		
14		
10		

Table 33.6: Output voltage, V_{OUT}, and the voltage at pin 9 with V_{CC} set for +12 volts and different sine wave amplitudes with a frequency of 500 Hz.

INPUT VOLTAGE VIN, VOLTS P-P	OUTPUT VOLTAGE VOUT, VOLTS D.C.	VOLTAGE AT PIN 9, VOLTS D.C.
1		
2		
4		

1.14 With V_{CC} set to +12 volts, and a sinusoidal waveform with a frequency of 500 Hz and an amplitude of 1 volt p-p applied to pin 1 of the FVC, use a digital voltmeter to monitor the output voltage, V_{OUT}. <u>Slowly</u> increase the D.C. offset of the sine wave until it just exceeds +0.5 volt. What happens to V_{OUT}? Record your observation in the space provided below.

Observation: _____

_____

1.15 With V_{CC} set to +12 volts, and a sinusoidal waveform with a frequency of 500 Hz and an amplitude of 1 volt p-p applied to pin 1 of the FVC, use a digital voltmeter to monitor the output voltage, V_{OUT}. <u>Slowly</u> decrease the D.C. offset of the sine wave until it just exceeds -0.5 volt. What happens to V_{OUT}? Record your observation in the space provided below.

Observation: _____

_____

1.16 Return the D.C. offset of the sine wave to 0.0 volt.
1.17 Turn off the function generator and the D.C. power supply. Replace the 0.01 µF capacitor (C1) with a 0.001 µF capacitor.
1.18 Turn on the D.C. power supply and the function generator. The D.C. power supply should be set for +12 volts and the function generator set for a sine wave output with an amplitude of 1 volt p-p.
1.19 Starting at a frequency of 100 Hz, measure V_{OUT} using a digital voltmeter. Repeat the voltage measurement for each of the frequencies shown in Table 33.7.
1.20 Return the frequency to 500 Hz. Use your oscilloscope to display the voltage waveform at pin 3 of the FVC. Record the resulting waveform on Graph 33.1.
1.21 Turn off the function generator and the D.C. power supply. Replace the 1µF capacitor (C2) with a 0.01 µF capacitor. Again use the oscilloscope to display the voltage waveform at pin 3 of the FVC. Record this waveform on Graph 33.1. Also use your digital voltmeter to measure the output voltage, V_{OUT}. Record V_{OUT}.

V_{OUT} at 500 Hz with C2 = 0.01 uF = _____

Table 33.7: Output voltage, V_{OUT}, as a function of frequency with C1 = 0.001 uF and a sine wave input.

INPUT VOLTAGE VIN, FREQUENCY, HERTZ	OUTPUT VOLTAGE VOUT, VOLTS D.C.
100	
200	
300	
400	
500	
1000	
2000	
3000	
5000	
7000	
9000	
10,000	
30,000	
50,000	

NOTE: For a more detailed description of the effect that the size of C2 has on the operation of the frequency-to-voltage converter, consult National Semiconductor's® **Linear Application Specific IC's Databook** or the **Linear Applications Handbook**.

Graph 33.1: Voltage waveforms at pin 3 with C1 = 0.001 µF and C2 = 1.0 µF then C2 = 0.01 µF.

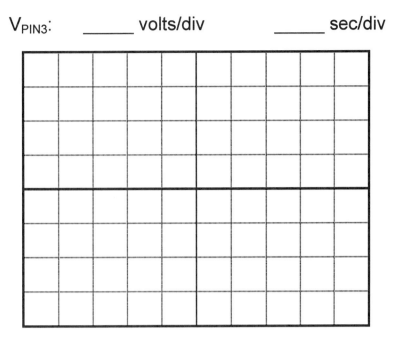

V_{PIN3}: _____ volts/div   _____ sec/div

PART 2: Application
Background:
This activity is intended to demonstrate how the FVC might be used in a control application. In this example, the output from the FVC will be used to vary the speed of a small fan. You will observe how the speed of the fan varies as the frequency of the waveform applied to the FVC varies. There are many industrial applications in which the speed of a process must be changed by varying the speed of a D.C. or A.C. motor. In situations such as these, the output from a transducer, counter, or some other type of sensor is used indirectly to regulate the process speed.

Procedure:
2.1 Construct the circuit shown in Figure 33.5. Adjust pot R4 for a resistance of 5 Kohms and pot R2 for 50 Kohms. You may use a 20-volt D.C. supply if a 24-volt source is not available.

2.2 Connect the output of your function generator to an oscilloscope. Adjust the function generator to provide a 1 volt peak-to-peak sine wave with a frequency of 50 Hz and no D.C. offset.

2.3 Apply power to your circuits. Connect the function generator output to the input of the FVC.

2.4 Test your circuit by varying the frequency of the sine wave. The speed of the fan should vary. If not, ask your instructor for help. Your circuit may be wired incorrectly, or you may need to change the resistance values in the interface circuit to achieve the desired current gain for your particular fan.

2.5 Vary the input frequency to determine the range of operation. Note what happens to the speed of the fan as the frequency of the sine wave is increased then decreased.

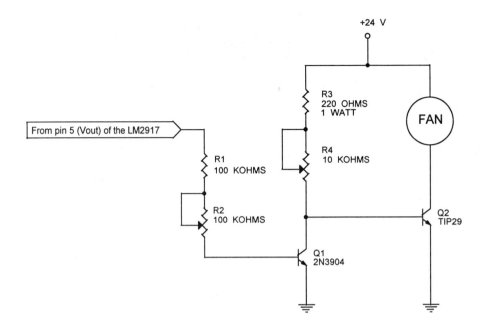

Figure 33.5: Fan circuit interface to the circuit shown in Figure 33.2.

PART 3: Questions

3.1 What is the significance of the results from steps 1.14 and 1.15 as it would apply to a control circuit that you might design in the future when using a frequency-to-voltage converter?

3.2 How would you eliminate the D.C. offset from a sine wave or a square wave signal that was to be applied to the frequency input of an FVC?

3.3 Referring to Figure 33.1, how do you think you could modify the circuit shown in Figure 33.2 to produce an output voltage from the FVC when the input voltage has a D.C. offset greater than the peak of the A.C. waveform? Try to come up with a solution different than your response to Question 3.2.

3.4 Using graph paper or some graphing software, plot V_{OUT} as a function of frequency for the data from Tables 33.2 and 33.7. For comparison purposes, use the same coordinate system for each set of data.

3.5 Referring to your graphs from Question 3.4, comment on the linearity of your results.

3.6 What effect does the value of C1 have on the magnitude of V_{OUT}? Refer to your data while answering this question. Be specific!

3.7 Compare the experimental results from Tables 33.3 and 33.4 to the data recorded in Table 33.2. What can you conclude about the operational characteristics of the FVC?

3.8 Compare the experimental results from Table 33.2 to the data recorded in Table 33.7. What can you conclude about the operational characteristics of the FVC?

3.9 Analyze the data recorded in Table 33.5. What can you conclude about the operational characteristics of the FVC?

3.10 How small do you think you could make V_{CC} in the circuit shown in Figure 33.2 without changing the results shown in Table 33.2?

3.11 Analyze the data recorded in Table 33.6. What can you conclude about the operational characteristics of the FVC?

3.12 What effect does increasing frequency have on the output of a FVC?

3.13 What effect does the value of C2 have on the output of a FVC?

3.14 In Part 2 of this experiment, what happened to the speed of the fan as the frequency of the sine wave applied to the FVC increased? Explain how this happened. Explain your answer in detail. In your answer, include what happens to the FVC output voltage as the input frequency increases and what would happen to Q1, Q2, and the fan in the circuit shown in Figure 33.5 as a result.

PART 4: Practice Problems

4.1 Use Formula 33.1 and the measured component values from Table 33.1 and the <u>measured</u> value for the voltage at pin 9 of the FVC to calculate the theoretical output voltage for an input voltage having a frequency of 500 Hz for a C1 = 0.01 µF and C1 = 0.001µF. If you used an LM2907, then use Formula 33.2

Calculations:

4.2 How do the results from Practice Problem 4.1 compare to the values for V_{OUT} at 500 Hz recorded in Tables 33.2 and 33.7?

PART 5: FVC Formulas

$$V_{OUT} = (V_{(+)}) \cdot (f_{IN}) \cdot (C1) \cdot (R1) \qquad \textbf{33.1}$$

$$V_{OUT} = (V_{CC}) \cdot (f_{IN}) \cdot (C1) \cdot (R1) \qquad \textbf{33.2}$$

Solid-State Relays: 34

INTRODUCTION:

In your studies to date, you have worked with a variety of switching devices, including mechanical switches, transistors, control relays, SCRs, triacs, optoisolators, and possibly others. Now you will study yet another switching device, the solid-state relay (SSR). The SSR combines the operating principles of the optoisolator and the triac or SCR. Like a control relay, SSRs are designed to turn on and off high-voltage (high-current) loads while being controlled (typically) by a low-voltage (low- current) input. However, most SSRs have only a single, normally-open set of contacts (SPST-NO), whereas most control relays have two or more sets of normally-open and normally-closed contacts (DPDT). Control relays are electromagnetic devices. SSRs, on the other hand, rely on an infrared emitter-detector pair to switch on and off the high-power output stage, which is built around a single triac or two inverse-parallel connected SCRs. Depending on the type of SSR, the load to be controlled can be either D.C. or A.C. powered, but not both. This is another difference between control relays and SSRs. The contacts of a relay can carry either D.C. or A.C. current. For D.C. loads, the output control device is often a power MOSFET. Both types of relays then provide for electrical isolation of the input from the output. Like control relays, the input or control voltage may be A.C. or D.C. depending on the type of SSR you choose for your application. One advantage of D.C. controlled SSRs is that the typical input voltage range is from 3 to 32 volts. This represents a much wider range of control voltage as compared to that for a control relay, which is designed to be energized by a fixed voltage level. Because the input impedance of the SSR is much higher than that of the control relay, the D.C. controlled SSR can be activated (driven) by TTL devices. Figure 34.1 shows how a NAND gate might be used to turn on a SSR. When the output of the NAND gate is active (low), the SSR will be turned on.

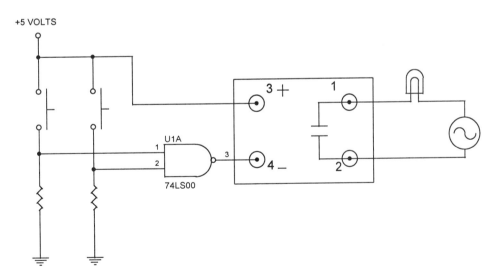

Figure 34.1: SSR triggered by a NAND gate.

SSRs can be categorized as either *zero voltage turn-on* or as *random voltage turn-on*. Random voltage turn-on SSRs operate very much like the triac circuit that you constructed in Experiment 21. In that circuit, the triac could be triggered into conduction at any point along the 60 Hz sine wave input. If the triac is triggered into conduction at or near the peak of the sinusoidal input, there will be a sudden in-rush of current when powering a low-resistance load. This is undesirable, because it will result in electromagnetic interference (EMI). Refer to Figure 34.2 for an example of the triac-controlled load voltage waveform that you produced in Experiment 21. This waveform caused noise (RFI) on your A.M. radio when you brought it close to that circuit. The noise occurs when the load waveform suddenly changes from 0 volts to its peak of approximately 175 volts. Zero voltage turn-on SSRs incorporate zero-crossover circuitry into the gate triggering circuit to prevent the occurrence of EMI and RFI. The zero-crossover circuitry senses the magnitude of the sinusoidal source at all times and will only allow the triac to switch on or off when the sine wave is at or near 0 volts. Figure 34.3 shows the output voltage waveform across a resistive load being controlled by a zero voltage turn-on SSR. Notice that there are no sudden vertical transitions, as is the case for the waveform shown in Figure 34.2. It is not always desirable to power a load with a zero voltage turn-on SSR. For some inductive loads, it is better to use a random voltage turn-on SSR. Consult the manufacturer for your particular application. Advantages offered by SSRs when compared to the conventional control relay include no contact bounce, no accidental switching due to mechanical vibration, no arcing of contacts, and no input coil to produce inductive kick from a collapsing magnetic field. In this experiment, you will work with a zero voltage turn-on SSR that is designed to operate from a D.C. source.

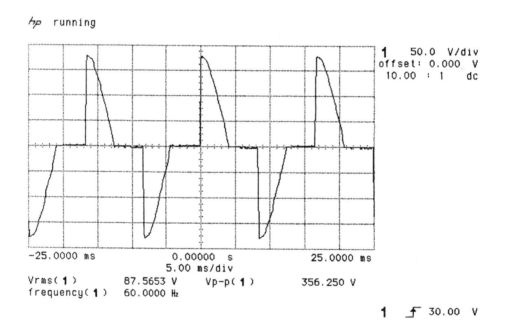

Figure 34.2: Voltage waveform across a load controlled by a triac. Note the sudden vertical transitions from 0 volts when the triac is open to the peak output voltage when the triac is triggered into conduction.

SAFETY NOTE:

When selecting an SSR for a new application, make sure that you consult the manufacturer's data sheet to make sure that it meets the voltage and current requirements of your application, including an appropriate factor of safety. SSRs that are operated by a D.C. input voltage are polarity sensitive. It is important not to inadvertently switch the input terminals. Some SSRs have internal circuitry to prevent damage due to incorrect wiring; others do not. Review Experiment 24 for ways to prevent accidental reverse biasing of LEDs and IREDs. You will be working with a 120 volts A.C. source. Take extra precaution to prevent electrical shock! Observe all safety precautions as outlined by your instructor. As always, be sure to wear safety glasses while working in the lab, remove jewelry from fingers and wrists, and have your instructor check your work before you energize your circuit and each time you make changes thereafter.

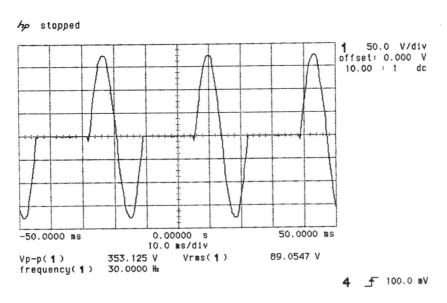

Figure 34.3: Voltage waveform across a load controlled by an SSR. Note the absence of the sudden vertical transitions associated with the load waveform of the triac-controlled load.

OBJECT:
Upon successful completion of this experiment and all reading assignments, the student should be able to:
- construct an operational control circuit using an SSR
- use an oscilloscope to observe the voltage waveform across a SSR-controlled load

MATERIALS:
- 1 - zero voltage turn-on SSR such as Potter & Brumfield's® SSR-240D50 or the equivalent
- 1 - 60 watt incandescent light bulb
- 1 - light bulb socket
- 2 - switches, SPST
- 1 - fuse, fast-blow, rated for 2 amps
- 1 - in-line fuse holder
- miscellaneous lead wires and connectors

EQUIPMENT:
- 1 - hand-held digital multimeter
- 1 - function generator
- 1 - frequency counter
- 1 - adjustable D.C. power supply
- 1 - isolation transformer or line-isolated 120 volts A.C. supply
- 1 - dual-trace oscilloscope
- 2 - oscilloscope probes
- 1 - inexpensive A.M. radio

PART 1: Operating Characteristics of the SSR

Background:
The first circuit you are to construct is shown in Figure 34.4. You must use a line-isolated A.C. supply to power the lamp shown in Figure 34.4. If you do not have a line-isolated supply, you may electrically isolate the lamp and SSR from a conventional 120 volt A.C. outlet by using an isolation transformer. Have your instructor explain the safety implications of this before beginning this experiment.

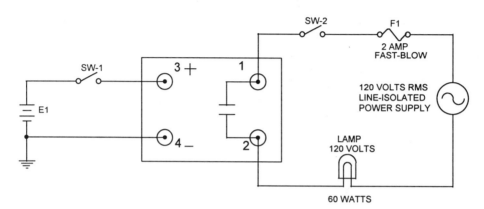

Figure 34.4: Experimental circuit to determine the operational characteristics of an SSR with a D.C. input.

Procedure:
1.1 Obtain all materials and equipment required to complete this experiment.
1.2 Construct the circuit shown in Figure 34.4. The D.C. power supply should initially be set for an output of 5.0 volts.
1.3 With SW-1 open, close SW-2. Use your handheld voltmeter to measure the lamp voltage. Record your measurement in Table 34.1.
1.4 Close SW-1. Again measure the lamp voltage. Record your measurement in Table 34.1.
1.5 Open both switches. Reduce the D.C. power supply output voltage to 0.0 volt. Close SW-2, then close SW-1. Use your oscilloscope to monitor the lamp's voltage.
1.6 Use your D.C. voltmeter to monitor the D.C. supply voltage. Slowly increase the D.C. supply voltage until the SSR is triggered into conduction. Note the D.C. voltage that just triggers the SSR into conduction. Record this value in Table 34.1.

Table 34.1: Experimental data for the SSR circuit shown in Figure 34.4.

LAMP VOLTAGE WITH SW-1 OPEN, VOLTS A.C.	
LAMP VOLTAGE WITH SW-1 CLOSED, VOLTS A.C.	
SSR D.C. THRESHOLD TRIGGER VOLTAGE, VOLTS D.C.	

1.7 Continue to increase the D.C. input voltage. Note the shape of the lamp voltage waveform. Does the shape or magnitude of the lamp voltage change any as the D.C. input voltage is increased? Record your observation(s) in the space provided below.

Observation:_____

1.8 Open both switches. Turn off both power supplies. Reverse the connections to the SSR at terminals 1 and 2. Adjust the D.C. power supply for 0.0 volt. Turn on both power supplies. Close both switches. Monitor the lamp voltage with your oscilloscope while slowly increasing the D.C. supply voltage until it reaches 12 volts. Did reversing the output terminals of the SSR change its operation in any way? Record your observation(s) in the space provided below.

Observation:_____

1.9 Open both switches. Turn off both power supplies. Disconnect the D.C. power supply from the SSR.

1.10 Adjust your function generator to produce a 5 volt peak-to-peak square wave with a +2.5 volt D.C. offset at a frequency of 5000 Hz. Turn off the function generator, connect the positive test lead from your function generator to terminal 3 of the SSR, and connect the negative (ground) lead of the function generator to terminal 4 of the SSR as shown in Figure 34.5.

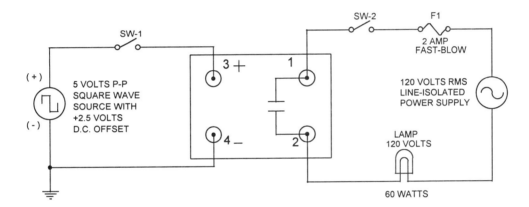

Figure 34.5: Experimental circuit to determine the operational characteristics of the SSR with a square wave input.

1.11 Turn on both power supplies. Close both switches. Monitor the lamp voltage with your oscilloscope and record the lamp voltage waveform on Graph 34.1.

1.12 Now reduce the frequency of the square wave input to 10 Hz. Display both the lamp voltage and the signal voltage, V_S, on the oscilloscope. Have your instructor assist you with this step. If you can get a good trigger, select a time base of 20 milliseconds/division. If not, try 5 or 10 ms/div. Examine the resulting waveforms closely. Draw both V_S and V_{LAMP} in proper time phase with each other on Graph 34.2. Completely label the vertical and horizontal axes with voltage and time base values. Draw each waveform neatly and accurately.

1.13 Increase the frequency of V_S to 20 Hz. Select a time base of either 5 or 10 ms/div depending upon which produces the most desirable results. Draw both V_S and V_{LAMP} in proper time phase with each other on Graph 34.3. Completely label the vertical and horizontal axes with voltage and time base values. Draw each waveform neatly and accurately.

1.14 Increase the frequency of V_S to 30 Hz. Select a time base of either 5 or 10 ms/div depending upon which produces the most desirable results. Draw both V_S and V_{LAMP} in proper time phase with each other on Graph 34.4. Completely label the vertical and horizontal axes with voltage and time base values. Draw each waveform neatly and accurately.

1.15 Increase the frequency of V_S to <u>exactly</u> 120 Hz. In the space that follows, describe in detail any and all observations you can make.

Observations:_____

_____

_____

_____

1.16 Decrease the frequency of V_S to 30 Hz. Adjust your A.M. radio to a local station. Bring the radio in close proximity to the SSR. Do you notice any RFI?

Observation:_____

1.17 Turn off both power supplies.

Graph 34.1: V_{LAMP} for the circuit shown in Figure 34.5 with a supply frequency of 5000 Hz.

V_{LAMP}: _____ volts/div _____ sec/div

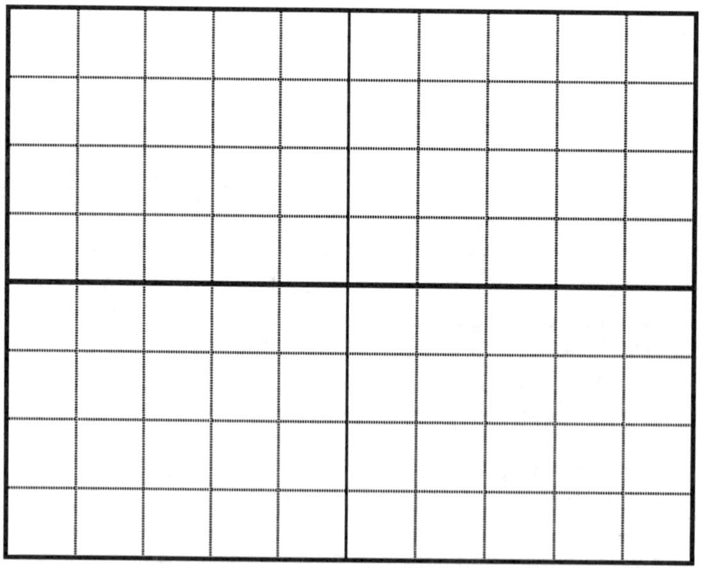

Graph 34.2: V_S and V_{LAMP} for the circuit shown in Figure 34.5 with a supply frequency of 10 Hz.

V_S: _____ volts/div _____ sec/div
V_{LAMP}: _____ volts/div _____ sec/div

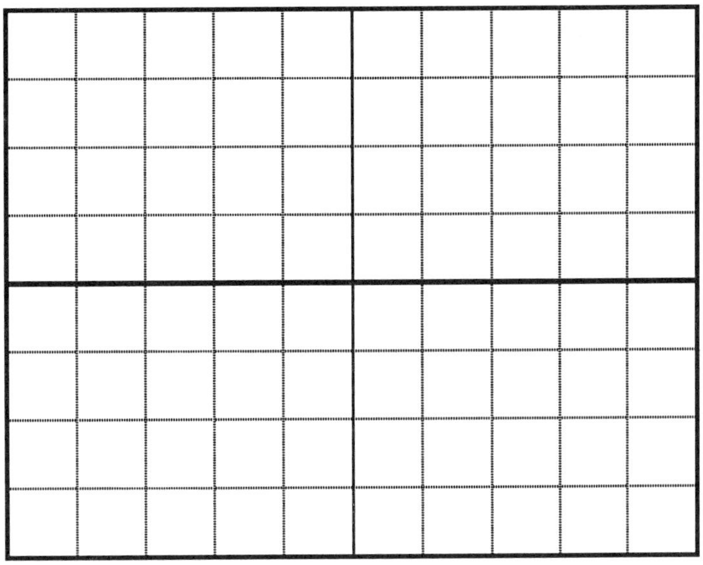

Graph 34.3: V_S and V_{LAMP} for the circuit shown in Figure 34.5 with a supply frequency of 20 Hz.

V_S: _____ volts/div _____ sec/div
V_{LAMP}: _____ volts/div _____ sec/div

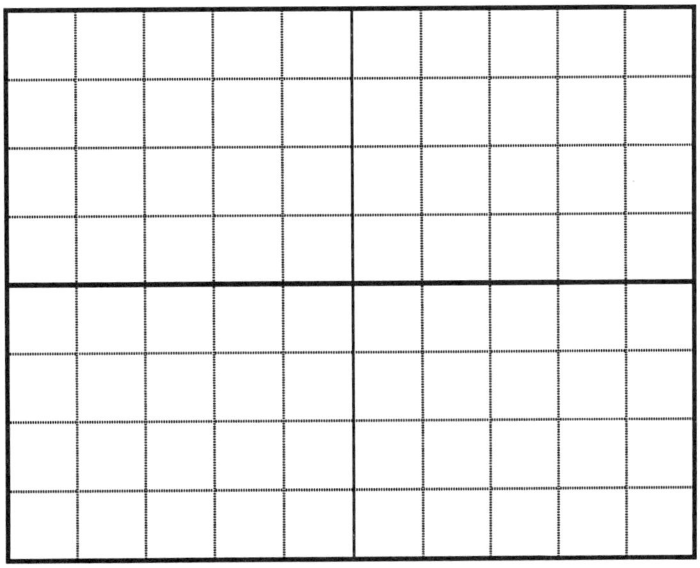

Graph 34.4: V_S and V_{LAMP} for the circuit shown in Figure 34.5 with a supply frequency of 30 Hz.

V_S: _____ volts/div   _____ sec/div
V_{LAMP}: _____ volts/div   _____ sec/div

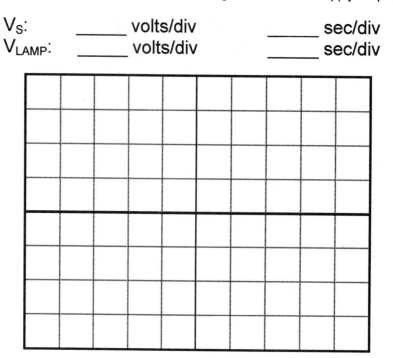

PART 2: Questions

2.1 How does the SSR threshold trigger voltage you recorded in Table 34.1 compare to that specified by the manufacturer of your SSR?

2.2 Compare the operation of your SSR to each of the following devices. Specifically, compare their operation when controlling an A.C. powered load. Include both similarities and differences. Refer to the data you collected today and to any of your experimental data from Experiments 3, 19, 20, and 21.

a) control relay

b) triac

c) SCR

2.3 What effect does the <u>magnitude</u> of the D.C. input voltage to the SSR have on the shape and magnitude of the lamp voltage in the circuit shown in Figure 34.4?

2.4 What effect does the <u>frequency</u> of the input voltage, V_s, to the SSR have on the shape and magnitude of the lamp voltage in the circuit shown in Figure 34.4?

2.5 What effect did reversing the connections to the output of the SSR in step 1.8 have on its operation?

2.6 Explain the cause of the phenomenon that you observed when you increased the frequency to 120 Hz in Step 1.15.

2.7 Explain why there was no noise picked up by your A.M. radio when performing step 1.16.

2.8 Explain the difference in the shape of the waveform drawn in Graph 34.1 when compared to the shape of the waveform drawn in Graph 34.4.

Solid-State Timers: 35

INTRODUCTION:
The subject of industrial electronics deals with the control of machinery and equipment in some type of manufacturing setting. This includes turning machines or electrical components on and off and/or varying the magnitude of the voltage and current to these machines. These processes involve the collection of information or data via some type of sensor or transducer, then comparing it to the desired value or level, and finally taking action based on the results. Again, the action consists of turning something on or off or increasing or decreasing the applied voltage. The input or stimulus that determines when an electrical switching or adjustment of some type is made can be load, temperature, flow rate, quantity (object count), time, or some combination of these or any number of other inputs. Here the focus will be on the time element. Specifically, the construction and application of time-delay relays will be covered. Timers can take a variety of forms. Prior to the evolution of electronics, several different mechanical and/or electromechanical devices were used to control the timing or sequence of events in an industrial process. This included the motor-driven, cam-operated, limit-switch controlled timer. (Whew! What a mouthful of words that is!) You may already be familiar with one version of this particular device. If your family has ever been on vacation for an extended period of time, you may have installed any number of security devices in your home to give it a lived-in look. Perhaps this involved some type of timer to control the turn on and off of lights in the house. This is quite often accomplished by plugging a motor-driven timer into a wall outlet and plugging the lamp into the timer. Switches on the timer are then positioned to determine when the light(s) will go on and off. In Experiment 32, you studied the 555 timer, an I. C. that can be used in a wide variety of time-related control applications. The study of electronic timers will be furthered in this experiment by first constructing a time-delay relay using discrete components. You will follow this by examining and wiring an actual off-the-shelf, time-delay relay. You will also be introduced to *timing diagrams*. These are pictorial descriptions of the sequence of events that take place in a circuit, whether controlled by an electronic timer or some other device(s).

SAFETY NOTE:
Familiarize yourself with the manufacturer's ratings for all devices and components to be used in this experiment before constructing any circuits. As always, wear safety glasses, and remove all jewelry from your hands and fingers before the start of the experiment.

OBJECT:
Upon successful completion of this experiment and all reading assignments, the student should be able to:
- construct an operational solid-state time-delay relay using discrete components
- draw from memory the schematic symbols for the coil and contacts of a time-delay relay
- given the schematic diagram of a control circuit incorporating a solid-state time-delay relay, construct the circuit and demonstrate it to your instructor
- given the schematic diagram of a control circuit incorporating a solid-state time-delay relay, draw the timing diagram for it and refer to it when explaining and demonstrating the operation of your completed circuit

REFERENCES:
Chapter 2 of Maloney's <u>Modern Industrial Electronics</u>

MATERIALS:
- 1 - fast-recovery diode, 1N914 or the equivalent
- 1 - rectifier diode, 1N4003 or the equivalent
- 1 - electrolytic capacitor, 50 μF, 25 WVDC
- 2 - incandescent lamps, 28 volt rating
- 2 - light bulb sockets
- 1 - normally-open pushbutton switch
- 1 - normally-closed pushbutton switch
  - miscellaneous lead wires and connectors
- 1 - 10 Megohms resistor or two series-connected 4.7 Megohms resistors, 1/4 watt
- 3 - single-pole, single-throw switches or one dip switch pack with eight SPST switches
- 1 - low-power SCR, MCR22-3
- 1 - UJT, 2N2646 or ECG6401
- 4 - 2.2 Megohms resistors, 1/4 watt
- 3 - 1 Kohms resistors, 1/4 watt
- 1 - 50 Kohms ten-turn trim pot
- 1 - 220 Ohms resistor, 1/2 watt
- 1 - 1 Megohms potentiometer
- 1 - solderless breadboard

EQUIPMENT:
- 1 - 12 volt D.C. power supply
- 1 - 24 volt D.C. power supply
- 2 - digital multimeters
- 1 - MIDTEX® control relay, Model 157-22B200 with 120 Ohms, 12 volt coil, DPDT contacts rated for 10 amperes, or the equivalent
- 1 - control relay socket
- 1 - MACROMATIC® programmable solid-state, time-delay relay, Model SS-60226 with 12 VAC/DC input and DPDT contacts rated for 10 amperes, or the equivalent
- 1 - time-delay relay socket

INSTRUCTOR/STUDENT NOTE: For most electronic circuits to operate properly, the components that they are constructed from must fall within some rather tight tolerances. For the circuit shown in Figure 35.1 to work properly, you must use the SK9123 UJT and the MCR22-3 SCR or components that have nearly identical manufacturer's ratings. If the ratings of your replacements deviate too much from those of the SK9123 and the MCR22-3, the SCR may not turn on at all or will turn on immediately without a time delay.

PART 1: UJT-Based Solid-State Timer
Background:
In Experiment 22 you used a UJT and an R-C network to create a relaxation oscillator. In Experiment 19, you studied the electrical characteristics of the SCR. In Experiment 3, you learned about the control relay. The knowledge you acquired in these previous experiments will be integrated in this experiment. Specifically, you will use the principles presented in those experiments to build an adjustable solid-state time-delay relay. The circuit you will build is shown in Figure 35.1. Note that the circuit has been broken up into blocks. The leftmost section is the relaxation oscillator. The output of the relaxation oscillator is used to trigger the SCR into conduction. Once latched on, the SCR turns on the control relay. The relay's contacts control the load--in this case, an incandescent light bulb.

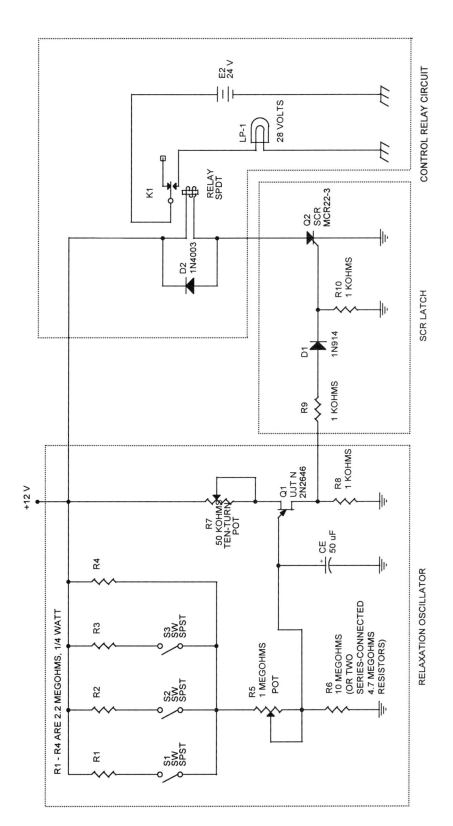

Figure 35.1: Solid-state, time-delay relay constructed from discrete components.

Procedure:

1.1 Obtain all materials and equipment required to complete this experiment. Use an ohmmeter to measure all resistors and ensure that they are within tolerance.

1.2 If your lab has one, use a capacitance tester to check the capacitance value of your capacitor. If it is out of tolerance, inform your instructor and obtain a replacement.

1.3 I suggest that you build and test this circuit in stages. First build the R-C network, turn on the power supply, and measure the voltage across C_E to ensure that it will charge up to its theoretical maximum, $E_{THEVENIN}$.

1.4 Once you have tested the R-C network by trying different switch combinations and adjusted the 1 Megohms potentiometer to ensure that the capacitor's charging does change with changing resistance, you may proceed to add the UJT, R_8, and the 50 Kohms trim pot. Once you have added these components, turn on the power supply and adjust the 50 Kohms potentiometer until the voltage across R_8 is 0.8 volt.

1.5 Turn off the power supply, and place a 220 Ohms, 1/2 watt resistor in parallel with the capacitor to discharge it. Once the capacitor is discharged, remove the 220 Ohms resistor.

1.6 Place a voltmeter in parallel with C_E and another in parallel with R_8. If your digital voltmeter has an autoranging feature, do **not** use it. If you do so, you may miss the fast-changing voltage across R_8. The capacitor voltage, V_{CE}, should be near 0.0 volt. If not, discharge the capacitor one more time. Turn on the power supply, and monitor V_{CE} and V_{R8}. V_{R8} should be nearly constant at approximately 0.8 volt, and V_{CE} should be increasing. Observe both voltages very closely, in particular V_{R8}. At some point, you will see a momentary voltage spike across R_8. At the same time, the value for V_{CE} will drop and the capacitor will start to charge again. This completes the relaxation oscillator stage of the circuit. This circuit operates very much like the relaxation oscillator that you constructed in Experiment 22. The only difference is that this circuit has a much longer R-C time constant.

1.7 Turn off the power supply, and complete construction of your circuit by adding the remaining components. Discharge the capacitor.

1.8 Adjust the 1 Megohms potentiometer for maximum resistance. For each of the switch settings shown in Table 35.1, measure and record the time delay or time lag from the moment the power supply is turned on until the lamp comes on. Each time you change the switch settings, turn off the power supply and discharge the capacitor.

1.9 Adjust the 1 Megohms potentiometer for minimum resistance. For each of the switch settings shown in Table 35.2, measure and record the time delay or time lag from the moment the power supply is turned on until the lamp comes on.

1.10 If time permits, try some other combination of switch and potentiometer settings to determine the effect that they have on the time delay.

1.11 Turn off the power supply.

Table 35.1: Time delays for two different switch settings for the circuit shown in Figure 35.1 with the 1 Megohms potentiometer set for maximum resistance.

SWITCH SETTINGS	TIME DELAY, SECONDS
ALL SWITCHES CLOSED	
ALL SWITCHES OPEN	

Table 35.2: Time delays for two different switch settings for the circuit shown in Figure 35.1 with the 1 Megohms potentiometer set for minimum resistance.

SWITCH SETTINGS	TIME DELAY, SECONDS
ALL SWITCHES CLOSED	
ALL SWITCHES OPEN	

PART 2: Basic Time-Delay Relay Circuits

Background:

The time delay circuit that you constructed in Figure 35.1 is an example of what is called a *delay on energize*, or a *delay on pull-in*, or *delay-on-operate*, or simply an *on-delay* time-delay relay. What this means is that there is a time delay from the moment the timer circuit is energized until the control relay is energized and its contacts change state--pull in. Some relays are designed such that the contacts will not change state until some time delay after the timer is de-energized. These time-delay relays are referred to as *delay on release*, *delay on de-energize*, *delay on drop-out*, or *off-delay*. An example of the latter is the timer circuit in many garage door openers. The security light stays on for a period of time after the garage door circuit and motor have turned off. The schematic symbol for the coil and contacts of an on-delay time-delay relay are shown in Figure 35.2 a), while the schematic symbol for the timer coil and contacts for an off-delay time-delay relay are shown in Figure 35.2 b). Some designers use the symbols for the contacts of a control relay when designing a circuit with a time-delay relay in it rather than using the unique symbols shown in Figures 35.2 a) and b). For this reason, it is important to label both the contacts and coil of a time-delay relay with some kind of identification that will distinguish them from a control relay coil and its contacts. For example, refer to the circuit shown in Figure 35.3. This circuit incorporates both a time-delay relay and a control relay. The control relay coil is identified by the label 1-CR and its contacts by 1-CR-A and 1-CR-B. Please note, there is no universally applied standard for labeling the coil and contacts. The drafting standards vary from one company to another (unfortunately). The coil of the time-delay relay is identified by the label 1-TDR and its contacts by 1-TDR-A. In this experiment, you will use an on-delay time-delay relay to construct the circuit shown in Figure 35.3.

a) DELAY ON PULL-IN   b) DELAY ON DROP-OUT

Figure 35.2: Schematic symbols for the input coil and contacts of delay on pull-in and delay on drop-out time-delay relays.

Procedure:

2.1 Obtain all of the components shown in Figure 35.3.
2.2 Adjust one of your power supplies for 12 volts and the other for 24 volts (you may use 20 volts if that is the maximum output of your power supply). Turn off both power supplies.

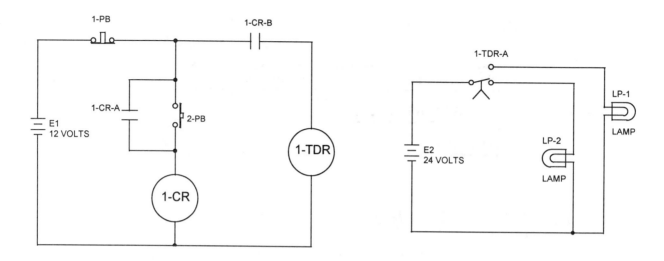

Figure 35.3: Control circuit using a time-delay relay and a control relay.

2.3 Construct the circuit shown in Figure 35.3. Approach this like you did the construction of the circuit shown in Figure 35.1. Build in stages; test in stages. A good plan would be first to build the holding relay circuit. If you have forgotten what this is, ask your instructor. Once built, test it. If it works, add on the timer. Lastly, wire the lamps to the 24-volt power supply through the time-delay relay contacts.

2.4 Set your timer for a delay of about 10 seconds. Turn on both power supplies. What happened? Record your observation.

Observation:_____

2.5 In this step you will need to measure the time in seconds from the moment switch 2-PB is pressed until lamp LP-1 comes on. Press 2-PB and release it. At the same moment, start your stop watch. Record your observation in the space below.

Observation:_____

_____

_____

2.6 Once on, how long does (will) lamp LP-1 stay on?

_____

2.7 Press 1-PB. What happens? Record your observation.

Observation:_____

2.8 With E2 and E1 on, press and release 2-PB. Wait until lamp LP-1 turns on. Turn off E2. Note what happened. Turn E2 back on. Note what happened. Record your observations and explain what happened.

Observations:_____

_____

Explanation:_____

_____

_____

_____

2.9 With E2 on, turn off E1. Note what happens. Turn E1 back on. Note what happens. Record your observations and explain everything that happened.

Observations:_____

_____

Explanation:_____

_____

_____

_____

PART 3: Timing Diagrams
Background:
A helpful troubleshooting guide or aid that can also be used to interpret how a circuit works is the *timing diagram*. A timing diagram is a pictorial representation of the state (activated or de-activated) of each component in a circuit during its operation with respect to time. For example, Figure 35.4 is the timing diagram for the circuit shown in Figure 35.3. The solid black bars indicate when each component is in its activated state. Now let's define the term *activated* as it applies to our timing diagram. For example, when either pushbutton is pressed, it is said to be activated. When the pushbutton is in its normal position, it is said to be de-activated. Note that when 1-PB is pressed (activated), no current flows through it. When 2-PB is pressed (activated), current does flow through it. The short black lines following 1-PB and 2-PB in Figure 35.4 indicate that the pushbuttons are only momentarily activated. The length of the black lines indicates the duration of time that each device is energized. A white or blank space indicates that a device is deactivated.

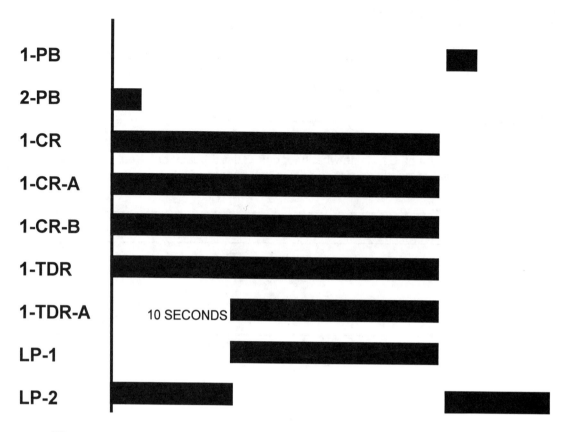

Figure 35.4: Timing diagram for the circuit shown in Figure 35.3.

Procedure:

3.1 In the space provided below, draw the timing diagram for the circuit shown in Figure 35.6.

Figure 35.5: Timing diagram for the circuit shown in Figure 35.6.

474

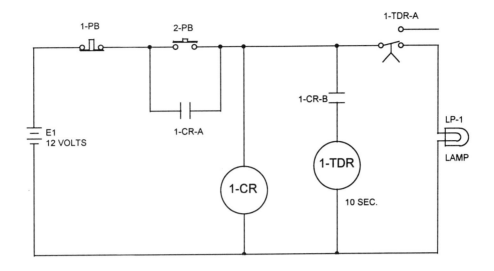

Figure 35.6: Practice circuit for constructing timing diagrams.

PART 4: Questions

4.1 The circuit shown in Figure 35.1 uses a relaxation oscillator circuit in which the capacitor continuously charges and discharges through the UJT. Why doesn't the control relay also turn on and off continuously?

4.2 Referring to your data, what are the maximum and minimum delay times for the circuit shown in Figure 35.1?

4.3 How many different <u>combinations</u> of time delays can be programmed into the time-delay relay circuit shown in Figure 35.1?

4.4 Once the control relay in Figure 35.1 turned on, what did you have to do to turn it off?

4.5 Explain in detail the operation of the circuit shown in Figure 35.1. Start with the closure of switch S1, and work your way from left to right as you explain the role each component plays in the circuit.

4.6 What is the purpose of 2-PB in Figure 35.3?

4.7 What is the purpose of 1-PB in Figure 35.3?

4.8 What is the purpose of 1-CR-A in Figure 35.3? In other words, why is it needed? Would the circuit work without it? Explain.

4.9 What is the purpose of 1-CR-B in Figure 35.3? In other words, why is it needed? How would the circuit's operation change if it were replaced by a short?

Programmable Controllers: 36

INTRODUCTION:

In Experiment 3, you constructed circuits using the control relay. In Experiment 34, you investigated the solid-state relay (SSR). And most recently, you studied the time-delay relay. These three devices have been, and continue to be, used in countless control applications throughout this country and the world. In modern manufacturing settings in the United States, machining and assembly lines have to be designed for flexibility. A manufacturing line that is making snow blowers one week may be switched over to make lawn mowers the next week. Since everyone knows that time is money, the process of converting over from one product to another must be done in as timely a manner as possible. Many of you have heard of the term *just-in-time (JIT) manufacturing*. In today's highly competitive marketplace, manufacturers are trying to increase the number of customers by offering a product of higher quality delivered when the customer wants it. Well, to accomplish the goal of timely product delivery, the manufacturer must minimize the downtime of production equipment. If plant electricians have to rewire control circuitry based on discrete components such as control relays and time delay relays, days or even weeks may be involved depending upon the size and complexity of the manufacturing operation.

American management and labor have found innovative ways to convert a production line from one product type or model to another in a matter of hours rather than days. One invention in particular has helped speed this process. That device is the *programmable controller*, also referred to as *programmable logic controller* (PLC). Simply put, the PLC is a computer-controlled control relay, timer, counter, and sequencer built into one package. Because the PLC is built around a microcomputer chip, it has the flexibility to be programmed and then reprogrammed when the need arises. Circuits that previously were controlled by relays and timers that would have to be manually rewired can now be replaced by a PLC. When the need to change the control function arises, a new program can be loaded into the PLC and a new function performed, often without changing as much as a single wire. It is therefore of utmost importance that today's maintenance technician know how to wire and program PLCs. A drawing of the physical layout of the Allen-Bradley SLC 5/04® PLC is shown in Figure 36.1.

SAFETY NOTE:

Read the operator's and/or programming manual for the PLC in your lab before using it. The terminals on your PLC input/output modules are very close together; therefore, make sure that any connections you make to the PLC are firmly connected and that no bare wire is exposed. Follow the directions provided by your instructor for choosing the proper size and color wire for wiring your PLC. As always, wear safety glasses, and remove all jewelry from your hands and fingers before the start of the experiment.

OBJECT:

Upon successful completion of this experiment and all reading assignments, the student should be able to:
- construct a logic-ladder diagram given the schematic for a control circuit that uses discrete components
- write, enter, and execute a program for your PLC given a ladder-logic diagram
- wire the input/output modules of your PLC given the wiring diagram for a particular control application
- draw the wiring diagram to implement a given control application for your PLC

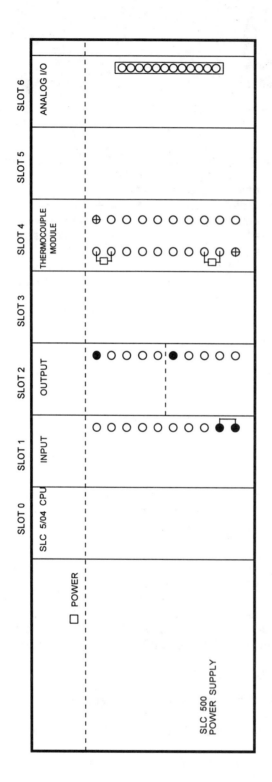

Figure 36.1: Allen-Bradley PLC with power supply, SLC 5/04® CPU, input module, output module, thermocouple module, and analog I/O module.

REFERENCES:
Chapter 3 of Maloney's Modern Industrial Electronics
www.ab.com

MATERIALS:
2 - incandescent lamps (the voltage rating will depend on the PLC that you are using)
2 - light bulb sockets
1 - normally-open pushbutton switch
1 - normally-closed pushbutton switch
- miscellaneous lead wires and connectors

EQUIPMENT:
1 - programmable logic controller with I/O modules,
1 - thermocouple module and analog I/O module are optional

PART 1: PLC Control Relay Functions
Background:
PLCs can perform a wide variety of functions. In this experiment, you will have the opportunity to investigate some of the more common and powerful functions performed by the PLC. This will include control relay functions, timers, and counters. In your first activity, you will write a short program and use the input and output modules of your PLC to replace a hard-wired control relay circuit. Figure 36.2 shows a simple start/stop circuit powered by a D.C. source. Take a moment to familiarize yourself with the operation of this circuit. This circuit has been redrawn in the form of a ladder diagram in Figure 36.3. Rails L1 and L2 represent either the (+) and (-) terminals of the D.C. supply shown in Figure 36.2, or each could represent A.C. line and neutral if the circuit were powered by an A.C. supply. The timing diagram for this circuit is shown in Figure 36.4.

The PLC was designed to allow a maintenance technician to wire switches to a common input module and permit a programmer to use software to determine the *logical* connection of the switches. The wiring diagram for a 24 VDC Allen-Bradley *input module* is shown in Figure 36.5. In a similar fashion, devices such as solenoids, relays, and magnetic contactors can be wired to an *output module* and the ladder logic program will determine when a given device will be activated. The wiring diagram for a 24 VAC Allen-Bradley *output module* is shown in Figure 36.6.

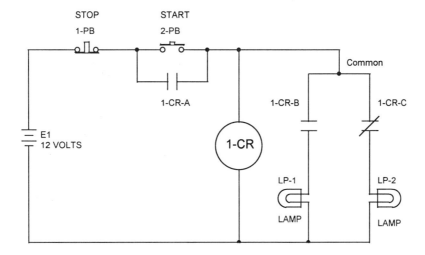

Figure 36.2: Basic start/stop circuit with D.C. source.

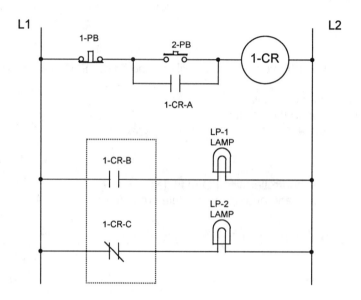

Figure 36.3: Ladder diagram equivalent of the circuit drawn in Figure 36.2.

Figure 36.4: Timing diagram for the circuit shown in Figure 36.3.

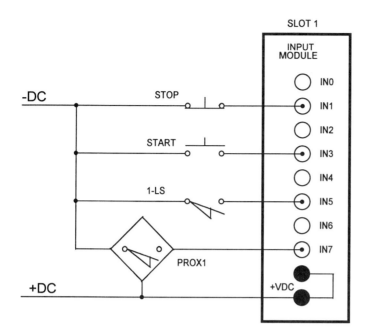

Figure 36.5: Wiring diagram for 24 VDC input module.

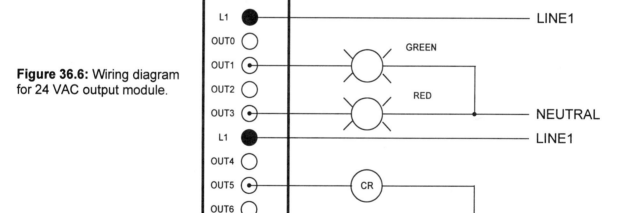

Figure 36.6: Wiring diagram for 24 VAC output module.

The ladder logic diagram that corresponds to the holding relay circuit of Figure 36.3 is shown in Figure 36.7. However, there is a problem with the logic of this diagram. If you program your PLC using this ladder logic, your circuit will not work properly. To get the control relay to energize, you would have to press and hold the stop button, 1-PB, the entire time. And if you released 1-PB, the circuit would become deactivated. This is the opposite of what should happen. This is one of the confounding issues about PLCs confronted by those just learning about PLCs for the first time.

As long as a closed switch is wired to Input 1 of the PLC input module, the CPU senses the presence of voltage and places a binary "1" in its memory location for that particular address--the address of Input 1. If the normally-open start pushbutton is pressed, a binary "1" is placed in the memory location associated with Input 3. The CPU decides what action to take on the basis of binary logic. Since both pushbuttons are "wired logically" in series, the CPU logically ANDs these two inputs together. If both memory locations have a "1", then the CPU will activate Output 0. With both switches closed, the CPU senses the application of voltage at Inputs 1 and 3 treating each as a logical or binary "1". However, the normally-closed contact *symbol* in the ladder logic diagram acts as a logical *inverter* and converts the binary "1" from the closed stop pushbutton to a binary "0". The CPU then ANDs a "0" with the "1" from the start pushbutton that has just been pressed. This results in a binary "0" causing Output 0 to be off.

So how do we remedy this problem? Well, panic stop pushbuttons such as that in Figures 36.2, 36.3, and 36.5 must be logically programmed as a set of normally open contacts. However, you must always make sure that safety switches and panic stop switches are wired normally-closed. Your instructor should discuss this safety topic with you at greater length. The correct ladder logic diagram is shown in Figure 36.8. Notice that two additional rungs of programming have been added for the two indicator lamps. This diagram has another improvement over that in Figure 36.7. The contacts that "seal in" or "latch" the coil are "binary bit" contacts in the CPU module rather than "physical" contacts on our output module. This is a more efficient use of our hardware resources.

Procedure:

1.1 Referring to the circuit shown in Figure 36.3, draw the corresponding PLC ladder-logic diagram.
1.2 Write a program for your PLC that will implement your ladder logic diagram.
1.3 Prepare a wiring diagram showing all connections to the input/output modules of your PLC. Label your wiring diagram with the appropriate module number and pin numbers that correspond to the program you wrote.
1.4 Have your instructor check your work for accuracy.
1.5 Enter your program into your PLC. Under the supervision of your instructor, test your program for any logic or programming errors.
1.6 Once your program is correct, turn off the PLC and wire your pushbutton switches and lamps to the PLC. Have your instructor check your work.
1.7 Once your wiring has been approved, turn on the PLC. Execute your program and note what happens.

Observation:_____

1.8 Press the start button (2-PB). Note what happens.

Observation:_____

1.9 Press the stop button (1-PB). Note what happens.

Observation:_____

1.10 Did your circuit perform as expected? If not, ask your instructor for assistance.

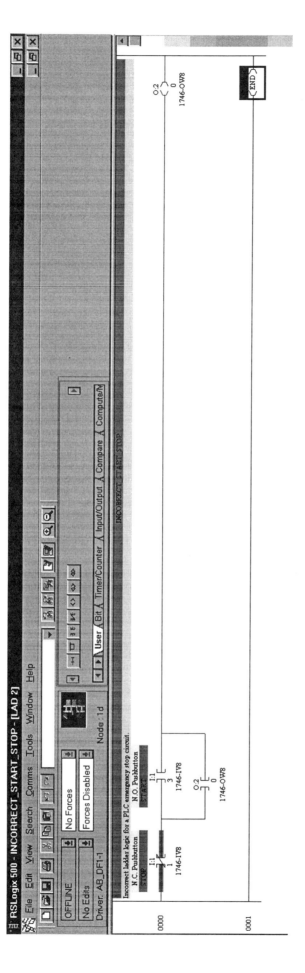

Figure 36.7: Incorrect ladder logic for basic start/stop circuit.

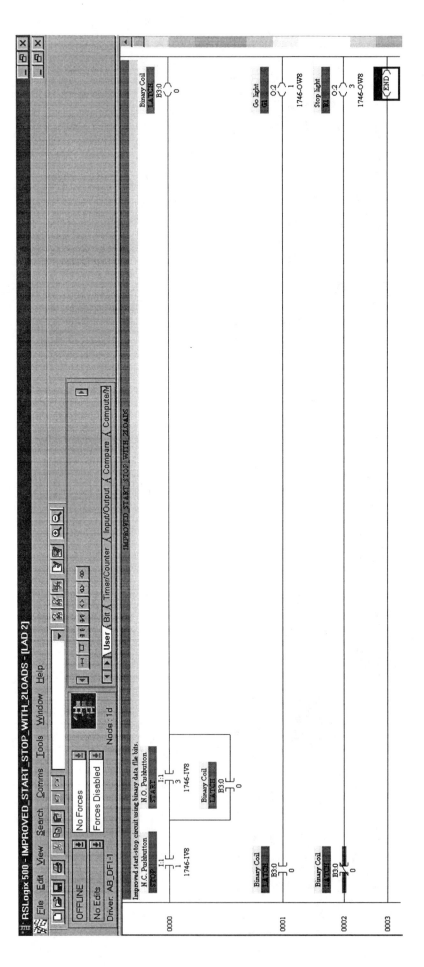

Figure 36.8: Correct ladder logic for basic start/stop circuit.

PART 2: PLC Timer Functions

Background:
The next circuit that you will implement with your PLC is shown in Figure 36.9. This is a time-delay relay circuit very similar to the one you constructed in Experiment 35. The only difference is that it is drawn in ladder logic form. Take a moment to familiarize yourself with the operation of this circuit. The timing diagram for this circuit is shown in Figure 36.10.

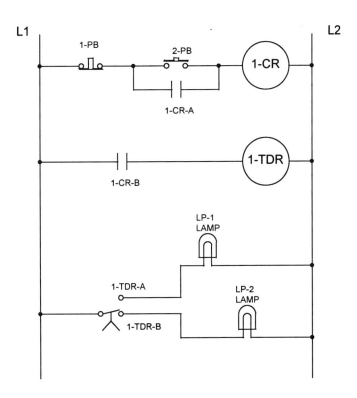

Figure 36.9: Control circuit using a control relay and a time-delay relay.

Procedure:
2.1 Referring to the circuit shown in Figure 36.9, draw the corresponding PLC ladder-logic diagram.
2.2 Write a program for your PLC that will implement your ladder logic diagram.
2.3 Prepare a wiring diagram showing all connections to the input/output modules of your PLC. Label your wiring diagram with the appropriate module number and pin numbers that correspond to the program you wrote.
2.4 Have your instructor check your work for accuracy.
2.5 Enter your program into your PLC. Under the supervision of your instructor, test your program for any logic or programming errors.
2.6 Once your program is correct, turn off the PLC and wire your push-button switches and lamps to the PLC. Have your instructor check your work.
2.7 Once your wiring has been approved, turn on the PLC. Execute your program and note what happens.

Observation:_____

2.8 Press the start button (2-PB). Wait a few seconds. Note what happens.

Observation:_____

2.9 Press the stop button (1-PB). Note what happens.

Observation:_____

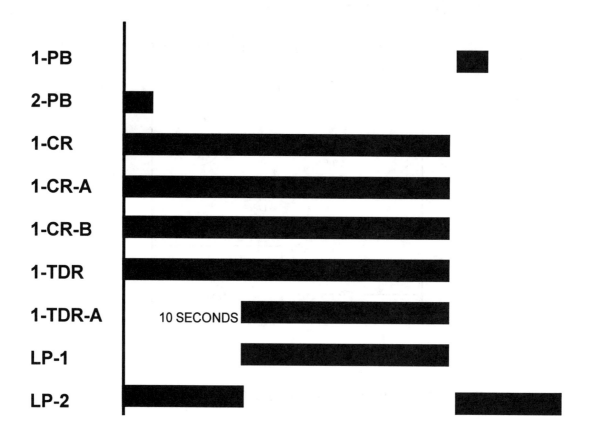

Figure 36.10: Timing diagram for the circuit shown in Figure 36.9.

PART 3: PLC Counter Functions
Background:
There are very few, if any, manufacturing operations that do not require a count of the number of objects passing through an assembly line. Computers or PLCs must keep track of the number of parts or finished products for purposes of inventory control and for the control and sequencing of machining, assembly, and packaging operations.

Procedure:
3.1 Draw the PLC ladder logic diagram for a circuit that will count to ten and on the tenth count cause an indicator lamp to light. Use one of your pushbuttons to act as the sensor input by

repeatedly pressing it. The opening and closing of the pushbutton should increment your PLC's internal counter. Use your second pushbutton to act as a manual reset to reset the counter to zero.

3.2 Write a program for your PLC that will implement your ladder logic diagram.

3.3 Prepare a wiring diagram showing all connections to the input/output modules of your PLC. Label your wiring diagram with the appropriate module number and pin numbers that correspond to the program you wrote.

3.4 Have your instructor check your work for accuracy.

3.5 Enter your program into your PLC. Under the supervision of your instructor, test your program for any logic or programming errors.

3.6 Once your program is correct, turn off the PLC and wire your pushbutton switches and lamp to the PLC. Have your instructor check your work.

3.7 Once your wiring has been approved, turn on the PLC. Execute your program. With your instructor's assistance, examine the contents of the PLC's internal counter. Press the pushbutton that is acting as the counter input. Does the internal counter increment? It should. If not, check your work. Continue pressing the pushbutton until the count reaches ten. Does your lamp turn on? If not, check your work. Continue pressing the pushbutton. Note what happens to the internal counter and the lamp.

Observation:_____

_____

3.8 Press the reset button. Note what happens to the internal counter and the lamp.

Observation:_____

_____

3.9 What happens if you press the reset button before the counter reaches the count of ten?

Answer:_____

3.10 Modify your ladder-logic diagram and program so that the counter will count to ten and <u>reset itself</u>. You should still provide for manual reset. Demonstrate the program to your instructor. What happens when the counter reaches a count of ten? Watch closely!

Observation:_____

_____

PART 4: Ladder Logic Design

4.1 Draw the PLC ladder logic diagram for the operation depicted by the timing diagram shown in Figure 36.11. Use one of your pushbuttons to start the operation. Write a program for your PLC that will implement your ladder logic diagram. Prepare a wiring diagram showing all connections to the input/output modules of your PLC. Label your wiring diagram with the appropriate module number and pin numbers that correspond to the program you wrote. Have your instructor check your work for accuracy. Enter your program into your PLC. Under the supervision of your instructor, test your program for any logic or programming errors. Once your program is correct, turn off the PLC and wire your pushbutton switch and lamps to the PLC. Have your instructor check your work. Once your wiring has been approved, turn on the PLC. Execute your program. Demonstrate your program to your instructor.

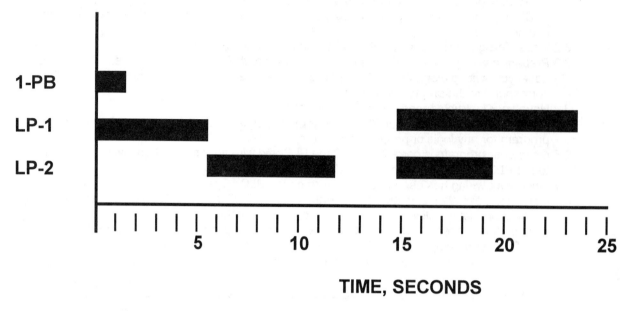

Figure 36.11: Timing diagram to be implemented using a PLC.

4.2 If your PLC has a thermocouple input module, design a PLC program that will turn on a heating element if the temperature of some process you are to control is below a given temperature. The heater should turn off if the temperature exceeds this set point. Daw the PLC ladder logic diagram for this operation. Prepare a wiring diagram showing all connections to the input/output modules of your PLC. Label your wiring diagram with the appropriate module number and pin numbers that correspond to the program you wrote.

4.3 If your PLC has an analog input module, design a PLC program that will turn on a control relay or magnetic contactor when the analog voltage (or current) from an industrial sensor reaches a certain set point (say +5 volts or 10 mA). Draw the PLC ladder logic diagram for this operation. Prepare a wiring diagram showing all connections to the input/output modules of your PLC. Label your wiring diagram with the appropriate module number and pin numbers that correspond to the program you wrote.

PART 5: Assignment

5.1 For each of the programs that you completed, submit a professional-looking copy of your ladder logic diagram, your program listing, and your wiring diagram.

Magnetic Proximity Switch: 37

INTRODUCTION:
In Experiment 26, you were introduced to the Hall-effect sensor. In that experiment, you constructed a circuit that could be used as a simple magnetic switch. Electronic switch manufacturers have incorporated the Hall-effect sensor into I. C. switch packages. In addition to the Hall-effect proximity switch, there are two other common types of magnetic proximity switches, the *reluctance proximity sensor* and the *eddy-current-killed oscillator (ECKO) sensor*. The Hall-effect sensor will switch states due to the presence of an external magnetic field. The reluctance and ECKO sensors will change states when a metal of sufficient mass is brought into close proximity of either of these switches. The reluctance sensor consists of a permanent magnet core around which are wrapped many turns of fine wire. When a magnetic material (iron or nickel-based) is brought close to the sensor, a low-reluctance path is created through the magnetic material. This causes the magnetic field flux density to change ($d\Phi/dt$). This momentary change will *induce* a voltage in the surrounding wire. This is why these devices are also referred to as *inductive* sensors. The small voltage is amplified and wave shaped so that it can be applied to external switching or counting circuitry. Even though external circuitry is added to the reluctance sensor, it can be considered an active transducer like a thermocouple or solar cell since it generates (produces) its own, but very small, voltage. On the other hand, the ECKO sensor is a passive transducer as it requires an external power supply to operate an oscillator circuit. The A.C. signal produced by the oscillator is applied to a coil to produce an electromagnetic field. When a metal target object passes close to the ECKO sensor, eddy currents are induced in the metal. If the eddy currents are sufficiently large, they will cause the magnetic field to collapse (hence the term *killed*). The collapse of the magnetic field will be sensed by the integrated circuitry causing the switching action to occur.

SAFETY NOTE:
Familiarize yourself with the manufacturer's ratings for your proximity switch before constructing any circuits. In particular, familiarize yourself with its supply voltage rating and current sinking rating. Of course, make sure you correctly identify all terminals (supply, ground, and output). As always, wear safety glasses, and remove all jewelry from your hands and fingers before the start of the experiment.

OBJECT:
Upon successful completion of this experiment and all reading assignments, the student should be able to:
- construct an operational magnetic proximity switch circuit
- experimentally determine the effect that distance between the target object and the sensor has on sensor performance
- experimentally determine the effect that target object composition has on sensor performance
- experimentally determine the effect that sensor supply voltage has on its sensitivity

REFERENCES:
Chapter 10 of Maloney's Modern Industrial Electronics

MATERIALS:
- 1 - proximity switch such as Agastat's® PCI12MFNA (3-wire DC NPN, N.O.) or Veeder-Root's® 653010-010 (3-wire DC NPN, N.O.) or the equivalent
- miscellaneous resistors
- miscellaneous ferromagnetic, metallic, and nonmetallic materials (steel, nickel, copper or brass, aluminum, wood, plastic, ceramic)
- miscellaneous lead wires and connectors
- 1 - 14-volt incandescent lamp and socket

EQUIPMENT:
- 1 - adjustable D.C. power supply
- 1 - digital multimeter
- 1 - oscilloscope
- 1 - oscilloscope probe

INSTRUCTOR/STUDENT NOTE: It is not necessary to use either of the proximity switches included in the preceding parts list. Any similar device will be sufficient. However, choose the externally-connected pull-up resistor so that the maximum sink-current rating of your particular proximity switch will not be exceeded.

PART 1: The Magnetic Proximity Switch
Background:
The schematic symbol for the proximity sensor is shown in Figure 37.1. Note that it is shown as a normally-open switch. This is typical for most magnetic proximity switches. However, consult the manufacturer's data sheet for your sensor to be certain that it is indeed normally-open and not normally-closed. Note also that a certain color-coding scheme is associated with the terminals of the sensor. The brown, black, and blue color typically (though not always) correspond to the supply voltage, output, and ground terminals, respectively, for this type of sensor. Again, consult the manufacturer's data sheet to confirm this. Figure 37.1 shows how Agastat's® PCI12MFNA (3-wire DC NPN, N.O.) or Veeder-Root's® 653010-010 (3-wire DC NPN, N.O.) proximity switch can be configured to produce a high output when no target object is present and a low output when a target object is present. In other words, when a target object is present, the switch will close; otherwise, it will be open. The Agastat® switch is rated for a supply voltage in the range of 10 to 60 volts D.C. and can sink a maximum of 400 mA. The Veeder-Root® switch is rated for a supply voltage in the range of 10 to 50 volts D.C. and can sink up to 200 mA.

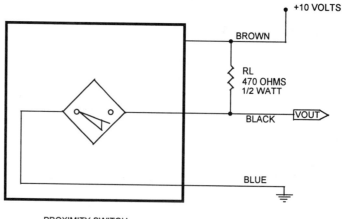

Figure 37.1: Schematic diagram of a proximity switch circuit.

Procedure:

1.1 Obtain all materials and equipment required to complete this experiment. Adjust your power supply to provide 10 volts D.C. Turn off your power supply.

1.2 Construct the circuit shown in Figure 37.1. If your proximity switch cannot sink 10 V/470 Ω = 21.3 mA, then modify your circuit accordingly. Turn on your power supply.

1.3 Use a digital voltmeter to measure the voltage V_{OUT} relative to ground. Record this value in Table 37.1.

1.4 Place a ferromagnetic material such as a steel bar in close proximity to the sensor. Again measure the value for V_{OUT} and record in Table 37.1.

1.5 Place a piece of copper or brass in close proximity to the sensor. Again measure the value for V_{OUT} and record in Table 37.1.

1.6 Place a piece of aluminum in close proximity to the sensor. Again measure the value for V_{OUT} and record in Table 37.1.

1.7 Place a piece of wood in close proximity to the sensor. Again measure the value for V_{OUT} and record in Table 37.1.

1.8 Place a piece of plastic in close proximity to the sensor. Again measure the value for V_{OUT} and record in Table 37.1.

1.9 Place a piece of ceramic material in close proximity to the sensor. Again measure the value for V_{OUT} and record in Table 37.1.

1.10 Hold a ferromagnetic material whose size is about one-half of the diameter of your proximity sensor about 1" from the sensor and slowly move the target object directly toward the sensor. Monitor V_{OUT}. Determine the distance between the magnet and the sensor at which the sensor output switches from high to low. Measure this distance in millimeters, and record your measurement in Table 37.2.

1.11 Repeat step 1.10 for a ferromagnetic material whose size is approximately equal in diameter to the proximity switch.

1.12 Repeat step 1.10 for a ferromagnetic object whose size is approximately twice the diameter of the proximity switch.

1.13 Repeat steps 1.10 through 1.12 for a copper or brass target object. Record your results in Table 37.2.

1.14 *Before you complete this step, ask your instructor if your proximity switch is designed for this higher voltage and current.* Increase the supply voltage to 20 volts D.C. Now repeat steps 1.10 through 1.13 at this higher supply voltage. Record your results in Table 37.3.

1.15 Now replace the 470 Ohms resistor with a 14-volt incandescent lamp and reduce the supply voltage to 12 volts as shown in Figure 37.2.

1.16 Move a metal target object close to the proximity switch. The lamp should turn on. Move the target object away from the switch. The light should turn off.

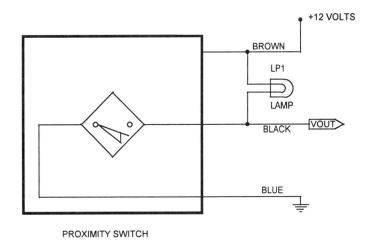

Figure 37.2: Schematic diagram of a proximity switch controlling an indicator lamp.

Table 37.1: V_{OUT} for target objects of varying composition.

TARGET OBJECT MATERIAL	NONE	STEEL	COPPER	ALUMINUM	WOOD	PLASTIC	CERAMIC
OUTPUT VOLTAGE VOUT, VOLTS D.C.							

Table 37.2: Switching distance between sensor and target object as a function of target object size and material composition with a supply voltage of 10 volts D.C.

TARGET OBJECT SIZE	TARGET OBJECT MATERIAL	SWITCHING DISTANCE BETWEEN SENSOR AND TARGET OBJECT, mm
SMALL	STEEL	
MEDIUM	STEEL	
LARGE	STEEL	
SMALL	COPPER	
MEDIUM	COPPER	
LARGE	COPPER	

Table 37.3: Switching distance between sensor and target object as a function of target object size and material composition with a supply voltage of 20 volts D.C.

TARGET OBJECT SIZE	TARGET OBJECT MATERIAL	SWITCHING DISTANCE BETWEEN SENSOR AND TARGET OBJECT, mm
SMALL	STEEL	
MEDIUM	STEEL	
LARGE	STEEL	
SMALL	COPPER	
MEDIUM	COPPER	
LARGE	COPPER	

PART 2: Questions

2.1 Based on your experimental results, is your sensor a reluctance proximity switch or an ECKO switch? Explain how you arrived at your answer.

2.2 Based on your results, explain the relationship between target size and the sensitivity of your proximity switch.

2.3 Based on your results, explain the relationship between the type of target material and the sensitivity of your proximity switch.

2.4 Based on your results, is there a relationship between the sensitivity of your proximity switch and the magnitude of the supply voltage? If so, what is it?

PART 3: Circuit Design

3.1 Connect your proximity switch to the input module of a programmable logic controller (PLC). Write and execute a PLC program that will count ten steel gears as they pass by your sensor on an automotive parts manufacturing line. After the count of ten, the PLC output module should energize a 24 volt solenoid that controls a pneumatic cylinder. The cylinder will push a box containing the gears onto another conveyor belt that will move the gears to the shipping department. Print a copy of your program and attach it to your report. Create a CADD drawing of your PLC wiring diagram showing the proximity switch and solenoid. Print a copy of your drawing and attach it to your report.

3.2 Modify your proximity switch circuit so that the output voltage will be applied to a counter circuit that will operate a seven-segment display capable of counting from 0 to 9. If time permits, construct this circuit and demonstrate it to your instructor. Create a CADD drawing of your circuit. Label all components. Print a copy of your schematic and attach it to your report.

3.3 Modify the circuit shown in Figure 37.2 so that the proximity switch will turn on and off a 12-volt D.C. control relay instead of the light. The contacts of the control relay should turn on a 120-volt lamp when a metal object passes in front of the proximity switch. If time permits, construct this circuit and demonstrate it to your instructor. Create a CADD drawing of your circuit. Label all components. Print a copy of your schematic and attach it to your report.

The D.C. Shunt Motor: 38

INTRODUCTION:
D.C. motors have been used for decades in a wide variety of applications. When you drove to school today, you used at least one D.C. motor. That was, of course, your car's starter motor. If your car has power windows or a power radio antenna, then several additional D.C. motors are at work in your car. If you own a portable cassette tape player/recorder, it is operated by a small D.C. motor. However, D.C. motors are by no means limited to these relatively low-power examples. Applications where a variable speed drive is needed are potential candidates for a D.C. motor. Steel mills that roll (squeeze) strips of steel into thin sheets require D.C. motors of several hundred horsepower to wind and unwind huge rolls of steel as the steel is being squeezed to a smaller gage. D.C. motors may be a permanent-magnet (P.M.) type, having a permanent magnet on either the armature (rotor) or the stationary field (stator). P.M. motors are typically limited to low-power applications. Higher-horsepower applications require motors having both a wound field and a wound armature to produce the required field strength and, hence, torque. Wound motors are available in three different configurations--series, shunt, and compound. Series-wound motors have their stationary field wired in series with the armature. Shunt motors have their field windings wired in parallel with the armature. Compound motors (like compound generators) have both series and shunt windings. The torque/speed requirements dictate which type of motor is chosen for a particular operation. In this experiment, you will work with the shunt motor. It is a very popular type of D.C. motor, because it can maintain a nearly constant speed over a wide range of loads.

SAFETY NOTE:
Motors such as the one you will be using in today's experiment represent many potential hazards, not the least of which is electrocution. Because motors have an output shaft rotating at high speed, they should be securely mounted and the motor/load connection covered with a steel enclosure. When working with rotating machinery, you must wear snug fitting clothing so that your clothes will not be caught on a moving piece of machinery and pull you into it. Long hair must be tied back! This author nearly lost his scalp once when his head (and long hair) got too close to the spindle of a rotating drill press. Before the start of this experiment, have your instructor review with you the proper operation and installation of your motor and load unit. Make sure that you never open the field of a D.C. motor that is lightly loaded or operating under no load! Have your instructor explain the hazardous consequences of doing so. As always, wear safety glasses, and remove all jewelry from your hands and fingers before the start of the experiment.

OBJECT:
Upon successful completion of this experiment and all reading assignments, the student should be able to:
- construct an operational D.C. shunt motor circuit
- interpret the nameplate data for a D.C. shunt motor
- calculate the speed regulation for a D.C. shunt motor
- explain how to reverse the direction of rotation of a D.C. shunt motor
- explain how to vary the speed of a D.C. shunt motor

REFERENCES:
Chapter 12 of Maloney's Modern Industrial Electronics
www.baldor.com

MATERIALS:
- 1 - 500 Ohms, 100 watt wire-wound power rheostat such as Ohmite's® RKS500 or the equivalent
- miscellaneous lead wires and connectors

EQUIPMENT:
- 1 - hand-held multimeter
- 1 - hand-held tachometer
- 1 - D.C. ammeter, 0 to 5 amperes
- 1 - D.C. power supply adjustable from 0 to 120 volts with circuit breaker
- 1 - dynamometer or D.C. generator to serve as a load for your motor
- 1 - D.C. motor up to 1/3 horsepower, rated for 120 volts

PART 1: D.C. Shunt Motor Winding Connections
Background:
The direction of rotation of a D.C. motor depends upon the direction of the magnetic field in the stator in relation to the magnetic field produced in the armature (rotor). For this part of the experiment, you will observe the effect that different field and winding connections have on the direction of rotation of a D.C. shunt motor.

INSTRUCTOR NOTE: Depending on the equipment in your lab, you may connect either a dynamometer or a D.C. generator to your motor to act as a load. If you use a D.C. generator, place a variable, high-power resistive load across its armature to vary the load to the motor. If your lab does not have either of these devices, you may still complete the other parts of the lab that do not require an adjustable load. However, make sure that the speed of your D.C. motors do not become excessive due to operating under too light of a load. Also make sure that the students select an appropriate current range on their ammeters to prevent excessive upscale deflection and the subsequent meter movement damage.

Procedure:
1.1 You will be working with high voltage throughout this experiment; therefore, do not make any connections with the power on. Do not leave any leads or connections exposed. Obtain all materials and equipment required to complete this experiment.
1.2 Adjust the output of your D.C. power supply for 0.0 volt. Turn off the power supply.
1.3 Construct the circuit shown in Figure 38.1 a).
1.4 Turn on the power supply. Slowly increase the supply voltage until the motor just begins to turn. Note the direction of rotation (clockwise or counterclockwise) of the motor's output shaft. Record your observation in Table 38.1. Reduce the supply voltage to 0.0 volt, and turn off the power supply.
1.5 Repeat Step 1.4 for the remaining configurations shown in Figure 38.1
1.6 Turn off the power supply.

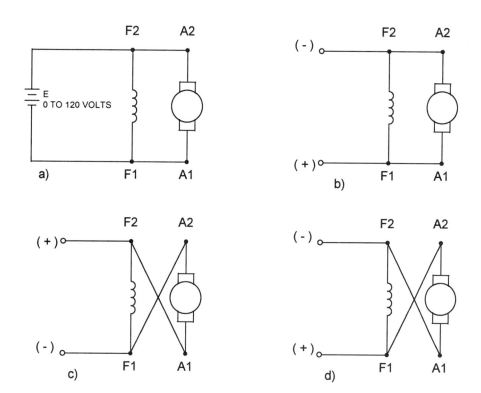

Figure 38.1: Circuits used to determine the effect that winding configuration and power supply polarity have on the direction of rotation of a D.C. motor's armature.

Table 38.1: Direction of motor rotation for the winding and power supply configurations shown in Figure 38.1.

CIRCUIT CONFIGURATION	DIRECTION OF ROTATION
FIGURE 38.1 a)	
FIGURE 38.1 b)	
FIGURE 38.1 c)	
FIGURE 38.1 d)	

PART 2: Speed Control of the Shunt Motor
Background:
While the direction of rotation of a D.C. motor depends on the relative direction of the magnetic fields in the armature and shunt field windings, the speed and torque of the motor depend on the strength (flux density) of the magnetic field produced by each of these windings. In this part of the experiment, you will use two different methods to vary the speed of your motor. In the first method, you will vary the

magnitude of the supply voltage. In the second, you will vary the motor's speed by varying the shunt field current.

Procedure:

    2.1 Adjust the output of your D.C. power supply for 0.0 volt. Turn off the power supply.
    2.2 Construct the circuit shown in Figure 38.1 a). Connect a dynamometer or D.C. generator to your motor in order to place a light load on it. This will prevent your motor from rotating at excessive speed when operated at rated voltage.
    2.3 Turn on the power supply. Increase the supply voltage in even increments until you reach rated voltage for the motor. The motor's rated voltage should be on the nameplate riveted to the motor. If not, have your instructor provide you with this information. For each voltage, use a handheld tachometer to measure the motor's speed. Record your results in Table 38.2. Reduce the supply voltage to 0.0 volt, and turn off the power supply.
    2.4 Install a rheostat in series with the shunt field winding as shown in Figure 38.2.
    2.5 Adjust the rheostat for 0.0 Ohms of resistance.
    2.6 Turn on the power supply, and increase the supply voltage to the motor's rated voltage.
    2.7 Measure the shunt field voltage. If the rheostat has been set for 0.0 Ohms, then the field voltage should equal the supply voltage. Measure and record the motor's speed in Table 38.3.
    2.8 Increase the resistance of the rheostat in even increments until it reaches a maximum of 500 Ohms. Measure and record the corresponding field voltages and motor speeds in Table 38.3. Monitor the supply voltage as you vary the field resistance. Make sure that the supply voltage is maintained at the motor's rated voltage.
    2.9 Reduce the supply voltage to 0.0 volt, and turn off the power supply.

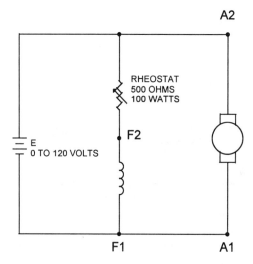

Figure 38.2: Experimental circuit to demonstrate speed control of a D.C. shunt motor by varying the shunt field voltage.

Table 38.2: D.C. shunt motor speed as a function of line voltage.

LINE VOLTAGE, VOLTS	MOTOR SPEED, RPM
0	0
RATED VOLTAGE =	

Table 38.3: D.C. shunt motor speed as a function of shunt field voltage.

SHUNT FIELD VOLTAGE, VOLTS	MOTOR SPEED, RPM

PART 3: Load Characteristics of the Shunt Motor

Background:
As you might imagine, the load operated by most motors can vary widely. Very few industrial operations involve controlling a piece of machinery that is under constant load. However, many industrial operations must be maintained at or near a constant speed. Recall that one of the measures of the quality of a power supply is its voltage regulation. In a similar fashion, a figure of merit for electric motors is speed regulation. While voltage regulation is a measure of a power supply's ability to maintain a constant voltage over a range of electrical loads, speed regulation (S.R.) is a measure of a motor's ability to maintain a constant speed over a range of mechanical loads. It is calculated using the following formula:

$$\% \text{ S.R.} = [(N_{NL} - N_{FL})/N_{FL}] \cdot 100 \qquad \textbf{38.1}$$

N_{NL} is the motor speed (RPM) under no-load conditions. N_{FL} is the motor speed (RPM) under full-load conditions. Full-load is the operating condition that occurs when a motor is producing rated torque, at its rated shaft speed, when operating at rated voltage and drawing rated line current. In this part of the experiment, you will investigate the performance of your motor under load.

Procedure:
3.1 Refer to the nameplate data for your motor, and complete Table 38.4. If any of the information is not available, write N/A.

Table 38.4: Shunt motor nameplate data.

MANUFACTURER'S NAME	
SPEC.	
RATED HORSEPOWER	
RATED LINE VOLTAGE	
RATE LINE CURRENT	
RATED SPEED	
SERVICE FACTOR	
CLASS OF INSULATION	

3.2 Construct the circuit shown in Figure 38.3. Increase the supply voltage to the motor's rated voltage. Adjust the field rheostat until the motor is operating at 1800 RPM or its rated no-load speed if different than this. The no-load speed should be about 3% to 7% greater than the full-load speed. If you do not know what the rated no-load speed is, adjust your rheostat to produce a no-load speed of 5% greater than the motor's rated full-load speed.

3.3 Increase the load on the motor until it draws rated line current. Make sure that your ammeter is set for a full-scale current that will prevent damage to your meter. Make sure that you maintain the supply voltage at the motor's rated voltage.

3.4 Measure the full-load motor speed, N_{FL}. Is the speed you measured equal to the rated motor speed recorded in Table 38.4? If it is close to this value, then record it as the full-load motor speed in Table 38.5. If not, adjust both the motor load and the field rheostat until the motor is operating at rated full-load speed and drawing rated line current.

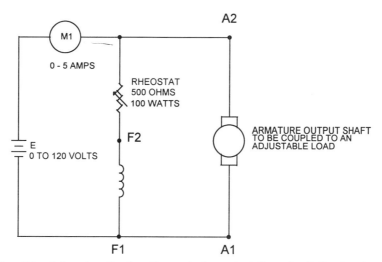

Figure 38.3: Circuit to determine the load/speed characteristics of a D.C. shunt motor.

3.5 If you are using a dynamometer as a load for the motor, record the full-load torque in Table 38.5. Reduce the motor load to 0.0 lb-in of torque. Measure and record the no-load speed in Table 38.5. Reduce the supply voltage to 0.0 volt. Turn off the power supply.

3.6 Use your data to calculate the percent speed regulation for your motor. Record in Table 38.5.

Calculations:

Table 38.5: Shunt motor no-load and full-load speed data.

NO-LOAD SPEED, RPM	FULL-LOAD SPEED, RPM	% SPEED REGULATION	FULL-LOAD TORQUE, #-IN

PART 4: Questions

4.1 Use your data from Table 38.2 to plot a graph of motor speed as a function of armature (line) voltage. Label both vertical and horizontal axes. Attach a copy of your graph to your lab report.

4.2 Use your data from Table 38.3 to plot a graph of motor speed as a function of shunt field voltage. Label both vertical and horizontal axes. Attach a copy of your graph to your lab report.

4.3 Judging from your graphs in Questions 4.1 and 4.2, which of the two methods (armature voltage or field voltage) of speed control would you judge to be more desirable? Explain your answer.

4.4 The formula for calculating the horsepower output of a rotating machine is

$$\text{Horsepower} = (2 \cdot \pi \cdot \tau \cdot N)/(33{,}000 \text{ lb-ft/min/hp}) \qquad \textbf{38.2}$$

In this formula $\pi = 3.14159$, τ is the motor torque in lb-ft, and N is the motor speed in RPM. Use this formula and your motor's <u>rated</u> output horsepower to calculate your motor's <u>rated</u> output torque in lb-ft and in lb-in.

4.5 The formula for efficiency (%) is

$$\text{Efficiency} = (\text{OUTPUT POWER}/\text{INPUT POWER}) \cdot 100 \qquad \textbf{38.3}$$

The conversion constant relating electrical power in watts and mechanical power in in horsepower is 746 watts = 1 hp. Use this conversion constant, the formula for electrical power ($P = E \cdot I$), and Formula 38.3 to calculate the efficiency of you motor when operating at rated load (rated output power).

4.6 Referring to the web site www.baldor.com, what is the catalog number for an SCR drive, permanent magnet DC, explosion-proof motor?

4.7 Referring to the Electrical Engineering Pocket Handbook, sketch and label the connection diagram for a series D.C. motor.

4.8 Referring to the Electrical Engineering Pocket Handbook, what is the full-load current that will be drawn by a 20 Hp D.C. motor when powered by a 240 VDC source?

PART 5: D.C. Shunt Motor Formulas

% S.R. = $[(N_{NL} - N_{FL})/N_{FL}] \cdot 100$ **38.1**

Horsepower = $(2 \cdot \pi \cdot \tau \cdot N)/(33{,}000 \text{ lb-ft/min/hp})$ **38.2**

Efficiency = (OUTPUT POWER/INPUT POWER) $\cdot$ 100   **38.3**

Stepper Motors: 39

INTRODUCTION:
In Experiment 38, you were introduced to D.C. motors having both a wound field and a wound armature. In applications where physical space is a premium and the power requirements are low, a permanent-magnet (P.M.) motor can be used. As discussed in Experiment 38, P.M. motors may have permanent magnets installed on either the stationary field or on the armature. While P.M. motors as a class are lower power than D.C. motors with a wound field and wound armature, they have the benefit of providing a constant flux density over their intended range of operation. P.M. motors having the permanent magnets on the armature have the added benefit of having no brushes. Remedying brush and commutator bar wear consume more of the electrical maintenance technician's time than any other motor problem. Eliminating the brushes and commutator bars reduces the overall length of the brushless D.C. motor. This type of motor can also be controlled (commutated) electronically. Power to the stationary field windings are switched on and off electronically by transistors, SCRs, or power MOSFETs. This provides the designer with the flexibility of choosing between an integrated circuit, a microcomputer, or a microcontroller to control the motor. In this experiment, you are going to be introduced to one type of brushless D.C. motor, the stepper motor. Stepper motors have been used in many low-power applications requiring precise positioning. For example, stepper motors have been used for many years to position the read/write heads in the disk drives of microcomputers. In this experiment, you will have the opportunity to construct a circuit that will allow you to digitally control the direction of rotation and speed of a stepper motor.

SAFETY NOTE:
The stepper motor you will use in this experiment is a relatively low-power device. However, that does not mean that safety considerations are thrown out the window. Before the start of this experiment have your instructor help you identify the leads to your stepper motor. Also familiarize yourself with the current and voltage ratings of your stepper motor in order to prevent damage to the motor due to the application of excessive voltage and/or current. Construct your circuit modularly. Construct one stage of your circuit at a time, and test each stage as you progress. As always, wear safety glasses, and remove all jewelry from your hands and fingers before the start of the experiment.

OBJECT:
Upon successful completion of this experiment and all reading assignments, the student should be able to:
- construct an operational stepper motor controller circuit
- explain how to vary the speed of a stepper motor
- explain how to change the direction of rotation of a stepper motor
- define the term *step angle*
- determine the shaft speed of a stepper motor for a given step angle and clock frequency

REFERENCES:
Chapter 13 of Maloney's Modern Industrial Electronics

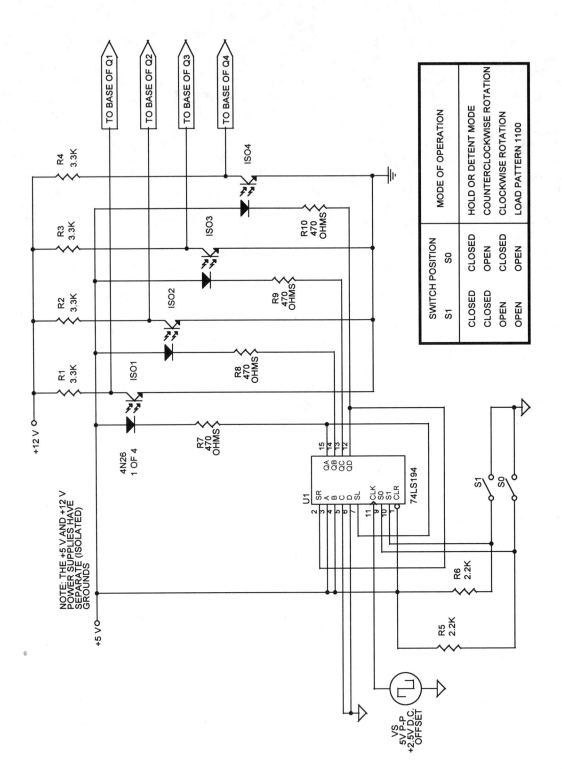

Figure 39.1 a): Stepper motor control circuit.

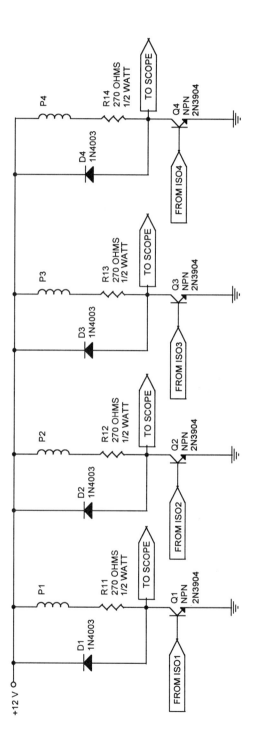

Figure 39.1 b): Stepper motor control circuit.

MATERIALS:

- 4 - SK2040 or 4N26 optoisolators or the equivalent
- 4 - diodes, 1N4003, or the equivalent
- 2 - SPST switches or one dip switch pack
- 1 - 74LS194 bidirectional universal shift register
  - miscellaneous lead wires and connectors
- 1 - four-pole stepper motor from a computer disk drive having a step angle of 1.8°/step
- 4 - 2N3904 transistors
- 4 - 470 Ohms, 1/4 watt resistors
- 4 - 3.3 Kohms, 1/4 watt resistors
- 4 - 270 Ohms, 1/2 watt resistors
- 2 - 2.2 Kohms, 1/4 watt resistors
- 1 - solderless breadboard

EQUIPMENT:

- 1 - digital multimeter
- 1 - handheld tachometer
- 1 - stop watch if a tachometer is not available
- 1 - function generator
- 1 - frequency counter
- 1 - dual-outlet D.C. power supply
- 1 - oscilloscope
- 2 - 10X oscilloscope probes

PART 1: Stepper Motor Operation

Background:

Stepper motors might very well be considered digital motors, since the power applied to each of the poles is a square wave. In the stepper motor control circuit shown in Figures 39.1 a) and b), a binary pattern of 1100 is loaded into the 74LS194 when switches S1 and S0 are open. For more details on the 74LS194, refer to a manufacturer's data manual of common TTL devices. Closing either S1 or S0 will allow this pattern to be shifted right or left. If S1 is closed and S0 is open, the pattern initially loaded will be shifted right on each rising edge of the square wave applied to the CLK input. In other words, the initial pattern of 1100 will shift right (recirculating the rightmost bit back to the leftmost position) on each ensuing clock pulse to become 0110, 0011, 1001, and back to 1100. Electrical isolation of the digital circuitry from the stepper motor's coils and current gain are being provided, respectively, by the optoisolators and 2N3904 NPN transistors. The shifting waveform, when applied to the stepper motor, effectively produces a rotating waveform that circulates either clockwise or counterclockwise around the circumference of the motor. This rotating waveform effectively "pulls" the magnetic rotor along with it. If the initial binary pattern is shifted left, it will change the phase relationship of the voltage waveforms applied to the poles of the motor such that the direction of rotation of the motor will be reversed. This is quite a convenience, because this could be performed under program control by a microcomputer. In the circuit that you are to build, the 4-bit output of the 74LS194 shift register could easily be replaced by the data bus of a microcomputer (with the proper signal conditioning of course).

Because current to the stepper motor field windings is a square wave, the stepper motor then does not make one smooth continuous motion; rather, it moves in small steps. Specifically, it rotates one step or increment for each clock pulse. The amount that the stepper motor rotates for a given clock pulse is referred to as the *step angle*. A common step angle for a stepper motor is 1.8°/step. For each clock pulse, a motor with this step angle would rotate 1.8°. Hence, it would take 360°/(1.8°/step) = 200 steps or clock pulses for this motor to make one revolution. For the stepper motor control circuit that you are to build today, the angular velocity of the stepper motor's armature in revolutions per minute can be found from the following formula:

$$\text{RPM} = (f_{CLK}) \cdot (1 \text{ rev}/360°) \cdot (\text{degrees/step}) \cdot (60 \text{ sec/min}) \quad \textbf{39.1}$$

In this formula, degrees/step represents the step angle of your stepper motor and f_{CLK} is the frequency of the square wave applied to the 74LS194 in Hz, cycles/sec, pulses/sec, or steps/sec. You should be able to find the step angle on the nameplate of your stepper motor. Before constructing the stepper-motor control circuit, it will be necessary to identify the lead wires coming out of your stepper motor. Most stepper motors have either three or four sets of windings (poles or phases) on the stator. The controller circuit shown in Figure 39.1 a) and b) was designed to operate a four-pole stepper motor. Stepper motors having four poles have either five or six wires exiting the body of the stepper motor. The two common internal configurations of a four-pole stepper motor are shown in Figures 39.2 a) and b). First count the number of different wires your motor has. Then connect an ohmmeter across the leads as shown in either Figures 39.3 a) or b) depending on how many wires your stepper motor has. Note that in either figure, the resistance reading of M1 will be twice that of M2. If you need more help with the lead wire identification process, just ask your friendly instructor. In this experiment you will be connecting a 12 volt D.C. source to the center or common tap(s) of the motor windings. The four remaining leads will be connected to the stepper motor controller circuit that will be switching the stator poles on and off.

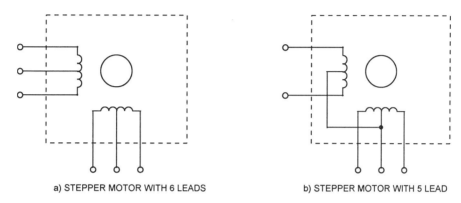

Figure 39.2: Two common internal wiring configurations of a four-pole (four-phase) stepper motor.

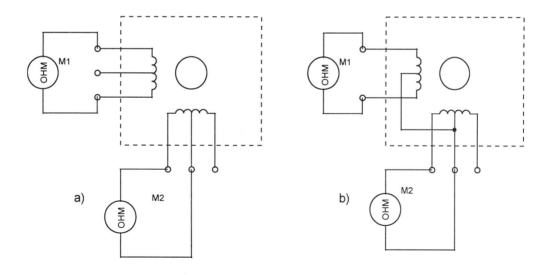

Figure 39.3: Using an ohmmeter to identify stepper motor lead wires.

Procedure:

1.1 Obtain all parts and equipment required to complete this experiment.

1.2 Once you have identified the leads of your stepper motor, construct the circuit shown in Figure 39.1 a).

NOTE: The successful operation of this circuit will depend on the current gain (transfer function) of the optoisolator. If the gain of your optoisolator is not very large, it may be necessary to increase the size of resistors R7, R8, R9, and R10 to as large as 22 Kohms.

1.3 Set your function generator for a 5 volt peak-to-peak square wave with +2.5 volts of D.C. offset at a frequency of 200 Hz. Connect your function generator to your frequency counter so that you may adjust the frequency to exactly 200 Hz.

1.4 Test the circuit shown in Figure 39.1 a) by applying power to the circuit and the square wave signal to the CLK input of the 74LS194.

1.5 Use your oscilloscope to display the voltage waveforms present at the collectors of each optoisolator. You should display the output of ISO1 on Channel 1 of your oscilloscope, trigger on this waveform, and display the remaining three waveforms in proper time phase with it. Have your instructor check the accuracy of your work.

1.6 Once the circuit shown in Figure 39.1 a) is working properly, you may construct the circuit in Figure 39.1 b) and attach it to the first stage.

1.7 Once both stages are complete, turn on all D.C. power supplies and your function generator.

1.8 Load the binary pattern 1100 into the 74LS194 by opening both S1 and S0. Now close S1. If the motor does not rotate, switch phases P2 and P3. If the motor still does not rotate, switch what is now phase P3 with phase P4. At this point, the motor should rotate. If it does not, the frequency of your clock may be too great, or you did not properly load the pattern 1100, or something else is wrong with your circuitry. If your motor still does not rotate, call your instructor for assistance.

1.9 Use your oscilloscope to display the voltage waveforms present at the collectors of each 2N3904 transistor. You should display the collector voltage for Q1 on Channel 1 of your oscilloscope, trigger on this waveform, and display the remaining three waveforms in proper time phase with it. Have your instructor check the accuracy of your work. Draw the four waveforms in the space provided on Graph 39.1. Draw all four waveforms on the same graph in proper time phase. Label the time base. Your waveforms should be drawn as neatly and as precisely as possible. Note the direction of rotation.

1.10 Now close S0. What happened to the stepper motor? What happened to the waveforms on your oscilloscope? Record your observations in the space provided below.

Observations:_____

_____

_____

_____

1.11 With S0 closed, open S1. What effect did changing the relative position of switches S1 and S0 have on the direction of motor rotation?

Observation:_____

1.12 Draw the resulting transistor voltage waveforms in the space provided on Graph 39.2. Make sure that you are still displaying the waveform for Q1 on Channel 1 of your oscilloscope and triggering on it.

Graph 39.1: Stepper motor controller voltage waveforms shown in proper time phase for the circuit of Figure 39.1 with S1 closed and S0 open.

_____ volts/div   _____ sec/div

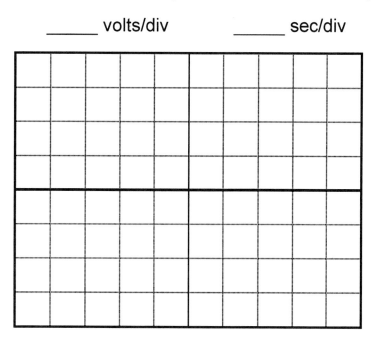

Graph 39.2: Stepper motor controller voltage waveforms shown in proper time phase for the circuit of Figure 39.1 with S1 open and S0 closed.

_____ volts/div   _____ sec/div

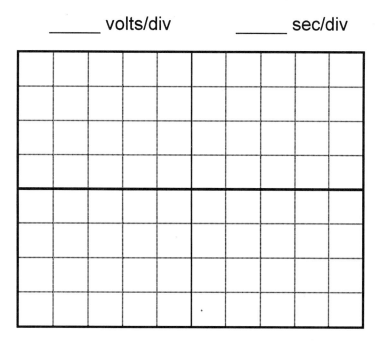

1.13 Adjust your frequency generator to produce each of the frequency values shown in Table 39.1. For each of these values, use your tachometer to determine the motor's RPM. If you do not have a tachometer, the motor should be rotating slow enough so that you can time the revolutions using a stop watch. Also use the period of each of the transistor waveforms to calculate the frequency of the waveforms being applied to the stepper motor. Record in Table 39.1.

Calculations:

Table 39.1: Stepper motor speed as a function of clock frequency.

f_{CLK}, Hertz	MOTOR SPEED, RPM	Q1 FREQUENCY, Hertz
100		
200		
300		
400		

1.14 Slowly increase the clock frequency until the motor stops rotating. Return the clock frequency to 200 Hz.

1.15 Turn off the function generator and both D.C. power supplies. Remove the diodes from your circuit. Turn on the D.C. power supplies and the function generator. Again load the pattern 1100. Close S1, and note the transistor waveforms on the oscilloscope. What effect did removing the diodes have on circuit operation? Record your observations in the space provided below.

Observations:_____

_____

_____

_____

1.16 Turn off the function generator and both D.C. power supplies. Remove the 270 Ohms resistors from your circuit. Reinstall the diodes in parallel with the motor windings. Turn on the D.C. power supplies and the function generator. Again load the pattern 1100. Close S1, and note the transistor waveforms on the oscilloscope. What effect did removing the resistors have on circuit operation? Record your observations in the space provided below.

Observations:_____

_____

_____

_____

1.17. Turn off the function generator and both power supplies.

PART 2: Questions

2.1 What is the relationship between the frequency of the clock input and the frequency of the 2N3904 transistor voltage waveforms? What multiple is one of the other?

2.2 What effect did changing the switch positions from S1 closed and S0 open to S1 open and S0 closed have on the phase relationship of the 2N3904 transistor voltage waveforms?

2.3 In order from left to right, write out the first four binary patterns from Graph 39.1, starting with 1100.

2.4 In order from left to right, write out the first four binary patterns from Graph 39.2, starting with 1100.

2.5 Why does closing both switches S1 and S0 cause the motor to stop rotating?

2.6 Why did the stepper motor stop rotating when the clock frequency was increased beyond a certain point?

2.7 What role do the diodes play in the operation of this circuit?

PART 3: Stepper Motor Formulas

RPM = (f_{CLK}) · (1 rev/360°) · (degrees/step) · (60 sec/min) **39.1**

Three-Phase Motors: 40

INTRODUCTION:
The three-phase motor is the most widely used type of motor in manufacturing settings today. Since three-phase power is not available in your home, it makes sense that all of your motor-driven electric appliances such as vacuum cleaners, dishwashers, clothes dryers, and can openers utilize single-phase motors. While single-phase motors do find use in industry, three-phase motors are more popular, because they run smoother and operate more efficiently than single-phase motors of equal horsepower rating. The applications for three-phase motors are endless. They can be used to operate hydraulic pumps, air compressors, milling machines, engine lathes, punch presses, conveyors, etc. Common three-phase motor configurations include the *wound rotor induction motor*, *squirrel cage induction motor*, and *synchronous motor*. The most common of these is the squirrel cage induction motor. It popularity is attributable to its simple, yet rugged construction and excellent full-load operating characteristics. In applications requiring motor start up under load, the wound rotor induction motor is often chosen. In applications requiring a constant speed over a wide range of loads, the synchronous motor should be chosen.

SAFETY NOTE:
Motors such as the one you will be using in today's experiment represent many potential hazards, not the least of which is electrocution. Because motors have an output shaft rotating at high speed, they should be securely mounted and the motor/load connection covered with a steel enclosure. When working with rotating machinery, you must wear snug-fitting clothing so that your clothes will not be caught on a moving piece of machinery and pull you into it. Long hair must be tied back! Before the start of this experiment, have your instructor review with you the proper operation and installation of your motor and load unit. Familiarize yourself with the full-load operating characteristics of your motor, in particular the full-load line currents. As always, wear safety glasses, and remove all jewelry from your hands and fingers before the start of the experiment.

OBJECT:
Upon successful completion of this experiment and all reading assignments, the student should be able to:
- construct an operational three-phase motor circuit with start/stop control
- calculate the percent slip for a three-phase motor
- explain how to reverse the direction of rotation of a three-phase motor
- explain how to vary the speed of a three-phase motor

REFERENCES:
Chapter 14 of Maloney's Modern Industrial Electronics
Electrical Engineering Pocket Handbook
www.baldor.com

MATERIALS:
- miscellaneous lead wires and connectors

EQUIPMENT:
- 1 - handheld multimeter
- 1 - handheld tachometer
- 3 - A.C. ammeters, 0 to 5 amperes
- 2 - single-phase wattmeters or 1 three-phase wattmeter
- 1 - A.C. power supply, 120/208, with circuit breaker
- 1 - dynamometer or D.C. generator to serve as a load for your motor
- 1 - three-phase induction motor up to 1/3 horsepower
- 1 - N.O. push button switch
- 1 - N.C. push button switch
- 1 - magnetic contactor
- 1 - thermal overload relay
- 1 - pilot lamp

PART 1: Three-Phase Motor Winding Connections
Background:
When power is applied to the field windings of a three-phase motor, a rotating magnetic field is created due to the continuously rising and falling three-phase currents that differ in phase by 120°. The speed of this rotating magnetic field is referred to as the *synchronous speed*, N_S. The theoretical value for the synchronous speed of a three-phase motor can be calculated using the following formula:

$$N_S = (120 \cdot f)/p \qquad \textbf{40.1}$$

In this formula f represents the frequency of the A.C. source in Hertz and p is the number of field poles per phase. In the United States, f will be 60 Hz. In other countries the line frequency may be 50 Hz. So there are two ways to change the speed of a three-phase motor, either change the supply frequency or select a motor with a different number of poles. While D.C. motors are still widely used in applications requiring a range of operating speeds, variable frequency drives are available that can be used to adjust the speed of three-phase motors. Common synchronous speeds for fractional horsepower three-phase induction motors installed in the U.S. are 900, 1200, and 1800 RPM. The rotating magnetic field induces a voltage and current in the rotor bars of an induction motor. As you may recall from an earlier A.C. circuits class, a current-carrying conductor in the path of a magnetic field will experience a mechanical force. The stronger the magnetic field and/or the greater the induced currents, the greater the mechanical force (or resultant torque in this case). The resultant force acts to pull the rotor in the same direction as the rotating magnetic field. The direction of rotation of the rotating magnetic field, and hence the direction of rotation of the motor's rotor, are determined by the order or sequence of the three voltages or phases. Changing the order or sequence of any two of the three phases will change the direction of the rotating magnetic field.

INSTRUCTOR'S NOTE: Depending on the equipment available in your lab, you may connect either a dynamometer or a D.C. generator to your motor to act as a load. If you use a D.C. generator, place a variable, high-power resistive load across its armature to adjust the load applied to the motor. Also make sure that the students select an appropriate full-scale value for their ammeters and wattmeters to prevent excessive upscale deflection and damage to the meter movement. Before starting Part 1 of this experiment, walk the students through the operation of the circuit shown in Figure 40.1. Review the connection of wattmeters for measuring three-phase power before the students start Part 2 of this experiment.

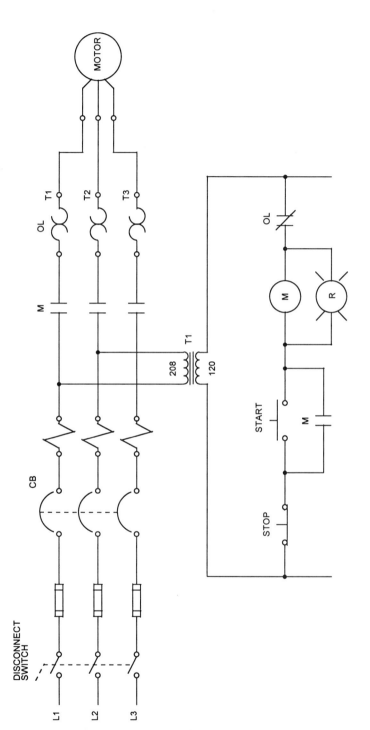

Figure 40.1: Basic start/stop control circuit used to operate a three-phase induction motor.

Procedure:

1.1 You will be working with high voltage throughout this experiment; therefore, do not make any connections with the power on. Do not leave any leads or connections exposed. Obtain all materials and equipment required to complete this experiment. Before constructing the circuit shown in Figure 40.1, have you instructor help you determine the current and voltage ratings of your pushbuttons, magnetic contactor, and connecting wires. Make sure that they are rated for the conditions under which they will be operating. Have your instructor help you adjust your thermal overload relay for an appropriate trip-out time and current.

1.2 Make sure that your power supply is turned off.

1.3 Construct the circuit shown in Figure 40.1. Build the circuit in stages by first constructing the start/stop control part of the circuit before connecting the motor.

1.4 Turn on the power supply. Press the start button. Does the pilot light turn on? Did you hear the contacts on the magnetic contactor close? If so, press the stop button. Did the pilot light go out? If so, proceed to the next step. If not, check your work with the assistance of your instructor. Turn off all power.

1.5 Now connect the motor to your circuit. Connect a dynamometer or generator to the motor in order to place a light load on it. Make sure that the dynamometer can be rotated both clockwise and counter-clockwise.

Warning: In the next step you will test your circuit by trying to start the motor. If the motor does not start when you press the start button or if there is a loud hum or buzz or you smell something getting hot at any point, immediately turn off all power!

1.6 Turn on the power supply. Press the start button. Does the motor shaft rotate? If not, immediately press the stop button and call your instructor for help.

1.7 Press the stop button. Note the direction of rotation (clockwise or counterclockwise) of the motor's rotor as it coasts to a stop. Record your observation in the space provided below.

Observation:_____

1.8 Turn off all power. Switch any two of the three lines going to your motor as shown in Figure 40.2.

1.9 Turn the power on again. Press the start button, and allow the motor to reach normal operating speed. Use a handheld tachometer to measure the motor's speed in RPM. Based on your reading, record the synchronous speed (900, 1200, or 1800) of your motor in Table 40.1.

1.10 Press the stop button. Note the direction of rotation of the motor as it coasts to a stop. Record your observation in the space provided below. Turn off all power.

Observation:_____

1.11 Use the value of the synchronous speed for your motor and Formula 40.1 to calculate the number of poles per phase on your motor. Record your result in Table 40.1.

Calculations:

Figure 40.2: Three-phase motor with two phases interchanged.

Table 40.1: Synchronous speed and number of poles per phase for your three-phase motor operated from a 60 Hz supply.

SYNCHRONOUS SPEED, RPM	NUMBER OF POLES PER PHASE

PART 2: Load Characteristics of the Three-Phase Motor
Background:
A three-phase induction motor operating under little load or no-load will rotate at its synchronous speed. As the load on the rotor is increased, the speed will start to decrease. As it does, the current in the rotor bars will increase, in turn increasing the motor torque. At full-load, an induction motor will rotate at a speed slightly less than its synchronous speed. Refer to the nameplate data on your motor for more detailed information. The difference between the synchronous speed and the speed at rated load is referred to as the *slip*. This parameter is often specified as a percent and is calculated as follows:

$$\% \text{ SLIP} = [(N_{NL} - N_{FL})/N_{NL}] \cdot 100 \qquad \textbf{40.2}$$

N_{NL} represents the no-load or synchronous speed (RPM) of the motor and N_{FL} is the full-load motor speed. Full-load is the operating condition that occurs when a motor is producing rated torque, at its rated shaft speed, when operating at rated voltage and drawing rated line current. In this part of the experiment, you will have the opportunity to explore the performance of your motor under load.

Procedure:
2.1 Refer to the nameplate data for your motor, and complete Table 40.2. If any of the information is not available, write N/A.

Table 40.2: Three-phase motor nameplate data.

MANUFACTURER'S NAME	
SPEC.	
RATED HORSEPOWER	
RATED LINE VOLTAGE	
RATE LINE CURRENT	
RATED SPEED	
SERVICE FACTOR	
CLASS OF INSULATION	

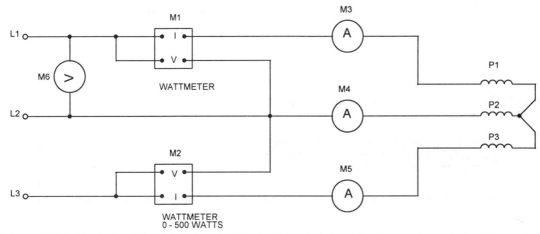

Figure 40.3: Test circuit to determine the load characteristics of a three-phase induction motor.

2.2 Install ammeters in your circuit as shows in Figure 40.3 to measure the line current drawn by each phase of your motor. You may use either in-line ammeters or clamp-on style ammeter(s). Install two single-phase wattmeters or one three-phase wattmeter to measure the total real power. Use a voltmeter to measure the line-to-line voltage. Make sure that the full-scale setting of your meters meets with the approval of your instructor. P1, P2, and P3 represent the wye-connected field windings of your motor. Before starting your motor, ask your instructor to tell you what direction the motor should be rotating when the load is applied by the dynamometer. Finalize your wiring at this time. Your dynamometer should be set for no load.

2.3 Turn on the power supply. At the moment you press the start button, observe what happens to the readings on your ammeters. Record your observation in the space provided below.

Observation:_____

2.4 Now that your motor is running, increase the load on your motor until it draws rated current. This represents 100% of rated load. Record all of the information specified in Table 40.3 for this load condition. Repeat your measurements for the remaining load conditions specified in Table 40.3.

2.5 Press the stop button. Turn off the power supply.

Table 40.3: Three-phase motor load data.

PERCENT OF RATED LOAD	TOTAL POWER WATTS	TORQUE LB-IN	V_{L1-L2} VOLTS	I_{LINE1} AMPS	I_{LINE2} AMPS	I_{LINE3} AMPS	SPEED RPM
0%							
25%							
50%							
75%							
100%							

PART 3: Questions

3.1 Explain how to reverse the direction of rotation of a three-phase motor.

3.2 In the space provided below, draw and completely label a timing diagram for the circuit shown in Figure 40.1.

3.3 When you turned on your motor in Step 2.3, why did the line current increase rapidly, then drop back off?

3.4 Referring to the data in Table 40.3, calculate the percent slip for your motor.

3.5 Referring to the data in Table 40.3, calculate each of the following for the no-load condition.

    a) apparent power =

    b) reactive power =

    c) power factor =

3.6 The formula for calculating the horsepower output of a rotating machine is

$$\text{Horsepower} = (2 \cdot \pi \cdot \tau \cdot N)/(33{,}000 \text{ lb-ft/min/hp}) \qquad \textbf{40.3}$$

In this formula, $\pi = 3.14159$, τ is the motor torque in lb-ft, and N is the motor speed in RPM. The formula for efficiency is

$$\text{Efficiency} = (\text{OUTPUT POWER}/\text{INPUT POWER}) \cdot 100 \qquad \textbf{40.4}$$

The conversion constant relating electrical power in watts and mechanical power in horsepower is 746 watts = 1 hp. Now, referring to the data in Table 40.3, calculate each of the following for the <u>full-load</u> condition.

    a) apparent power =

b) reactive power =

c) power factor =

d) horsepower =

e) efficiency =

3.7 Referring to the Electrical Engineering Pocket Handbook, sketch and label the terminal markings and connections for a three-phase, single-speed, single-voltage, wye-connected motor.

3.8 Referring to the Electrical Engineering Pocket Handbook, sketch and label the terminal markings and connections for a three-phase, single-speed, single-voltage, delta-connected motor.

3.9 Referring to the Electrical Engineering Pocket Handbook, what is the minimum size NEMA motor starter you would specify to start a 460 volt, 3-phase, 40 Hp motor?

3.10 Referring to the Electrical Engineering Pocket Handbook, what is the line current that would be drawn by a 460 volt, 3-phase, 40 Hp, squirrel cage motor operated at full load?

3.11 Referring to the web site www.baldor.com, what is the catalog number of a Baldor three-phase, rigid base motor rated for 10 Hp, when operated at a rated speed of 1725 RPM and a line frequency of 60 Hz having a service factor of 1.0?

3.12 Referring to the web site www.baldor.com, what is(are) the rated line voltage(s) and rated line current(s) for the motor in Question 3.11?

3.13 Referring to the web site www.baldor.com, what is the efficiency and power factor for the motor in Question 3.11?

PART 4: Circuit Design

4.1 With assistance from your instructor, design a start/stop cicuit that will be operated by a programmable controller (PLC). Use an output module of your PLC that is rated for 120 VAC loads to turn on and off your magnetic contactor. Connect your start and stop buttons to an input module of appropriate voltage rating. Write and execute a PLC program that will allow you to accept inputs from the start and stop pushbuttons. The program should then direct the output module to turn the magnetic contactor on or off. If time permits, construct this circuit and connect the motor to your magnetic contactor, then run the program. Print a copy of your program and attach it to your report. Create a CADD drawing of your PLC wiring diagram showing the pushbuttons and magnetic contactor. Print a copy of your drawing and attach it to your report.

PART 5: Three-Phase Motor Formulas

$N_S = (120 \cdot f)/p$ **40.1**

$\% \text{ SLIP} = [(N_{NL} - N_{FL})/N_{NL}] \cdot 100$ **40.2**

$\text{Horsepower} = (2 \cdot \pi \cdot \tau \cdot N)/(33{,}000 \text{ lb-ft/min/hp})$ **40.3**

$\text{Efficiency} = (\text{OUTPUT POWER}/\text{INPUT POWER}) \cdot 100$ **40.4**

Appendix

MATERIALS and EQUIPMENT LISTS

MISCELLANEOUS -

4 - switches, SPST
1 - light bulb, 120 volts, 60 watts
1 - light bulb socket for 120-volt lamp
1 - 14 volt lamp such as Radio Shack's Archer® Catalog No. 272-1118 rated for 200 mA
1 - bayonet-style lamp holder such as Radio Shack's Archer® Catalog No. 272-355
2 - light bulbs, 28 volt rating
2 - light bulb sockets
2 - normally-open push button switches
1 - normally-closed push button switch
1 - 120:12.6 V_{RMS} center-tapped transformer
1 - choke coil in the range of 4 to 12 Henrys
1 - small bar magnet (approximately 0.25" in diameter and no more than 0.5" in length)
1 - variable-speed, handheld electric drill
1 - 1000 mL or 2000 mL beaker
1 - test tube
2 - NTC thermistors
1 - PTC thermistor
  - miscellaneous thermocouples
1 - silicon solar cell 2 X 4 cm or larger, Radio Shack® Catalog No. 276-124A or the equivalent
1 - photoconductive cell, VACTEC® VT-204, VT935G, VT90D782G, or the equivalent
1 - rigid, opaque, nonconductive tube with an inner diameter of approximately 0.5" and a length of at least 7" to 8"
1 - zero voltage turn-on SSR such as Potter & Brumfield's® SSR-240D50 or the equivalent
1 - fuse, fast-blow, rated for 2 amps
1 - fuse, slow-blow, rated for 1/8 amp
1 - MIDTEX® control relay, Model 157-22B200 with 120 Ohms, 12 volt coil, DPDT contacts rated for 10 amperes, or the equivalent
1 - control relay socket
1 - MACROMATIC® programmable solid-state, time-delay relay, Model SS-60226 with 12 VAC/DC input and DPDT contacts rated for 10 amperes, or the equivalent
1 - time-delay relay socket
1 - programmable logic controller
1 - proximity switch such as Agastat's® PCI12MFNA (3-wire DC NPN, N.O.) or Veeder-Root's® 653010-010 (3-wire DC NPN, N.O.) or the equivalent
1 - pressure transducer, rated 0 – 500 psig, designed for hydraulic circuits
1 - four-pole stepper motor from a computer disk drive having a step angle of 1.8°/step
1 - dip switch pack with eight SPST switches
  - miscellaneous lead wires with alligator-style terminals
  - miscellaneous lead wires and connectors
  - miscellaneous switches and control devices
1 - in-line fuseholder
1 - solderless breadboard

ELECTRONIC DEVICES -

- 4 - NPN transistors, 2N3904 or the equivalent
- 1 - PNP transistor, 2N3905 or the equivalent
- 1 - NPN transistor, TIP29 or the equivalent
- 1 - N-channel, enhancement-mode MOSFET, SK9155 or a MOSFET of equal or higher power (if you do not have a low power MOSFET, you may substitute a high-power MOSFET such as the SK9502)
- 1 - diac, ST2, or SK3523, or the equivalent
- 1 - UJT, 2N2646, or SK9123, or the equivalent
- 1 - SCR, MCR218-6FP or the equivalent
- 1 - triac, MAC210A8 or the equivalent
- 1 - phototransistor, ECG3031, or SK2031, or the equivalent
- 1 - low-power SCR, MCR22-3 or the equivalent

FIXED RESISTORS -

- 4 - 470 Ohms, 1/4 watt
- 5 - 1 Kohms, 1/4 watt
- 2 - 1.2 Kohms, 1/4 watt
- 2 - 2.2 Kohms, 1/4 watt
- 1 - 3.3 Kohms, 1/4 watt
- 1 - 5.6 Kohms, 1/4 watt
- 2 - 10 Kohms, 1/4 watt
- 1 - 12 Kohms, 1/4 watt
- 1 - 15 Kohms, 1/4 watt
- 1 - 47 Kohms, 1/4 watt
- 1 - 100 Kohms, 1/4 watt
- 1 - 270 Kohms resistor, 1/4 watt
- 1 - 470 Ohms resistor, 1/4 watt
- 4 - 2.2 Megohms resistors, 1/4 watt
- 1 - 10 Megohms resistor or two series-connected 4.7 Megohms resistors, 1/4 watt
- 1 - 100 Ohms, 1/2 watt
- 1 - 220 Ohms, 1/2 watt
- 4 - 270 Ohms, 1/2 watt
- 1 - 330 Ohms, 1/2 watt
- 1 - 470 Ohms, 1/2 watt
- 1 - 560 Ohms, 1/2 watt
- 2 - 1 Kohms, 1/2 watt
- 3 - 1.2 Kohms, 1/2 watt
- 1 - 1.5 Kohms, 1/2 watt
- 1 - 22 Kohms, 1/2 watt
- 1 - 220 Ohms, 1 watt
- 1 - 470 Ohms, 1 watt
- 1 - 560 Ohms, 1 watt
- 1 - 1 Kohms, 1 watt
- 2 - 1.5 Kohms, 1 watt
- 1 - 220 Ohms, 2 watts
- 1 - 470 Ohms, 2 watts
- 1 - 1.5 Kohms, 25 watt power resistor
- 3 - 50 Ohms, 225 watt wire-wound power resistors
- 8 - 330 Ohms resistors or resistor pack, 1/4 watt

POTENTIOMETERS -

- 1 - 200 Ohms, ten-turn potentiometer
- 1 - 1 Kohms potentiometer
- 3 - 2 Kohms, ten-turn potentiometer
- 2 - 5 Kohms, ten-turn potentiometers
- 1 - 10 Kohms ten or twenty-turn trim pot
- 1 - 50 Kohms ten-turn trim potentiometer
- 1 - 100 Kohms ten-turn, trim pot
- 1 - 100 Kohms potentiometer, 2 watts
- 1 - 1 Megohms potentiometer
- 1 - 500 Ohms, 100 watt wire-wound power rheostat such as Ohmite's® RKS500 or the equivalent

DIODES -

- 2 - 1N914 diodes
- 4 - 1N4003 rectifier diodes or the equivalent
- 1 - bridge rectifier, such as the SK3985 or SK5042
- 1 - zener diode, 1N4742, 1 watt
- 6 - rectifier diodes, 1N5407, or SK9009, or the equivalent
- 8 - LEDs or 1 ten-segment bar display
- 1 - common anode, seven-segment display
  - miscellaneous red, green, and amber LEDs

CAPACITORS -

- 1 - 150 pF
- 1 - ceramic or tantalum capacitor, 0.001 µF
- 1 - ceramic or tantalum capacitor, 0.01 µF
- 1 - 0.047 µF
- 2 - 0.1 µF capacitors, 400 WVDC
- 2 - 0.1 µF capacitors, ceramic disc
- 1 - ceramic or tantalum capacitor, 0.1 µF
- 1 - 0.33 µF
- 1 - 1 µF aluminum or tantalum electrolytic
- 1 - 1 or 10 µF tantalum capacitor
- 2 - 10 µF electrolytic capacitors, 150 WVDC
- 1 - electrolytic capacitor, 50 µF, 25 WVDC or higher

INTEGRATED CIRCUITS -

- 1 - fixed-output, I-C voltage regulator, LM7812CT, or SK3592, or the equivalent
- 1 - adjustable-output, I-C voltage regulator, LM317T, or SK9215, or the equivalent
- 1 - LM334Z three-terminal adjustable current source
- 1 - Hall-effect switch such as Micro Switch's® 8SS7, or 65SS4, or the equivalent
- 1 - digital-to-analog converter, DAC0800, or the equivalent
- 1 - analog-to-digital converter, ADC0803, or ADC0804, or the equivalent
- 1 - temperature sensor, LM35CZ or the equivalent
- 1 - temperature sensor, LM335Z or the equivalent

2 - op-amps, LM741
1 - optoisolator (light-activated SCR), H11C3, or SK4929, or the equivalent
1 - optoisolator (light-activated triac), MOC3021, or SK2048, or the equivalent
4 - optoisolators, 4N26, or SK2040, or the equivalent
1 - 74LS08
1 - 74LS240
1 - LM555 or the equivalent
1 - LM2917, or SK9209, or the equivalent
1 - 74LS194 bidirectional universal shift register or the equivalent
1 - 74LS26
1 - instrumentation amplifier, AD620, or its equivalent

EQUIPMENT -

1 - dual-outlet D.C. power supply, 0 to 24 volts (or 0 – 20 volts)
1 - 5-volt D.C. power supply
2 - digital multimeters
1 - analog or digital curve tracer (if available)
1 - dual-trace oscilloscope
2 - oscilloscope probes
1 - handheld multimeter
1 - single-phase or three-phase, variable A.C. power supply capable of 0 to 120 volts RMS with circuit breaker
1 - in-line A.C. ammeter, 0 to 8 amps or hand-held clamp-on ammeter
3 - wattmeters, 0 to 500 Watts
1 - single-phase, 120 volt A.C. split-phase motor or equivalent inductive load
1 - three-phase, variable A.C. power supply capable of 0 to 120 volts RMS single-phase and 0 to 208 volts line-to-line voltage with circuit breaker
1 - three-phase, analog wattmeter, 0 to 600 watts
1 - isolation transformer
1 - function generator (capable of 16 volts p-p with D.C. offset capability)
1 - frequency counter (optional)
1 - electric hot plate
1 - thermometer or electronic temperature probe
1 - inexpensive A.M. radio
1 - hand-held tachometer
1 - D.C. ammeter, 0 to 5 amperes
1 - dynamometer or D.C. generator
1 - D.C. motor up to 1/3 horsepower, rated for 120 volts
3 - A.C. ammeters, 0 to 5 amperes
1 - three-phase induction motor up to 1/3 horsepower
1 - magnetic contactor
1 - thermal overload relay
1 - pilot lamp
1 - stopwatch